Cardiomyocytes – Active Players in Cardiac Disease

Klaus-Dieter Schlüter
Editor

Cardiomyocytes – Active Players in Cardiac Disease

Springer

Editor
Klaus-Dieter Schlüter
Institute of Physiology
Justus-Liebig-Universität Giessen
Giessen
Germany

ISBN 978-3-319-31249-1 ISBN 978-3-319-31251-4 (eBook)
DOI 10.1007/978-3-319-31251-4

Library of Congress Control Number: 2016943117

Printed on acid-free paper

This Springer imprint is published by Springer Nature
The registered company is Springer International Publishing AG Switzerland

Preface

In this book, scientists well established in the field describe the fascinating biology of cardiomyocytes, cells that keep us alive and that are perfectly adapted to perform regulatory contractile activity throughout every minute, every hour, every day, and year by year. Failure of regular heartbeat will necessarily lead to death of the whole organism. Therefore, the understanding of the biology of cardiomyocytes, that means the understanding of their capacity to adapt to different workload, their ability to use different substrates to generate energy that allows them to keep contracting, their ability to restructure their contractile elements, and their ability to withstand various stressors is the basis to improve our clinical options with respect to heart failure, one of the biggest challenges in medicine. That is why this book is intended to give an overview about the current understanding of the biology of cardiomyocytes.

Getting the possibility to isolate and culture terminal-differentiated cardiomyocytes is and was the key step to learn and study precisely the biology of cardiomyocytes in its pure sense. Chapter 1 will briefly review the process that has been developed to reach this aim and gives a methodological overview how we can analyze basal functions of these cells. Chapter 2 recapitulates the ontogenetic history of cardiomyocytes and explains their heterogeneity in the heart and the specific function of cardiomyocytes in the adult heart. Chapter 3 introduces the principles of electromechanical coupling with a strong focus on the unique electrophysiological properties of these cells. In Chap. 4, we will learn how these cells can use different energy sources and how they adapt these mechanisms due to alternations in workload and substrate availability. In Chap. 5, it is outlined in great detail how cardiac function can be adapted to acute changes in workload. Cellular molecules are identified that are targeted by neurohumoral factors to respond to increased workload. Mechanisms different from acute adaptations are required if workload remains high. The initiation of such processes, summarized by the term cardiac hypertrophy, will be described in Chap. 6. Proteins have a distinct halftime. Therefore, cardiomyocytes must be able to degrade their proteins and if required must regulate the function of their protein degradation machinery. This is certainly another prerequisite for cardiac adaptation and remodeling. Chapter 7 will introduce all aspects of protein degradation. Cardiomyocytes normally require oxygen to generate energy. However, they have established strategies that allow them to withstand at least small periods of ischemia. How cardiomyocytes react to ischemia and how they can deal with this challenge are discussed in Chap. 8. Even cardiomyocytes die. They have established several pathways to induce apoptosis, necrosis, and necroptosis to avoid further damage to their neighboring cells. Such mechanisms are explained in Chap. 9. Finally, although oxygen is required for cardiac function, it is also toxic. Cardiomyocytes must develop strategies to protect themselves against oxidative stress but at the same time, they have learned to use such molecules as signals. These mechanisms will be summarized in Chap. 10.

I have to thank all the authors for their contribution to this book. Without their input, this description about the biology of cardiomyocytes would be incomplete. Many thanks to all of them! I also have to thank the editors who gave me the motivation to start this project. I have learned a lot during the process of editing this book about these fascinating cells. I hope that the reader will share our enthusiasm about this interesting field of biology.

Giessen, Germany K.-D. Schlüter

Contents

Part I **Cardiomyocytes: Function and Regeneration**

1 **Ways to Study the Biology of Cardiomyocytes** 3
Klaus-Dieter Schlüter

2 **Cardiomyocytes: Function and Regeneration** 25
Marten Szibor

3 **Excitation–Contraction Coupling of Cardiomyocytes** 67
Jens Kockskämper

4 **Cardiac Metabolism and Energetic Control** 97
Susanne Rohrbach and Bernd Niemann

5 **Endogenous Mechanisms for Regulating Myocardial Contractility** 135
Rolf Schreckenberg

Part II **Cardiomyocytes in Ageing, Disease, and Protection**

6 **Growth Regulation of Cardiomyocytes: Control of Cell Size and Its Role in Cardiac Hypertrophy** 167
Klaus-Dieter Schlüter

7 **Protein Degradation in Cardiomyocytes: Target Proteins and Clinical Consequences** 191
Oliver Drews

8 **Ischemia and Reperfusion** 223
Jacqueline Heger

9 **Mechanisms of Cardiac Cell Death** 247
Gerhild Euler

10 **Oxidative Stress and Nitrosative Stress** 267
Rainer Schulz and Fabio Di Lisa

Contributors

Oliver Drews Division of Cardiovascular Physiology, Institute of Physiology and Pathophysiology, Heidelberg University, Heidelberg, Germany

Gerhild Euler Institute of Physiology, Justus Liebig University, Giessen, Germany

Jacqueline Heger Institute of Physiology, Justus-Liebig-Universität Gießen, Giessen, Germany

Jens Kockskämper Institut für Pharmakologie und Klinische Pharmazie, Philipps-Universität Marburg, Marburg, Germany

Fabio Di Lisa Department of Biomedical Science, University of Padova, Padova, Italy

Bernd Niemann Klinik für Herz-, Kinderherz und Gefäßchirurgie Justus-Liebig-Universität Gießen, Giessen, Germany

Susanne Rohrbach Institute of Physiology, Justus-Liebig-Universität Gießen, Giessen, Germany

Klaus-Dieter Schlüter Institute of Physiology, Justus-Liebig-Universität Gießen, Giessen, Germany

Rolf Schreckenberg Institute of Physiology, Justus-Liebig-Universität Gießen, Giessen, Germany

Rainer Schulz Institute of Physiology, Justus-Liebig-Universität Gießen, Giessen, Germany

Marten Szibor Institute of Biotechnology, University of Helsinki, Helsingin Yliopisto, Finland

Abbreviations

AAV	Adeno-associated virus
ACE	Angiotensin converting enzyme
ACh	Acetylcholine
ADP	Adenosine diphosphate
ADR	Adrenaline
AIF	Apoptosis-inducing factor
Alk	Activin receptor-like kinase
AP	Action potential
AP-1	Activator protein-1
ATP	Adenosine triphosphate
AM	Acetoxymethyl
AMP	Adenosine monophosphate
AMPK	AMP kinase
Ang	Angiotensin
ANP	Atrial natriuretic peptide
ANT	Adeninin nucleotide translocase
AT	Angiotensin receptor
Atg	Autophagy related
AV	Atrioventricular (= node of Aschoff and Tawara)
BAT	Baroreceptor activation therapy
BCAA	Branched-chain amino acids
BCAT	Branched-chain aminotransferase
BCKA	Branched-chain keto acids
BCKAD	Branched-chain keto acid dehydrogenase complex
BDM	Butanedione monoxime
bHLH	Basic helix-loop-helix
BMP	Bone morphogenic protein
BNP	Brain natriuretic peptide
CaMK	Calcium-calmodulin-dependent protein kinase
cAMP	Cyclic adenosine monophosphate
cGMP	Cyclic guanosine monophosphate
CASA	Chaperone-assisted selective autophagy
CAT	Carnitine acylcarnitine translocase
CaT	Calcium transients

CD	Cluster of differentiation
CHD	Congenital heart disease
CHF	Chronic heart failure
CICR	Calcium-induced calcium release
CK	Creatine kinase
CNC	Cardiac neural crest
CNCC	Cardiac neural crest-derived cells
CPT	Carnitine palmitoyltransferase
CT	Central
CTP	Carnitine Palmitoyltransferase
CVLM	Caudal ventrolateral medulla
DAD	Delayed afterdepolarization
DAG	Diacylglycerol
DGS/VCFS	DiGeorge or velo-cardio-facial syndrome
DISC	Death-induced signalling complex
Drp	Dynamin-related protein
EC	Excitation-contraction
EndoG	Endonuclease G
EOMES	Eomesodermin
EPAC	Exchange protein directly activated by cAMP
EPDC	Epicardium-derived cells
EMT	Epithelial to mesenchymal transcription
ER	Endoplasmatic reticulum
ERK	Extracellular responsive kinase
ET	Endothelin
ETC	Electron transport chain
ETF:Q	Electron transfer flavoproptein-ubiquinone
FAT	Fatty acid translocase
FABP	Fatty acid-binding protein
FACS	Fatty acetyl-CoA synthase
FAO	Fatty acid oxidation
FATP	Fatty acid transport proteins
FFA	Free fatty acids
FGF	Fibroblast growth factor
FHC	Familiar hypertrophic cardiomyopathy
FHF	First heart field
FKBP	FK506-binding protein
FoxH	Forkhead box H
FWHA	Full width at half amplitude
GDP	Guanosine diphosphate
GFP	Green fluorescence protein
GPCR	G-protein-coupled receptors
Gpx	Glutathione peroxidase
GRK	G-protein-coupled receptor kinase
GTP	Guanosine triphosphate

HCN	Hyperpolarization-activated and cyclic-nucleotide-regulated channel
HDAC	Histone-deacetylase
HIF	Hypoxia-inducible factor
HT	Histamine
HtrA	High temperature requirement protein A
Hz	Hertz
IAP	Intracellular caspase inhibitor
IGF	Insulin-like growth factor
IHD	Ischemic heart disease
IMS	Intermembrane space
IP	Inositol phosphate
I/R	Ischemia/reperfusion
Isl	Islet
JAK	Janus kinase
JNK	c-Jun N-terminal Kinase
K2P	K-2-Pores
Kir	K-inward rectifying
LAMP	Lyosome-associated membrane protein
LC	Light chain
LCFA	Long-chain fatty acid
LDH	Lactate dehydrogenase
lncRNA	Long non-coding RNA
LO	Lipoxygenase
LTCC	L-type calcium channels
LVAD	Left ventricular assist device
LVEDP	Left ventricular end-diastolic pressure
MAP	Mitogen-activated protein
MCU	Mitochondrial uniporter
Mef	Myocyte-specific enhancer factor
MEK	Mitogen-activated protein kinase kinase
Mesp	Mesoderm posterior
MHC	Myosin heavy chain
MI	Myocardial Infarction
Mito-Q	Mito-quninone
MPTP	Mitochondrial permeability transition pore
MOMP	Mitochondrial outer membrane permeabilisation
mRNA	Messenger RNA
miRNA	MicroRNA
mTOR	Mammalian target of rapamycin
MuRF	Muscle ring finger
MyBP-C	Myosin-binding protein-C
NCX	Na-Ca-exchanger
Nec	Necrostatin
NFAT	Nuclear factor of activated T cells
NHE	Na-H-exchanger

NKA	Na-K-ATPase
NO	Nitric oxide
NOR	Noradrenaline
NOS	Nitric oxide synthase
NOX	NADPH oxidase
NYHA	New York Heart Association
NTS	Nucleus tractus solitarii
OxPhos	Oxidative phosphorylation
PAH	Pulmonary arterial hypertension
PCI	Percutaneous coronary intervention
PE	Phosphatidylethanolamine
PLC	Phospholipase C
PCr	Phosphocreatine
PDE	Phosphodiesterase
PDGF	Platelet-derived growth factor
PDH	Pyruvate dehydrogenase complex
PDP	Pyruvate dehydrogenase complex kinase
PDP	Pyruvate dyhdrogenase phosphatase
PFK	Phosphofructokinase
PGC	PPAR-gamma-coactivator
PI3K	Phophoinositide 3 kinase
PLB	Phospholamban
POLG	Polymerase gamma
PKA	Protein kinase A
PKC	Protein kinase C
PP	Protein phosphatase
PPP	Pentose phosphate pathway
PPAR	Peroxisome-proliferator-activated receptor
PRC	Polycomb-repressive complex
PTM	Post-translational modification
RAAS	Renin-angiotensin-aldosterone-system
RIPK	Receptor interacting protein kinase
RISC	RNA-induced silencing complex
RIS	Reactive inflammatory species
RISK	Reperfusion injury salvage kinase
ROS	Reactive oxygen species
rhNRG	Recombinant human neuregulin
RNS	Reactive nitrogen species
RyR	Ryanodine receptor
RVH	Right ventricular hypertrophy
RVLM	Rostral ventrolateral medulla
SA	Sinoatrial
SERCA	SR-calcium-ATPase
SHF	Second heart field
SK	Small conductance

SNS	Sympathetic nervous system
SOD	Superoxide dismutase
SR	Sarcoplasmatic reticulum
SRF	Serum response factor
SSM	Subsarcolemmal mitochondria
TAC	Transverse aortic constriction
TAG	Triacylglycerol
Tbx	T-box transcription factor
TGF	Transforming growth factor
Tm	Tropomyosin
TnC	Troponin C
TNF	Tumor necrosis factor
TNFR	Tumor necrosis factor receptor
TnI	Troponin I
TnT	Troponin T
TRPC	Transient receptor potential channels
TTP	Time-to-peak
TUNEL	Terminal deoxynucleotidyl transferase assay
UCP	Uncoupling protein
ULK	Autophagy-activating kinase
UPS	Ubiquitin-proteasome system
V	Volume
V_{max}	Maximal velocity
VCAM	Vascular cell adhesion molecule
VEGF	Vascular endothelial growth factor
Wt	Wilms' tumor
XO	Xanthine oxidase

Part I

Cardiomyocytes: Function and Regeneration

1 Ways to Study the Biology of Cardiomyocytes

Klaus-Dieter Schlüter

Abstract

This chapter explains how experiments can be designed to investigate and quantify the biological function of cardiomyocytes. Cardiomyocytes are highly and terminally differentiated cells that are closely connected to each other in tissue. Isolation of cells requires disruption of cell-cell contacts without damaging these cells. This is performed by a transient depletion of extracellular calcium making reexposing of cardiomyocytes to a challenging procedure with slight methodological differences for cardiomyocytes from different species and parts of the heart. Under culturing conditions, cardiomyocytes rapidly adapt the specific conditions. The lack of mechanical load and loss of contractile activity leads to degradation of contractile units that requires specific attempts to analyze the behavior of such cells. This can be performed by mechanical load, electrical pacing, or induction of remodeling. Function of cardiomyocytes is mostly characterized by load-free cell shortening with remarkable reproducible results between cardiomyocytes from different species. Molecular aspects of cardiac hypertrophy can be analyzed by quantification of protein synthesis, protein degradation, and cell sizes. Although cardiomyocytes can be isolated and cultured from many species, the majority of researchers focused on small rodents, preferentially rats. These have a surprisingly strong comparability with other species in many aspects but not in electrophysiological aspects.

K.-D. Schlüter
Institute of Physiology, JLU Giessen, Aulweg 129, Giessen 35392, Germany
e-mail: Klaus-Dieter.Schlueter@physiologie.med.uni-giessen.de

K.-D. Schlüter (ed.), *Cardiomyocytes – Active Players in Cardiac Disease*,
DOI 10.1007/978-3-319-31251-4_1

1.1 How Are Cardiomyocytes Organized in the Tissue?

Cardiomyocytes are one of the most fascinating cells in highly developed organisms. With growing levels of evolutional development, these cells are getting more and more specialized. In most vertebrates these cells lose the ability to complete a cell division cycle. They are still able to undergo karyogenesis, but they do not complete cytokinesis. One of the big advantages of a tissue, in which mass is built up to 80 % of nonproliferating cells, is that it barely develops any type of tumors. One of the big disadvantages is, however, that these cells cannot be replaced by adjacent cardiomyocytes once they are damaged or lost. In the last decade, this traditional view of the heart as a tissue that consists of nonproliferating cells with low renewing capacity has been challenged. As an alternative hypothesis, researchers stated that cardiomyocytes can be renewed during aging by reactivation of tissue-specific progenitor cells or by recruiting circulating stem cells or progenitor cells. However, using highly sophisticated methods that allow determining the average age of cardiomyocytes in a given tissue, it turned out that myocyte regeneration is a process that is extremely rare (Bergmann et al. 2009). In fact, most of cardiomyocytes act for years as stable cells. They do their job, that is, to contract in a regular way approximately 60 times per minute in humans and 400–600 times per minute in rodents, day by day and night by night. As the maximal life-span of humans is approximately 120 years, this cell contraction will be repeated 3.8×10^9 times. This requires a coordinated way of contractile responsiveness between all cells that build up a ventricle, and this is done by building a functional syncytium between thousands of cardiomyocytes. This functional syncytium is formed by cell-cell contacts between individual heart cells. These cell-cell contacts are known as gap junctions and easily visualized as *Glanzstreifen* in histological preparations (Fig. 1.1). If we want

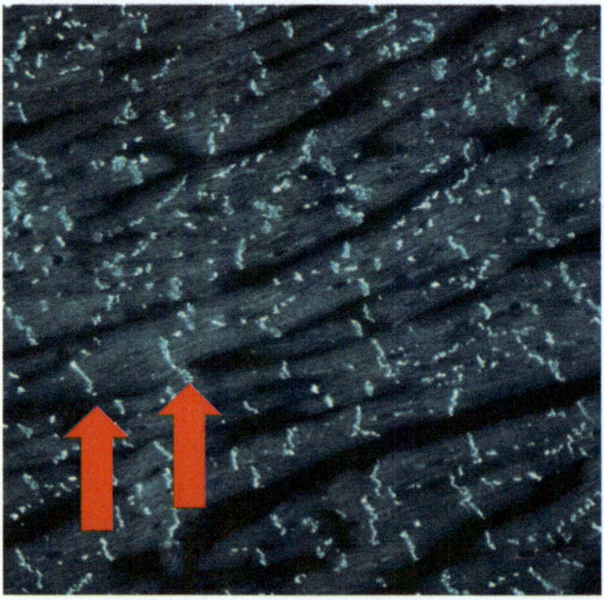

Fig. 1.1 Immunohistological staining of intact viable myocardium in which connexin 43 is stained (*red arrows*). Note the longitudinal shape of cardiomyocytes and the strong staining of connexin 43 at the co-called *Glanzstreifen*

to study the biology of these cells and characterize their ability to respond to electrical signals, to mechanical load, and to neurohumoral stimulation, it is required to isolate these cells from their cellular network.

As mentioned above, cardiomyocytes were getting more and more specialized during evolution. The evolutional process of heart development is recapitulated in embryogenesis. Therefore, it should be clearly mentioned at the beginning of this chapter that fetal cardiomyocytes, neonatal cardiomyocytes, and cardiomyocytes in the adult heart differ significantly in their biological characteristics. Just to mention some of these differences: Neonatal and adult cardiomyocytes differ in receptor density and expression, the rate of protein synthesis, the type of protein isoforms that are expressed, the stability of membrane potential, the type of metabolism, and the ability to complete cell division. Collectively, it is clear from these examples that a characterization of cardiomyocytes responsible for proper heart function in adult hearts requires the isolation and cultivation of cardiomyocytes in their terminally differentiated form.

The view of the heart as an organ in which individual muscle cells are connected to each other was not accepted until 1965 when Muir was the first to separate individual cardiomyocytes and analyze them under the microscope (Muir 1965). Until that time, it was known that individual myocytes are formed during embryogenesis, but it was considered that these cells form a skeletal-type syncytium during maturation. Muir successfully isolated cells by breaking down cell-cell connections of the *Glanzstreifen* by calcium depletion. Unfortunately, researchers were unable to reexpose cardiomyocytes to physiological concentrations of extracellular calcium at that time, a phenomenon named "calcium paradox." Therefore, although it was the first time that it was clearly shown that the heart consists of millions of individual cells, it was impossible to perform any physiological experiments with these cells. Meanwhile, the strong coupling of individual cardiomyocytes by cell-cell connections was identified as gap junctions built up mainly from hexamers of connexin 43. These cell-cell connections allow the heart to act as a "functional syncytium," although it is organized in a completely different way than the skeletal muscle where individual cells form an anatomical syncytium by cell fusion.

1.2 Isolation and Cultivation of Cardiomyocytes

Key characteristics of cardiomyocytes such as metabolism, electromechanical coupling, responsiveness to hormones, and other aspects of cellular biology cannot be studied in tissue preparations of hearts due to the strong interference between different cell types, such as endothelial cells, neurons, and cardiac fibroblasts. To study the responsiveness of cardiomyocytes to hormones and other factors, it is required to separate these cells from all other cells of the heart and avoid the barrier function of vessels that limits free exchange of factors in a heart. At the same time, isolation of cardiomyocytes is a double-edged sword. On the one hand, it allows researchers to study the biological adaptation of cardiomyocytes to definitive responsiveness to agonists under fully defined conditions, but on the other hand, information about cell-cell interactions that occurs in a given heart is lost. Nevertheless, without analyzing separated myocytes, we cannot learn anything about the interindividual variability of cells and the cellular participation to adaptations leading to heart failure. As heart failure is still the leading cause of death at least in higher industrialized countries, methods to study the biology of cells are indeed essential to develop future strategies to avoid the consequences of heart failure. For that reason, finding ways to isolate and culture cardiomyocytes was an important aim of intensive research for many years. Today, researchers are lucky enough to have established protocols for isolation and cultivation of cardiomyocytes in a reproducible way that allows them to study the biological responsiveness of cardiomyocytes to various challenges, such as hypertrophic stimulation, ischemia, and reperfusion, or to study molecular aspects of electromechanical coupling.

Historically, the first successful isolation of spontaneously beating cardiomyocytes was described approximately 100 years ago by Burrows. He separated these cells mechanically from embryonic chicken hearts (Burrows 1910). At the middle of the last century, the efficiency of this procedure was significantly approved by replacing mechanical dissection by proteolytic digestion. Using this technique based on trypsin digestion, spontaneously beating heart cells were isolated from chick embryos and neonatal rats (Cavanaugh 1955). Ten years later, it was found that the dense cell-cell contacts between individual cardiomyocytes in adult hearts can be separated by calcium depletion (Muir 1965). However, although these procedures now allowed the isolation of terminally differentiated cardiomyocytes, a key step in the progression of the analysis of cardiomyocyte's biology, the cells were not calcium tolerant. This means that the cells could not be reexposed to physiological calcium concentrations, a prerequisite to study their physiological behavior.

The breakthrough in isolation procedures of adult terminally differentiated cardiomyocytes was performed, when cells were initially isolated at low calcium concentrations (<100 μM but >50 μM) and thereafter exposed to increasing calcium concentrations in three to four steps until the physiological extracellular calcium concentration of 1–2 mM was reached again (Powell and Twist 1976). Furthermore, enzymatic digestion was optimized, and finally most researchers use a crude preparation of collagenase that contains low amounts of other proteolytic enzymes as well (Kono 1969). Unfortunately, crude collagenase preparations differ in quality, and therefore, a specific batch has to be identified. Moreover, different batches of collagenase have to be adapted to individual protocols. As a consequence of this, there is no general rule about the concentration of collagenase used to isolate myocytes and about the treatment time. On average, a mean concentration of 400–500 mg/l collagenase and a mean perfusion time of 20–30 min are commonly used. However, as mentioned already before, the exact perfusion time and concentration of collagenase depend strongly on the batch of collagenase, the age of the animal, and the sex of the animal.

An example of a protocol to isolate calcium-tolerant cells from rats is given in detail elsewhere (Schlüter and Piper 2005). In brief, hearts are rapidly removed from rats and connected to a standard Langendorff system, allowing perfusion of the heart with standard buffer, i.e., Powell medium. This removes blood and blood cells from the tissue. Thereafter, the perfusion system is switched to a nominal calcium-free collagenase-containing buffer, and the heart is continuously perfused with this buffer for 20–30 min until the tissue is soft. Then the heart is removed from the Langendorff system and the atria, and if requested, the left and right ventricles are separated. Next, this soft tissue will be cut into small pieces, i.e., by using a tissue chopper allowing to cut the heart into 0.7 mm long pieces. In the next step, the tissue is incubated with the same nominal calcium-free collagenase-containing buffer for up to 5 minutes, and the tissue will be mechanically softened during this time. Thereafter, the whole material is filtered through a nylon mesh (mesh size 200 μm), and the filtered cell containing material is transferred to a centrifugation step (25 g, 3 min). The supernatant will be removed and the sediment resuspended into a collagenase-free buffer containing 2% (vol/vol) $CaCl_2$ (stock solution 100 mM). This step will be repeated three times, and with each step, calcium concentration increases until it reaches a final amount of 1.0 mM. These cells are then plated onto culture dishes precoated with laminin (5 μg/ml). After 2 h the cell culture medium is renewed, and only calcium-tolerant rod-shape myocytes are left on the culture dish. On average, 2.8×10^6 cells/heart can be isolated and cultivated (plating density approximately 1.5×10^4 cells/1000 mm^2). Most of all researchers who aimed to analyze the biology of cardiomyocytes via isolation and cultivation of these cells were able to establish such isolation and cultivation procedures for cardiomyocytes isolated from rats; only few researchers were able to isolate cardiomyocytes from mice. Initially, the protocol of cell isolation for rats was modified to isolate cardiomyocytes from mice by replacing collagenase by protease extracts and adding butanedione monoxime (BDM) to perfusion buffers. The main effect of BDM is that it blocks the binding of the heads of myosin heavy chain molecules to actin filaments and thereby blocks any contractile activity. This avoids mechanical stress from the cells during the isolation procedure. Mechanical stress during the isolation process produces small membrane defects. These can be repaired in a calcium-dependent way. Most likely, this is the reason that a minimum of calcium is needed during the isolation procedure, although depletion of calcium is required to break down cell-cell contacts. It can be speculated that mouse cardiomyocytes are more sensitive than rat myocytes to membrane defects and therefore benefit from a reduction of mechanical stress. Meanwhile, methods that reestablish physiological values of calcium concentration more carefully have been used and allow the use of a similar protocol for isolation of mouse myocytes as for rat myocytes without the need of BDM (Schlüter and Schreiber 2005).

The next critical step is to get cardiomyocytes in culture for a couple of days. The entire size and protein expression of isolated cardiomyocytes is the consequence of direct or indirect mechanical stress to cardiomyocytes in a constitutively beating heart. As soon as calcium-tolerant cardiomyocytes are isolated from hearts, they stop beating (see above). This means that there is neither mechanical load on these cells nor contractile activity in a meaning of regular and repeated calcium transients. As a result of this, cardiomyocytes start to degrade their contractile proteins. This process can be significantly activated by pro-hypertrophic stimulation and can be considered as an accelerated form of cellular adaptation to hypertrophy that also

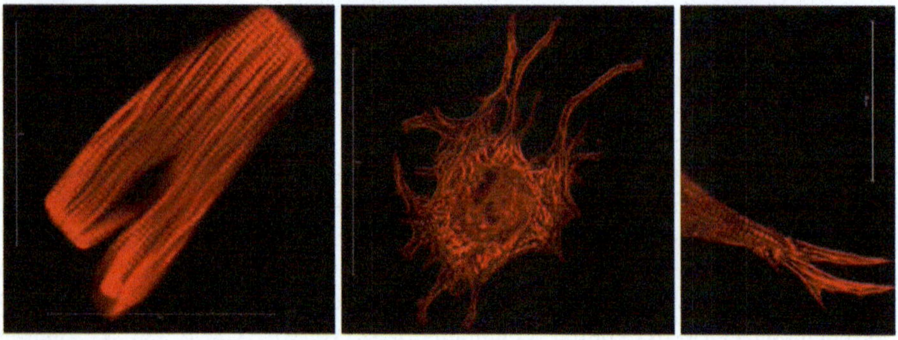

Fig. 1.2 Staining of actin filaments in isolated cultured cardiomyocytes. In the presence of 20 % (v/v) fetal cal serum, cells lose their initial striation pattern, build a new stress fiber network in the perinuclear region, and finally build new sarcomeres as indicated by the reappearance of striated structures at later time points

occurs in vivo. Mechanisms dealing with the regulation of protein degradation and protein synthesis leading to cardiac hypertrophy will be discussed in detail in Chaps. 6 and 7. Cardiac hypertrophy is not simply synthesis of more proteins increasing the numbers of sarcomere units leading to thicker cells but also characterized by replacing isoforms of contractile proteins that are specific for adult cardiomyocytes by those specifically expressed in neonatal cardiomyocytes (phenotype differentiation). Under culture conditions, the process of cellular differentiation can be activated by addition of excessive concentrations of fetal calf serum (up to 20 % v/v) as first observed in 1977 by Jacobson. Under such conditions cardiomyocytes rapidly start to degrade their contractile machinery. This is necessary for cultured cardiomyocytes to survive for longer time periods. In a second phase of adaptation to culture conditions, cardiomyocytes start to reform sarcomere-like structures (Fig. 1.2). However, at this stage they lose their rod-shape morphology. Instead of this they start to rebuild new cell-cell contacts. Under these optimized conditions, cardiomyocytes can be cultivated for months (Liu 2013). However, the final cell culture represents cardiomyocytes that share characteristics of neonatal cardiomyocytes (i.e., isoform expression of sarcomere proteins) with those of fetal-type cardiomyocytes (i.e., spontaneously beating) and those of adult cardiomyocytes (i.e., lack of cytokinesis). Without this excessive stimulation of protein degradation, in other words under serum-free conditions, cardiomyocytes also lose their contractile machinery step by step and getting thinner. However, this process is much slower than that under serum cultured conditions. There is also a constitutive loss of T tubules, again similar to adaptations seen in vivo during heart failure (Louch et al. 2004). Under serum-free conditions, cardiomyocytes remain more or less stable and maintain their membrane potential up to 3 days until most of these cells die. Some researchers tried to overcome these problems by continuous electrical stimulation during the entire cultivation period (Kuramochi et al. 2003). This was performed by biphasic electrical stimulation, similar to that used to study cell shortening. Although some promising results were initially reported, this procedure did not make the transition to standard procedures because the effect was rather limited. Chronic pacing during cultivation did not significantly increase any survival markers and did not activate any stress pathways. However, chronic pacing sensitized the cells toward hormonal stimulation. A possible reason that cultured myocytes did not strongly differ from non-paced myocytes may be that this type of pacing induces a type of load-free cell shortening not generating force to the cells and thereby lacks a main aspect of mechanical load exposed to cardiomyocytes in vivo. As an alternative approach, cells were attached to flexible matrices that can be extended to various levels in a repeated way (Shyu et al. 1995). This allows passive extension of cells during cultivation and generates a passive force to cells. However, as adult cardiomyocytes did not firmly attach to the flexible matrix material, this was used only for neonatal and fetal cardiomyocytes.

The aforementioned procedures of cell cultivation depend on cell adhesion to culture dishes that can be achieved by pre-cultivation of culture dishes with either precoating culture dishes with 4 % (v/v) fetal calf serum or with laminin. Other potential attachment substrates that have been tested such as polyamino acids or extracellular matrix molecules such as collagen and fibronectin were less suitable (Jacobson and Piper 1986). Studies investigating the biological behavior of cardiomyocytes under cultured conditions over a longer time period are nearly exclusively performed with cardiomyocytes isolated from rats. For reasons not obvious, mouse cardiomyocytes are less suitable for long-term cultivation. On the other side, cardiomyocytes isolated from guinea pigs seem to be less prone to remodeling processes of contractile structure than those of rats (Horackova and Byczko 1997), while isolation of cardiomyocytes from larger animals such as cats, rabbits, and pigs is expensive and also less prominent for ethical reasons and therefore rarely reported in the literature. When compared directly, cat and rat myocytes behave similarly under cultured conditions, making rat myocytes a suitable model to study remodeling processes in great detail as it occurs in the heart in situ under hypertrophic conditions (LoRusso et al. 1997). Similarly, cultivation of dog myocytes in serum-containing media resulted in rounding of cells (Sakai et al. 1989). As a consequence of this, it should be clearly stated that most of the current knowledge about the detailed biology of cardiomyocytes is related to that of rats. A MEDLINE search (current standing 12.08.2015) displayed 932 publications for cultured rat myocytes, 411 for cultured mouse myocytes, 33 for cultured rabbit myocytes, 11 for cultured dog myocytes, and five for cultured cat or frog-derived myocytes.

Another aspect of cultivation of cardiomyocytes is linked to cell culture media that are required for establishing cardiomyocytes cultures. Cell culture media allowing these terminally differentiated cells to survive are complex. In most cases they are based on medium M199, one of the most complex media among commercially available cell culture media, and M199 requires further additions such as creatinine, carnitine, and taurine. Further substitutions regularly added to culture media are antibiotics and β-arabinofuranoside to attenuate proliferation of potentially contaminating non-myocytes.

1.3 Methods to Study the Contractile Activity of Cardiomyocytes

Once terminally differentiated cardiomyocytes have been isolated and successfully been reexposed to physiological concentrations of calcium, the next important step is to analyze their main function, and that is their contractile activity. In a total heart, cell stimulation will be performed by generating an action potential in pacemaker cells that will then be distributed along the conduction system to the myocardium. When terminally differentiated cardiomyocytes are isolated and cultured as individual cells, they lose their cell-cell connection. They are no longer conducted to a pacemaker cell, and therefore, they should remain quiescent without any contractile activity. Normally, cells are considered as calcium tolerant if they do not beat spontaneously; it means that less than 10 % of cells show spontaneous calcium release and contractile waves and that this will not happen more than twice a minute. Nevertheless, it happens that in isolated cells, ryanodine channels in the sarcoplasmic reticulum (SR) open spontaneously, generating calcium waves that distribute through the cell (Lopez et al. 1995). These spontaneous releases of calcium from the SR depend on luminal and cytosolic calcium concentrations, and it is considered that these events contribute to the arrhythmic potential of cardiomyocytes (Landgraf et al. 2004). However, spontaneous calcium waves do not necessarily lead to contraction waves.

The process of electromechanical coupling can be studied if cardiomyocytes are stimulated by extracellular electrodes, i.e., exposed to biphasic electrical stimuli. In that case the membrane potential will be depolarized to values sufficient to reach the trigger potential of voltage-dependent sodium channels. At that point cardiomyocytes develop a membrane potential that initiates a calcium-dependent calcium release from the SR and activates the contractile machinery similarly to their in vivo behavior. This leads to a rapid cell shortening and subsequent relengthening of cardiomyocytes. When studied on the level of cultured cardiomyocytes, this type of contractile activity is termed load-free cell shortening and mimics a situation of isotonic contraction because cardiomyocytes do not generate any type of force under such conditions. This is not necessarily comparable with physiological types on contractile activity that are normally auxotonic contractions. However, mechanisms that activate load-free contractile activity reflect the situation of the heart in situ quite nicely. Moreover, changes in load-free cell shortening can be used to monitor changes in cell function similar to that observed in heart failure. A timely discussion about technical aspects of monitoring cell shortening is given elsewhere (Delbridge and Roos 1997).

Terminally differentiated cardiomyocytes have been isolated from cardiac tissue of various species, and load-free cell shortening has been analyzed thereafter. Table 1.1 summarizes the data from 304 publications found by a MEDLINE research using the terms cardiomyocytes and cell shortening (studies published until 2012). There are no systematic differences between the contractile parameters among different species. However, the majority of experiments are performed on myocytes isolated from small rodents like rats and mice.

Table 1.1 Species-dependent comparison of load-free cell shortening determined by percent shortening amplitude normalized to diastolic cell length (ΔL/L) and expressed as %

	n	ΔL/L (%)	Temp (°C)	Frequency (Hz)	Ca concentration (mmol/l)
Rat	213	7.77 ± 3.32	28.6 ± 6.0	1.3 ± 1.6	1.5 ± 0.8
Mouse	83	6.47 ± 3.13	27.6 ± 7.2	1.5 ± 1.6	1.6 ± 1.1
Dog	17	7.52 ± 5.07	34.2 ± 4.0	0.6 ± 0.3	1.6 ± 0.4
Rabbit	43	8.57 ± 4.67	30.2 ± 6.9	0.8 ± 0.5	2.0 ± 1.1
Guinea pig	26	10.01 ± 5.31	31.9 ± 6.1	0.9 ± 0.6	2.1 ± 1.2
Ferret	4	9.40 ± 1.02	27.2 ± 3.4	0.4 ± 0.1	1.9 ± 0.5
Cat	2	5.90 ± 4.10	28.7 ± 6.1	0.5 ± 0.3	1.4 ± 0.6
Pig	1	6.30	24.0	1.0	n.g.
Human	6	6.63 ± 2.76	32.0	0.7 ± 0.7	2.6 ± 2.2
Sheep	2	5.55 ± 0.85	23.0	0.25	1.8
Trout	2	8.30 ± 1.10	15.0	0.8 ± 0.6	2.5
Hamster	5	9.30 ± 3.97	33.0 ± 6.3	0.5 ± 0.0	1.9 ± 0.2

Data are means ± SD

n number of experiments that report about ΔL/L (%), *n.d.* not given in publication, *Temp.* temperature used to determine cell shortening, *Frequency* beating frequency rate where cell shortening was monitored, *Ca concentration* indicates the concentration of extracellular calcium

The data described in Table 1.1 do not take into account the different methods by which cell shortening can be analyzed. In general it should be noted that different systems are used to quantify cell shortening. Most researches report about studies in which cell shortening is quantified by cell-edge-detection assays. This means that the system recognizes cell boundaries along the long axis of cardiomyocytes and records continuously the cell length. Normally two different methods are available: systems that use a video-based detection system that allows recording cell shortening with a reading frame of 50 Hz and systems that use a line camera connected to an oscilloscope allowing reading frames of 500 Hz. The difference is the more accurate quantification of contractile dynamics, i.e., contraction and relaxation velocity, by using a higher reading frame. Furthermore, this type of analysis gives also information about total cell lengths that may differ when cardiomyocytes are isolated from different strains of animals or from animals undergoing myocardial hypertrophy. An alternative method is to analyze the sarcomere lengths by laser diffraction in cells that will also shorten during cell contraction (Kent et al. 1989). From a technical point of view, this requires very sensitive technique to monitor small changes in length. Cardiomyocytes have an average cell length of about 100 μm. If the cells shorten by approximately 10 %, it means that cell lengths at the magnitude of the systolic cell shortening are on average 10 μm smaller. Sarcomere length in cardiomyocytes is approximately 2.0 μm. This means that the same amount of cell shortening induces a shortening of sarcomere lengths of only 200 nm. Thus sensitivity of the cell recoding must be 50 times higher. Another variation between data recoded by various researchers comes from the temperature and from the beating frequency. Load-free cell shortening can easily be monitored and used to quantify

Fig. 1.3 Representative single cell recording of the length of a ventricular cardiomyocyte paced at 2 Hz

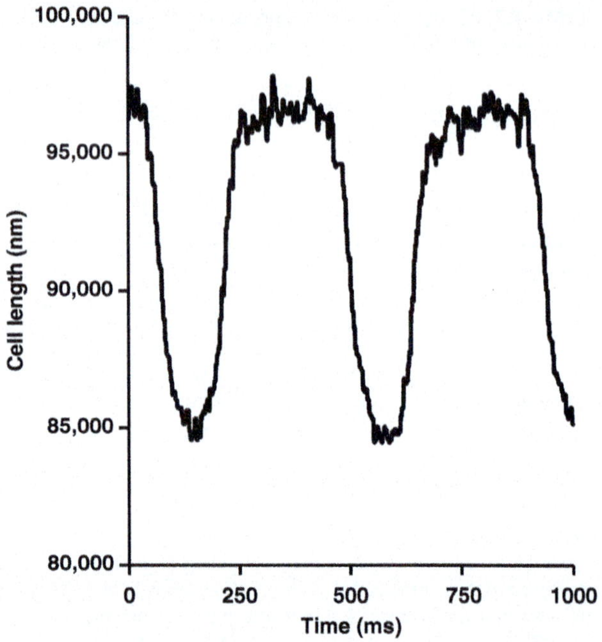

maximal load-free shortening, a parameter normally expressed as maximal shortening amplitude (ΔL) normalized to diastolic cell length (L) and expressed as %. Furthermore, the following kinetic parameters can be monitored from such recordings: the time to reach the peak of contraction (time to peak (TTP), given in ms), the time to reach 50 or 90% of relaxation (R50, R90, given in ms), the maximal contractile velocity (given as μm/s), and the maximal relaxation velocity (given as μm/s). Another parameter often given to characterize contractile behavior is the full width at half amplitude (FWHA, given in ms). The meaning of these determinates differs a little bit. ΔL/L (%) characterizes the overall contractility defined as cell shortening. TTP and FWHA are more closely related to calcium sensitivity of troponin C. Maximal contraction and relaxation velocities are related to the velocity of calcium release from SR and the subsequent refilling of SR and normalization of cytoplasmatic calcium, mainly depending on SERCA activity. Figure 1.3 gives an example of such a recording.

A low temperature (normally room temperature) favors a strong contractile responsiveness (Table 1.2a). Possible explanations for the more effective contractility at lower temperatures are the effect of temperature on calcium transport proteins favoring calcium load of SR under conditions of lower temperature. However, this seems not to be the case. Data on rat cardiomyocytes clearly show that an increase in temperature significantly increases relaxation velocity and contraction velocity most likely because diffusion of calcium ion is a temperature-dependent effect, and activity of SERCA is also temperature dependent. However, cell shortening amplitude is more related to myosin ATPase activity, and this enzyme has an optimal

Table 1.2 Influence of various parameters on load-free cell shortening of adult ventricular rat cardiomyocytes

(a) Effect of temperature

	Temperature (°C)	*n*	ΔL/L (%)	Con-Vel (μm/s)	Rel-Vel (μm/s)
Room temperature	22–25	78	8.21 ± 2.81	136 ± 48	130 ± 59
Moderate temperature	28–32	34	7.59 ± 2.69	150 ± 44	89 ± 39
Body temperature	35–39	60	7.38 ± 3.62	207 ± 80	179 ± 92

(b) Effect of pacing frequency

	Frequency (Hz)	n	ΔL/L (%)	Con-Vel (μm/s)	Rel-Vel (μm/s)
Low frequency	0.5	74	7.41 ± 3.07	149 ± 72	126 ± 62
Moderate frequency	1.0	48	8.30 ± 3.51	143 ± 69	114 ± 91
High frequency	2.0	25	7.66 ± 2.65	148 ± 43	141 ± 67

(c) Effect of calcium concentration

	Calcium (mM)	n	ΔL/L (%)	Con-Vel (μm/s)	Rel-Vel (μm/s)
Low Concentration	1.0	64	6.73±2.90	122±53	97±64
Small Concentration	1.2	31	8.03±3.86	164±57	116±61
Normal Concentration	1.5	6	8.45±4.50	145±28	145
High Concentration	1.8	43	8.61±2.21	171±29	160±46
Super physiological	2.0	10	9.30±3.16	68±36	73±32

temperature at room temperature (de Tombe and ter Keurs 1990). Consequently, load-free cell shortening is inversely correlated to temperature (Fig. 1.4). In contrast, pacing frequencies have minor effect on cell shortening. In general, low beating frequencies can easily be recorded and up to 2 Hz a frequency at which cardiomyocytes follow the external stimulation without induction of arrhythmias. However, when individual authors compare the effect of beating frequencies on load-free cell shortening, the data for rat cardiomyocytes show either a negative contractility-frequency relationship, a biphasic relationship, or positive relationships. As data from literature do not give a unique picture, it seems that the effect of pacing frequency is less important than other variables. The relationship between beating frequencies and load-free cell shortening are given in Table 1.2b. Of note, neither room temperature nor low beating frequencies (0.5–2.0 Hz) are normal physiological conditions. However, if isolated cardiomyocytes are used to address the question whether a specific signal or event modifies the entire contractile activity of cardiomyocytes, it is an advantage to analyze the cells under the best shortening conditions. Therefore it is highly justified to study contractile responsiveness of cardiomyocytes under such conditions (low frequency, low temperature). Another variable that limits comparability between different studies is the composition of the buffer. The impact of calcium concentration may be highlighted in this context (Table 1.2c). As shown in Table 1.2c, the concentration of extracellular calcium

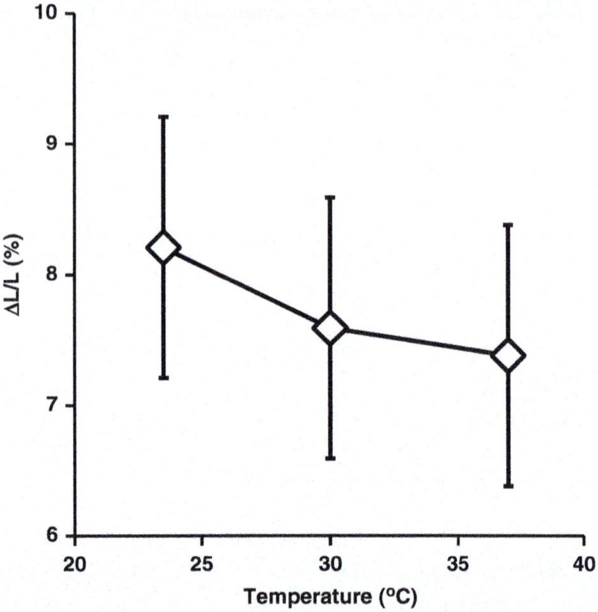

Fig. 1.4 Effect of temperature on load-free cell shortening of cultured cardiomyocytes paced at 2 Hz. Data are means ± SD from n = 36 cells

correlates with shortening amplitudes. The relationship between extracellular calcium concentration and shortening kinetics as shown in Table 1.2 must be considered with care because at some concentrations, only few (two to three) reports are given in the literature in which always shortening kinetics was recorded with video recording (low reading frame) or no statements are given about the methodological aspects. As mentioned above, video reading frames are sufficient to record shortening amplitudes but limited at high velocities and may therefore report nonphysiological values.

In many cases it will be necessary to determine load-free cell shortening of cells exposed to different stimuli that change the expression of proteins in these cells. This requires a long-term cultivation of cardiomyocytes, leading to studies investigating cell shortening after 24 h or later time points. Basically, cardiomyocytes remain more or less in a stable rod-like shape for 48 h, but due to the reduction in cell structures such as T tubuli, the question occurred whether this impacts a possible analysis of cell shortening. Data that compare cardiomyocytes that are analyzed within 4–6 h after the isolation with those analyzed 24–48 h later do not show major differences in percent cell shortening (Table 1.3).

As mentioned above, systems that allow to monitor cell shortening judge the activity of cardiomyocytes on the basis of load-free cell shortening. The question occurs of course whether additive information can be obtained from studies when cardiomyocytes are used to study isometric contraction or auxotonic contraction. The latter one can be analyzed if the viscosity of the culture is increased, i.e., by addition of methylcellulose at different molecular weights (Kent et al. 1989). Although administration of methylcellulose to culture media allows to increase afterload, the

Table 1.3 Effect of serum-free cultivation on load-free cell shortening of adult rat ventricular cardiomyocytes

	N	L diast (μm)	ΔL/L (%)	Con-Vel (μm/s)	Rel-Vel (μm/s)	TTP (ms)
Day 0	250	104.61 ± 18.38	9.84 ± 3.44	220 ± 86	209 ± 102	133 ± 37
Day 1	45	94.21 ± 15.97	8.99 ± 2.46	152 ± 51	153 ± 90	161 ± 38
Day 2	45	97.86 ± 15.04	8.72 ± 3.03	164 ± 59	163 ± 78	137 ± 21

Data are means ± SD

n number of cells under investigation, *L diast* = diastolic cell length, *ΔL/L* percent shortening amplitude normalized to diastolic cell length M, *Con-Vel* contraction velocity, *Rel-Vel* relaxation velocity, *TTP* time to peak of contraction

effect of changes in preload, as they occur by Frank-Starling mechanisms, cannot be investigated with such setups. To analyze the effect of preload, cardiomyocytes have to be attached to glass micro-rod coated with biological adhesive such as MyoTak, and these have to be connected to a force transducer (Posser et al. 2011). An advantage of this method is that preload can be altered because cell length can passively be manipulated. A major disadvantage is that only very few cells (one to two) can be analyzed within one preparation, and it remains unclear whether these cardiomyocytes are representative for the average cell population.

1.4 Methods to Discover the Electromechanical Coupling of Cardiomyocytes

In the last paragraph, the most important methods have been described that allow measuring contractile activity of cardiomyocytes. The activation of the contractile response consists of three different steps: first, building of an action potential (AP); second, influx of extracellular calcium; and third, calcium-induced calcium release from the SR. The ability to release calcium from the SR depends on the calcium filling of the SR and the activation of ryanodine receptors that release calcium. Therefore, the calcium-induced calcium release is a process that depends on diastolic function (refilling of SR) and systolic function (activation of ryanodine receptors). Analysis of membrane potential in cardiomyocytes is a difficult task, and although some reports in the literature give data about channel conductivity for myocytes from rats, the majority of reports are performed on cardiomyocytes isolated from guinea pigs. The reason is simply that cardiomyocytes from guinea pigs are more resistant to standard techniques of electrophysiology such as patch clamp techniques. For that reason an alternative method to monitor changes in membrane potential is the use of fluorescence dyes, such as RH237, di-4-ANEPPS, or JC-1 (Fujiwara et al. 2008; Windisch et al. 1995; Abdallah et al. 2011). With these dyes, changes in membrane potential can be analyzed; however, time resolution is a problem by this technique. As changes in intracellular calcium are the main triggers of cardiomyocyte activation, fluorescence dyes monitoring changes in calcium concentrations are commonly used. These can be admitted to cells as acetoxymethyl (AM)-ester-tagged molecules such as fluo-4 and Fura-2. Within the cell the AM part is enzymatically cleaved, and the fluorescence dye is stored in the cell. Variations in fluorescence dye uptake can easily be corrected by normalization of the specific signal to the isospectic wavelength (i.e., 340 nm and 380 nm). Normally, calcium concentrations are simply given as a ratio of fluorescence intensities of the two wavelengths. In some cases researchers finally equilibrated the fluorescence signal by comparison to standardized concentrations. In this case cells are finally permeabilized and calibrated with ionomycin. The disadvantage of monitoring calcium transients by fluorescence dyes is that the calcium indicator binds calcium at least in part, and this part is no longer able to bind to physiological binding partners such as troponin C. Therefore, load-free cell shortening and calcium transients should not be measured at the same time but rather in different approaches. Another important physiological parameter that can be measured by administration of fluorescence dyes is the pH that can be monitored by BCECF (Ladilov et al. 2003).

1.5 Analysis of the Activity of Subcellular Organelles in Cardiomyocytes

Calcium is by far the most important signaling factor in cardiomyocytes. Therefore, there is a need to quantify the relative calcium filling of the three most important calcium storing organelles: SR, mitochondria, and nucleus. Calcium load of the SR can be measured indirectly by the release of calcium from the SR, i.e., by adding thapsigargin to the cells. In this case the rapid increase of intracellular calcium as determined by calcium-binding fluorescence dyes corresponds to the pre-filling of the SR. Ryanodine can be used to block the SR calcium release. In the case of mitochondria and/or nuclei, the quantification is more complex because total calcium signals have to be negotiated. In the case of mitochondria, this can be performed by using Fura-2 as calcium-sensitive dye after quenching the cytosolic compartment with $MnCl_2$ (Abdallah et al. 2011). In the case of nuclei, this is performed by mag-fluo-4 AM, a low-affinity calcium indicator (Ljubojevic et al. 2014). Mitochondrial activity can also be determined indirectly by their metabolic activity. In this case, either mag-Fura-2 can be used as a fluorescence dye that is sensitive to Mg^{2+} which reflects changes in ATP concentration. Alternatively, time-resolved spectroscopy of endogenous NAD(P)H fluorescence can be used because it correlates with metabolic activity (Chorvatova et al. 2012).

1.6 Genetic Modification of Cultured Adult Cardiomyocytes

Any attempts that allow researchers to modify the function of cardiomyocytes by genetic manipulation require cultivation of cardiomyocytes. As cultivation of cardiomyocytes is a time-limited process (see above), this means that any attempts to downregulate proteins by genetic manipulation depend on the half-life of the respective target protein. As contractile proteins are in the focus of cardiac function, this is often a problem because contractile proteins have normally long half-lives.

Similarly, attempts to overexpress proteins of the contractile machinery have to replace native target proteins, again a time-consuming process. In conclusion, this limits the number of potential proteins that can be targeted by genetic manipulation. Nevertheless, adenoviruses, adeno-associated virus (AAV) transduction, and lentivirus-derived approaches have been successfully used to transduce adult terminally differentiated cardiomyocytes (Louch et al. 2011). The main advantage of AAV-based transfection methods is that these vectors preferentially infect nonproliferating cells (Zacchigna et al. 2014). In many of these cases, the genetic manipulations are induced in vivo by generation of transgenic animals, and subsequently the cardiomyocytes are isolated from these animals and analyzed ex vivo. A current problem in the field is that most of these transgenic animals are mice and terminally differentiated cardiomyocytes from mice can hardly be cultivated. This is certainly a limitation.

The easiest way to manipulate the protein expression in cardiomyocytes is to downregulate proteins by administration of small interfering RNA (siRNA) molecules to the culture medium. siRNA molecules are mono- or double-strain ribonucleic acids that form together with specific proteins, the RNA-induced silencing complex (RISC). RISCs use one of the two RNA strains that is complementary to the mRNA of a specific protein. Using this technique, molecules that display a high turnover, such as the calcium-sensing receptor, can be effectively downregulated (Schreckenberg et al. 2014). Alternatively, adenoviral constructs have been developed to transfect cardiomyocytes as performed by dominant negative constructs of Kir channel subunits in rabbit cardiomyocytes (Zobel et al. 2003). Control transfections can be performed with green fluorescence protein (GFP), allowing identification of transfected cardiomyocytes.

1.7 Ways to Challenge Isolated Cardiomyocytes with Ischemia and Reperfusion

A strong focus of cardiovascular research in general is the understanding of the response of cardiomyocytes to ischemic energy depletion and subsequent reoxygenation. It is generally accepted that targeting successfully this process in cardiomyocytes will protect the heart from the subsequent consequences of myocardial infarction. To analyze this pathophysiological scenario, it is required to establish experimental conditions that mimic exactly the conditions that occur during ischemic periods. During ischemia oxygen supply to cardiomyocytes is reduced or in the case of total vessel occlusion totally lost. This leads to a change in cellular metabolism from oxidative metabolism to non-oxidative metabolism such as glycolysis. As a result of this, metabolites of the glycolytic pathway such as lactate are released by the cells and accumulate in their surroundings. This leads to an increase in extracellular osmotic pressure and acidosis. Therefore, experimental setups that are aimed to mimic ischemic conditions do not only require hypoxia but also initiating slight acidosis. Standard protocols that consider ischemia use media that are made anoxic ($pO_2 < 1$ mmHg as indicated by polarographic oxygen sensors) and acidotic (pH 6.4) and have glucose deprivation (deoxy-glucose). Initially, oxygen in media was simply replaced with nitrogen by flushing the supernatant of the media. Oxygen depletion can be accelerated if media are autoclaved or heated. However, energy depletion in non-beating cells is a very slow process and normally increase in diastolic cell calcium due to energy-dependent loss of pump function of Na-K-ATPase and the subsequent rigor contraction cannot be monitored earlier than 30–40 min of simulated ischemia. In contrast to tissues in which large cardiomyocytes are packed in a dense way with little extracellular space, the situation under culture conditions is completely different. Here cells are far from each other, they do not beat and do not generate force, that means they have only a very low energy demand, and the extracellular space is large. Therefore, acidosis and accumulation of metabolites are also low. Cardiomyocytes can be exposed to ischemia in a much better way when the cells are overflowed by anoxic media rather than incubated with anoxic media. In this case, however, the continuous flow of media removes metabolites released by the cells.

Energy depletion of cardiomyocytes can easily be observed under a microscope because ATP depletion led to rigor contraction of cardiomyocytes. Typically, cardiomyocytes reduce their cell length by approximately 20 % due to rigor contraction. The effect of reperfusion can also be monitored in an easy way just by replacing the medium used to simulate ischemic conditions by normal oxygenated culture medium. Typically, the damage of cells, called reperfusion injury, can easily be monitored by hypercontraction of cells, leading to round cardiomyocytes with lot of blebs at the sarcolemmal membrane or by quantification of the release of enzymes into the media, such as lactate dehydrogenase (LDH), creatinine kinase (CK), and others. The release of cytoplasmatic proteins during reoxygenation of cultured cardiomyocytes is not simply due to hypercontraction but also due to increased osmotic fragility due to mechanical stress during ischemia and reoxygenation (Schlüter et al. 1996).

1.8 Ways to Quantify Cardiac Hypertrophy on the Cellular Level

Cardiomyocytes have a large ability to adapt to different levels of mechanical load and to change their cell shape and intracellular configuration. The pure meaning of the term hypertrophy is an increase in cell size. A stable cell size requires that the rate of protein synthesis is balanced by the rate of protein degradation. Cardiac hypertrophy requires an imbalance between protein synthesis and protein degradation favoring protein synthesis. This means that the efficiency of ribosomal activity is increased, whereas the half-life of proteins is prolonged. Overall, an increase in cell size should be the result of these processes. Rates of protein synthesis can be quantified by determination of the incorporation of radioactive-labeled amino acids into cell proteins. Normally this is achieved by either ^{3}H- or ^{14}C-labeled phenylalanine or lysine. It is important for the quantification of the rate of protein synthesis by radiolabeled amino acids into proteins that a stable concentration of these precursors is achieved. This can be done by long-term exposure of amino acids to cardiomyocytes. A small amount of radiolabeled amino acids is normally substituted by non-radiolabeled amino acids to minimize variation in the specific activity of the precursor pool responsible for protein synthesis. Under such conditions, the rate of incorporation of radioactive-labeled phenylalanine into acid-insoluble material was stable for 48 h (Schlüter and Piper 1992). Phenylalanine and lysine are normally used because these amino acids are incorporated into most of all proteins. Experiments revealed that the mean rate of protein synthesis can be increased under optimal conditions by approximately 20–25 %. Alternatively, radioactive-labeled amino acids can be added at certain time points for a small time period, i.e., after 12 h, 24 h, and 36 h for 60 min, and radioactive labeling of acid-insoluble material is quantified as described above. Under these conditions, time-dependent changes in the rate of protein synthesis can be monitored in contrast to the mean rate of protein synthesis. However, as the rate of protein synthesis is relatively constant, this has no specific advantages. To quantify the rate of protein degradation requires a pulse-chase experiment in which radioactive-labeled amino acids are added for an initial pulse and then replaced by non-labeled amino acids. The remaining amount of radioactive-labeled proteins is related to their half-life. Experiments have revealed that cardiomyocytes can increase protein half-time by 35–40 % (Millar et al. 1994).

Cell sizes can be monitored by microscopic techniques. It must be taken into account that terminally differentiated cardiomyocytes are cells that have an irregular cylindric shape. An approximation of their volume is given by the equitation: $l \times r^2 \times \pi$, where l is the length of the cells and r the radius of the cells normally measured at the middle of the length of rod-shaped cardiomyocytes. Using these techniques it has been shown that cardiomyocytes can increase their volume within 24 h by approximately 20–25 % when exposed to endothelin-1 (Hinrichs et al. 2011).

1.9 Species-Specific Aspects in Studies Related to the Biology of Cardiomyocytes

Cardiac function differs between various species, but species-dependent variations in electrophysiological aspects are more common than differences in mechanical aspects (Milani-Nejad and Janssen 2014). Heart rate differs by a factor of ten between humans and mice, but blood pressure is quite comparable. Furthermore, differences exist in the way of growth regulation as well, in particular in respect to the ability to perform cytokinesis and to respond to specific hypertrophic stimuli. Obviously, a big difference exists in the electrophysiological properties of cardiomyocytes of small rodents and higher vertebrates. The small size of rodent hearts leading to small stroke volumes (within the range of μl) requires a corresponding high beating frequency to compensate for the small stroke volume. It is easy to understand that this leads to strong differences in the duration of action potentials. Cardiomyocytes from mice lack a prominent plateau phase in their action potential in comparison to human cardiomyocytes. This is of course related to a different specification of ion channels between species. Therefore, an analysis of electrophysiological aspects in mice has a limited insight if the findings should be transferred to human aspects. On the one hand, blood pressure and therefore mechanical load of the heart is quite comparable between a broad spectrum of species such as rats and humans. Many aspects of growth regulation are quite similar on the cellular levels between cardiomyocytes from different species. On the other hand, neurohormonal regulation of cardiac hypertrophy, this means the outside-inside signaling, may differ significantly between species. For example, the pro-hypertrophic responsiveness of cardiomyocytes to catecholamines differs significantly between rat and mice, as in rat cardiomyocytes, α-adrenoceptor stimulation causes a strong hypertrophic response and β-adrenoceptor stimulation blocks α-adrenoceptor-dependent hypertrophy, while in mouse cardiomyocytes, β-adrenoceptor stimulation triggers hypertrophic responsiveness with no coupling of α-adrenoceptors to hypertrophy (Schlüter et al. 1992; Schäfer et al. 2001; Sabri et al. 2000).

1.10 Specific Aspects of Cardiomyocytes from Different Parts of the Heart

In most studies published on isolated cardiomyocytes, researchers used ventricular cardiomyocytes. This means that in these studies, differences between the right and left ventricle are ignored. As the majority of cells are derived from the thick left ventricle, data that arose from a ventricular preparation normally represent the behavior of left ventricular myocytes. The data can be improved if the thin right ventricle is separated from the left ventricle before the isolation of the cells. In this case, cardiomyocytes are completely derived from the left ventricle. The question whether cardiomyocytes from the left and right ventricle differ in physiological behavior has not yet been completely worked out. However, significant differences exist for the response of cardiomyocytes to similar electrical stimulation (Table 1.4). Furthermore, it was recently shown that right ventricular cardiomyocytes are unable to respond to oxidative stress in a similar way than cardiomyocytes from the left ventricle (Schreckenberg et al. 2015). The biological background of such a difference may lay in the ontogenetic origin.

Few studies have focused on the analysis of atrial myocytes. At first, due to the small amount of atrial tissue in relation to ventricular tissue, the amount of cells generated by isolation procedures is small. At second, protocols established to isolate myocytes from the atria differ from those to isolated cardiomyocytes from

Table 1.4 Comparison of load-free cell shortening between left (LV) and right (RV) ventricular cardiomyocytes

	N	L diast (μm)	TTP (ms)	Con-Vel (μm/s)	Rel-Vel (μm/s)	ΔL/L (%)
LV	71	99.19 ± 15.14	142 ± 13	139 ± 31	131 ± 69	7.38 ± 0.97
RV	42	98.61 ± 22.60	133 ± 17	127 ± 45	126 ± 62	6.48 ± 1.25
p-Value		0.885	0.008	0.113	0.677	0.018

Data are means ± S.D

L diast = diastolic cell length, *TTP* time to peak of contraction, *Con-Vel* contraction velocity, *Rel-Vel* relaxation velocity, *ΔL/L (%)* shortening amplitude normalized to diastolic cell length ventricles. Atrial myocytes have been analyzed preferentially with respect to electrophysiological characteristics because atrial fibrillation rather than atrial hypertrophy or impaired inotropy is the major pathophysiological complication under clinical conditions related to the atrium.

1.11 Conclusive Remarks and Future Perspectives

In conclusion, terminally differentiated cardiomyocytes represent a fascinating cell type that requires specific methods to analyze its functional and metabolic behavior. Most importantly, the specific differentiation of adult and terminally differentiated cardiomyocytes in comparison to fetal and neonatal cardiomyocytes requires the use of adult cardiomyocytes to understand the behavior of these cells under physiological and pathophysiological conditions. Future research must have a stronger focus on species differences, as the vast majority of our knowledge about cardiomyocytes depends on specific aspects of the heart of small rodents, and future research must consider more intensive differences between cardiomyocytes from various parts of the heart, such as right ventricle and atria.

References

Abdallah Y, Iraqi W, Said M, Kasseckert S, Shahzad T, Erdogan A, Neuhof C, Gündüz D, Schlüter KD, Piper HM, Reusch HP, Ladilov Y (2011) Interplay between Ca^{2+} cycling and mitochondrial permeability transition pores promotes reperfusion-induced injury of cardiac myocytes. J Cell Mol Med 15:2478–2485

Bergmann O, Bhardwaj RD, Bernard S, Zdunek S, Barnabe-Heider F, Walsh S, Zupicich J, Alkass K, Buchholz BA, Druik H, Jovinge S, Frisen J (2009) Evidence for cardiomyocytes renewal in humans. Science 324:98–102

Burrows MT (1910) The cultivation of tissues of the chick embryo outside the body. JAMA 55:2057–2058

Cavanaugh MW (1955) Pulsation, migration and division in dissociated chick embryo heart cells in vitro. J Exp Zool 128:573–589

Chorvatova A, Elzwiei F, Mateasik A, Chorvat D (2012) Effect of ouabain on metabolic oxidative state in living cardiomyocytes evaluated by time-resolved spectroscopy of endogenous NAD(P)H fluorescence. J Biomed Opt 17:101505

De Tombe PP, ter Keurs HE (1990) Force and velocity of sarcomere shortening in trabeculae from rat heart. Circ Res 66:1239–1254

Delbridge LM, Roos KP (1997) Optical methods to evaluate the contractile function of unloaded isolated cardiac myocytes. J Mol Cell Cardiol 29:11–25

Fujiwara K, Tanaka H, Mani H, Nakagami T, Takamatsu T (2008) Burst emergence of intracellular Ca^{2+} waves evokes arrhythmogenic oscillatory depolarization via the Na^{+}-Ca^{2+} exchanger. Circ Res 103:509–518

Hinrichs S, Heger J, Schreckenberg R, Wenzel S, Euler G, Arens C, Bader M, Rosenkranz S, Caglayan E, Schlüter KD (2011) Controlling cardiomyocyte length: the role of renin and PPAR-γ. Cardiovasc Res 89:344–352

Horackova M, Byczko Z (1997) Differences in the structural characteristics of adult guinea pig and rat cardiomyocytes during their adaptation and maintenance in long-term cultures: confocal microscopy study. Exp Cell Res 237:158–175

Jacobson SL (1977) Culture of spontaneously contracting myocardial cells from adult rats. Cell Struct Funct 2:1–9

Jacobson SL, Piper HM (1986) Cell cultures of adult cardiomyocytes as models of the myocardium. J Mol Cell Cardiol 18:661–678

Kent RL, Mann DL, Urabe Y, Hisano R, Hewett KW, Loughnane M, Cooper G 4th (1989) Contractile function of isolated feline cardiomyocytes in response to viscous loading. Am J Physiol 257:H1717–H1727

Kono T (1969) Roles of collagenase and other proteolytic enzymes in the dispersal of animal tissues. Biochim Biophys Acta 178:397–400

Kuramochi Y, Lim CC, Guo X, Colucci WS, Liao R, Sawyer DB (2003) Myocyte contractile activity modulates norepinephrine cytotoxicity and survival effects of neuregulin-1β. Am J Physiol Cell Physiol 286:C222–C229

Ladilov Y, Efe Ö, Schäfer C, Rother B, Kasseckert S, Abdallah Y, Meuter K, Schlüter KD, Piper HM (2003) Reoxygenation-induced rigor-type contracture. J Mol Cell Cardiol 35:1481–1490

Landgraf G, Gellerich FN, Wussling MH (2004) Inhibitors of SERCA and mitochondrial Ca-uniporter decrease velocity of calcium waves in rat cardiomyocytes. Mol Cell Biochem 256–257:379–386

Liu SJ (2013) Characterization of functional capacity of adult ventricular myocytes in long-term culture. Int J Cardiol 168:1923–1936

Ljubojevic S, Radulovic S, Leitinger G, Sedej S, Sacherer M, Holzer M, Winkler C, Pritz E, Mittler T, Schmidt A, Sereinigg M, Wakula P, Zissimopoulos S, Bisping E, Post H, Marsche G, Bossuyt J, Bers DM, Kockskämper J, Pieske B (2014) Early remodeling of perinuclear Ca^{2+} stores and nucleoplasmic Ca^{2+} signaling during the development of hypertrophy and heart failure. Circulation 130:244–255

Lopez JR, Jovanovic A, Terzic A (1995) Spontaneous calcium waves without contraction in cardiac myocytes. Biochem Biophys Res Commun 214:781–787

LoRusso S, Rhee D, Sanger JM, Sanger JW (1997) Premyofibrils in spreading adult cardiomyocytes in tissue culture: evidence for reexpression of the embryonic program for myofibrillogenesis in adult cells. Cell Motil Cytoskeleton 37:183–198

Louch WE, Bito V, Heinzel FR, Macianskiene R, Vanhaecke J, Flameng W, Mubagwa K, Sipido KR (2004) Reduced synchrony of Ca^{2+} release with loss of T-tubules-a comparison to Ca2+ release in human failing cardiomyocytes. Cardiovasc Res 62:63–73

Louch WE, Seehan KA, Wolska BM (2011) Methods in cardiomyocyte isolation, culture, and gene transfer. J Mol Cell Cardiol 51:288–298

Milani-Nejad N, Janssen PM (2014) Small and large animal models in cardiac contraction research: advantages and disadvantages. Pharmacol Ther 141:235–249

Millar BC, Schlüter KD, Zhou XJ, McDermott BJ, Piper HM (1994) Neuropeptide Y stimulates hypertrophy of adult ventricular cardiomyocytes. Am J Physiol 266:C1271–C1277

Muir AR (1965) Further observations of the cellular structure of cardiac muscle. J Anat 99:27–46

Posser BL, Ward CW, Lederer WJ (2011) X-ROS signaling, rapid mechano-chemo transduction in heart. Science 333:1440–1445

Powell T, Twist VW (1976) A rapid technique for the isolation and purification of adult cardiac muscle cells having respiratory control and a tolerance to calcium. Biochem Biophys Res Commun 72:327–333

Sabri A, Pak E, Alcott SA, Wilson BA, Steinberg SF (2000) Coupling function of endogenous alpha(1)- and beta-adrenergic receptors in mouse cardiomyocytes. Circ Res 86:1047–1053

Sakai S, Tokimasa T, Nohara M, Koga Y, Akasu T, Toshima H (1989) Electrophysiological properties of cultured dog myocytes obtained by endomyocardial biopsy. Circ Res 64:203–212

Schäfer M, Pönicke K, Heinroth-Hoffmann I, Brodde OE, Piper HM, Schlüter KD (2001) Beta-Adrenoceptor stimulation attenuates the hypertrophic effect of alpha-adrenoceptor stimulation in adult rat ventricular cardiomyocytes. J Am Coll Cardiol 37:300–307

Schlüter KD, Piper HM (1992) Trophic effects of catecholamines and parathyroid hormone on adult ventricular cardiomyocytes. Am J Physiol 263:H1739–H1746

Schlüter KD, Piper HM (2005) Isolation and culture of adult ventricular cardiomyocytes. In: Dhein S, Mohr FW, Delmar M (eds) Practical methods in cardiovascular research. Springer, Berlin/Heidelberg/New York, pp 557–567

Schlüter KD, Schreiber D (2005) Adult ventricular cardiomyocytes: isolation and culture. Methods Mol Biol 290:305–314

Schlüter KD, Jakob G, Ruiz-Meana M, Garcia-Doardo D, Piper HM (1996) Protection of reoxygenated cardiomyocytes against osmotic fragility by nitric oxide donors. Am J Physiol 271:H428–H434

Schreckenberg R, Dyukova E, Sitdikova G, Abdallah Y, Schlüter KD (2014) Mechanisms by which calcium receptor stimulation modifies electromechanical coupling in isolated ventricular cardiomyocytes. Pflugers Arch. doi:10.1007/s00424-014-1498-y

Schreckenberg R, Rebelo M, Deten A, Weber M, Rohrbach S, Pipicz M, Csonka C, Ferdinandy P, Schulz R, Schlüter KD (2015) Specific mechanisms underlying right heart failure: the missing upregulation of superoxide dismutase-2 and its decisive role in antioxidative defense. Antioxid Redox Signal. doi:10.1089/ars.2014.6139

Shyu KG, Chen JJ, Shih NL, Chang H, Wang DL, Lien WP, Liew CC (1995) Angiotensinogen gene expression is induced by cyclical mechanical stretch in cultured rat cardiomyocytes. Biochem Biophys Res Commun 211:241–248

Windisch H, Ahammer H, Schaffer P, Müller W, Platzer D (1995) Optical multisite monitoring of cell excitation phenomena in isolated cardiomyocytes. Pflugers Arch 430:508–518

Zacchigna S, Zentilin L, Giacca M (2014) Adeno-associated virus vectors as therapeutic and investigational tools in the cardiovascular system. Circ Res 114:1827–1846

Zobel C, Cho HC, Ngyuen TT, Pekhletski R, Diaz RJ, Wilson GJ, Backx PH (2003) Molecular dissection of the inward rectifier potassium current (IK1) in rabbit cardiomyocytes: evidence for heteromeric co-assembly of Kir2.1 and Kir2.2. J Physiol 550:365–372

2 Cardiomyocytes: Function and Regeneration

Marten Szibor

Abstract

In vertebrate embryonic development, the heart is the first organ to shape and start functioning. In higher developed vertebrates, it is organized by four chambers (two atria, two ventricles) that have distinct differentiation during embryonic development and either derives from the first heart field, the second heart field, or the cardiac neural crest. Interestingly, at least cardiomyocytes lose their ability to proliferate early after birth in higher developed vertebrates. Key factors triggering cardiac differentiation have been identified, such as Nkx2-5, GATA4, Tbx, Mef, Islet-1, and others. Although they are not strictly specific for cardiogenesis, they play a prominent role in cardiac differentiation. Recent advantages in the field have identified specific types of miRNA to orchester cardiac differentiation. The regulation of cardiac differentiation will be addressed in more detail in this chapter.

2.1 Introduction

In vertebrate embryonic development, the heart is the first organ to shape and start functioning. This is to match the rapidly increasing needs for oxygen and nutrients of the growing embryo as the diffusion distances become too long. In human this occurs roughly in the 3rd week of gestation. Although the work unit of the heart is the cardiomyocyte, the developing and adult heart consists of numerous highly specialized cells of different embryonic origins that are essential for its function. Despite the tremendous progress made in recent years, it is still not fully understood

M. Szibor
Institute of Biotechnology, FinMIT Cluster of Excellence, University of Helsinki, Viikinkaari 5, Helsinki FI-00790, Finland
e-mail: marten.szibor@helsinki.fi

K.-D. Schlüter (ed.), *Cardiomyocytes – Active Players in Cardiac Disease*,
DOI 10.1007/978-3-319-31251-4_2

how the different sources synchronize the assembly of their derivatives in to a fully functional cardiovascular system. Interestingly, the complex signaling cascades and the chronological sequence of events, i.e., migration, proliferation, and differentiation of cell populations, seem evolutionarily highly conserved. Because of its complexity, it is not surprising that any disturbance in this process can compromise the viability of the embryo, causing premature abortion or lead to cardiac abnormalities such as congenital heart disease (CHD), which accounts for almost one-third of all congenital abnormalities with a prevalence of up to 1 % in the newborn (reviewed in van der Linde et al. 2011).

Recent advances in the genetic manipulation of the mammalian genome made it possible to delineate the hierarchical order of events in a spatiotemporal resolution that was previously impossible to obtain. A better understanding of how cardiomyocytes mature and how the mammalian heart shapes is of importance beyond that of evolutionary and developmental interest. It helps to understand mechanisms leading to CHD as it is fundamental for the development of novel therapies in regenerative medicine, and it helps to explain the cardiac stress response under which adult cardiomyocytes show fetal characteristics in contractility and metabolism (reviewed in Nemer 2008; Ávila et al. 2014; Murry and Keller 2008; Szibor et al. 2014).

2.2 Shaping the Four-Chambered Heart

In the early phase of embryonic development, the single-layered blastula reorganizes itself by folding into the three-layered gastrula. The three embryonic layers are the ventral endoderm, the dorsal ectoderm, and the embedded mesoderm, each giving rise to different organs. During the gastrulation, heart field precursors migrate in an anterior-lateral direction from the primitive streak to form the bilateral cardiogenic mesoderm (reviewed in Harvey 2002; Moorman and Christoffels 2003; Buckingham et al. 2005; Rana et al. 2013; Paige et al. 2015). From there the precursors spread out further in a cranial and cranio-lateral direction to form the cardiac crescent, which is a curved sickle-shaped region underneath the head folds. The cardiac crescent is also referred to as the first heart field (FHF), and the progenitor population comprising it is the first to show the expression of myocardial transcription factors. After a rapid migration of progenitors from the cardiac crescent toward the midline and subsequent fusion, the linear heart tube is formed, which is elongated at both poles by a cell population originating from the second heart field (SHF). The heart tube becomes segmented and now consists of the truncus arteriosus, the bulbus cordis, the ventricle, an atrium, and the sinus venosus, from cranial to caudal. At this stage, the primary myocardium shows the generation and slow propagation of electrical waves, which results in a peristaltic movement (reviewed in Jensen et al. 2013). This cardiac peristaltic is the motor for the first directed blood flow from the venous to the arterial pole and is a crucial event for cardiogenesis and the embryonic development in general (reviewed in Paige et al. 2015). The heart tube starts then a complex transformation by looping rightward and a process called

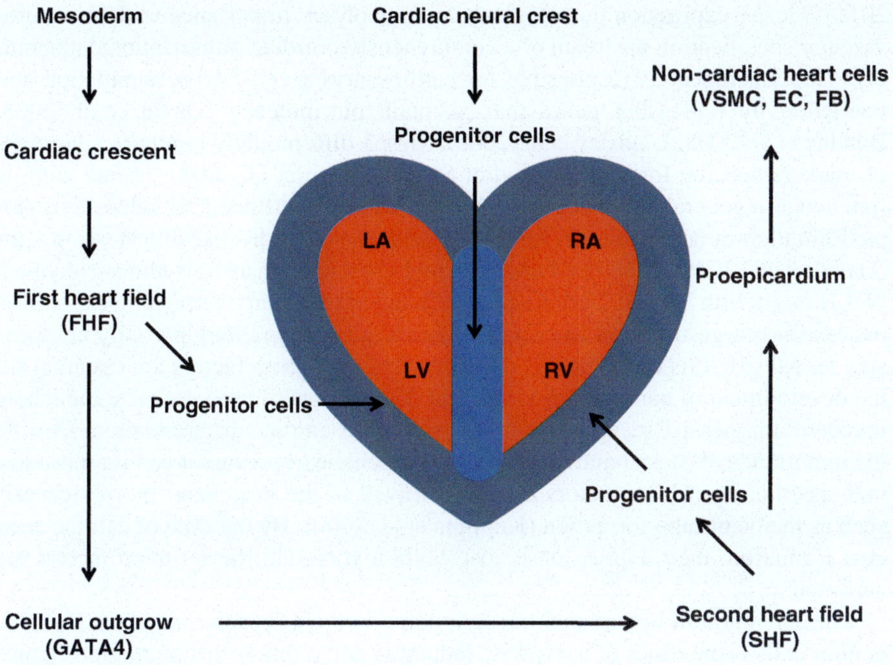

Fig. 2.1 Schematic overview about the different sources of cardiomyocytes that built the left ventricle, the right ventricle, and the septum

ballooning of the ventricle and atrium at the outer curvature, while the inner curvature remains tubular. The developing atria migrate cranially, while a series of septations and fusion complete the partition of the four chambers and the outflow tracts. At birth, this complex shaping of the heart with its selective proliferation, determination, differentiation, and partition is completed, and cardiomyocytes lose their proliferative capacity shortly thereafter (Fig. 2.1).

2.3 Determination of Cell Fate

Mesodermal precursors express the T-box transcription factor T (Brachyury), which has conserved roles in defining the midline of bilaterian organisms and mesoderm determination (Lartillot et al. 2002; Marcellini et al. 2003). After migration from the primitive streak, these precursors form bilateral cardiogenic regions. An important player in the specification of the cardiogenic region is the T-box transcription factor Eomesodermin (Eomes), which is essential as well for other embryonic events such as epithelial-to-mesenchymal transition and mesoderm migration. In principle, Eomes expression defines the early cardiac mesoderm and promotes formation of cardiovascular progenitors by activating another transcription factor, namely, the bHLH (basic-helix-loop-helix) mesoderm posterior 1 (Mesp1) (Costello et al.

2011). Mesp1 expression during gastrulation is of key importance to drive cardiovascular specification upstream of a comprehensive cardiac transcriptional machinery. This machinery is responsible for cardiovascular cell-fate determination and restriction by repressing genes that maintain pluripotency (David et al. 2008; Bondue et al. 2008; Lindsley et al. 2008). Mesp1 differentially regulates a large set of markers specific for the FHF and/or SHF (Bondue et al. 2008). In line with its function as a general coordinator, Mesp1 expression is sufficient to induce the ectopic formation of heart tissue in vertebrates and causes cardiovasculogenesis in vitro (David et al. 2008). Another family member, Mesp2, shares an almost identical bHLH motif with Mesp1. Both transcription factors have an overlapping expression pattern during gastrulation and somitogenesis, and Mesp2 may partially compensate for Mesp1 if its expression is deficient. Although both factors are essential for the development of cardiac mesoderm and descendants of both colonize the whole myocardium, Mesp1 seems to be more critical for cardiac differentiation. Despite the importance of Mesp1 and Mesp2 in the nascent heart precursor cell for mesodermal specification, these factors are not involved in the later heart morphogenesis such as the heart tube formation (Kitajima et al. 2000). By the time of cardiac crescent formation, their expression is lost while myocardial transcription factors are activated.

The activation of myocardial transcription factors in predetermined cardiac progenitor cells is the result of a stepwise induction and requires differentiation signals that arise from the neighboring tissue such as the primitive streak and the visceral embryonic endoderm (Arai et al. 1997). At this stage cardiac progenitors are exposed to a multitude of both activators and repressors of cardiogenesis. Proper development makes it necessary to integrate different, often opposing, signaling cascade in a delicate spatiotemporal manner. A fundamental and very complex signaling cascade is the so-called Wnt signaling pathway. Wnt proteins comprise a diverse family of secreted glycoproteins that by binding to Frizzled family receptors pass the signal to the recipient cell. Wnt signals regulate either gene transcription by activation of beta catenin, coined as canonical pathway, or the cytoskeleton or calcium homeostasis, the so-called noncanonical pathway (reviewed in Clevers and Nusse 2012; Willert and Nusse 2012). Expression of the noncanonical Wnt member Wnt-11 is required for cardiogenesis (Pandur et al. 2002). Also bone morphogenetic proteins (Bmp) are locally expressed and select cells that enter the cardiac myocyte lineage (Schultheiss et al. 1997; Andrée et al. 1998). Fibroblast growth factors (FGFs) cooperate with Bmp signaling, as well as Hedgehog proteins (Alsan and Schultheiss 2002; Dyer et al. 2001; Zhang et al. 2001). To restrain uncontrolled cardiogenesis, opposing signals arise such as canonical Wnt signals secreted from the neural tube (Tzahor and Lassar 2001) or the Bmp inhibitors, noggin and chordin (Schultheiss et al. 1997). But these repressing signals as well are opposed, specifically by secretion of canonical Wnt inhibitors from the endoderm such as dickkopf and the frizzled-related protein crescent (Marvin et al. 2001; Schneider and Mercola 2001). Taken together, at this early stage the cardiac progenitor cells are exposed to a multitude of signals and expression levels and exact timing of each signal is crucial for every step of cardiogenesis (Tzahor 2007) (Fig. 2.2).

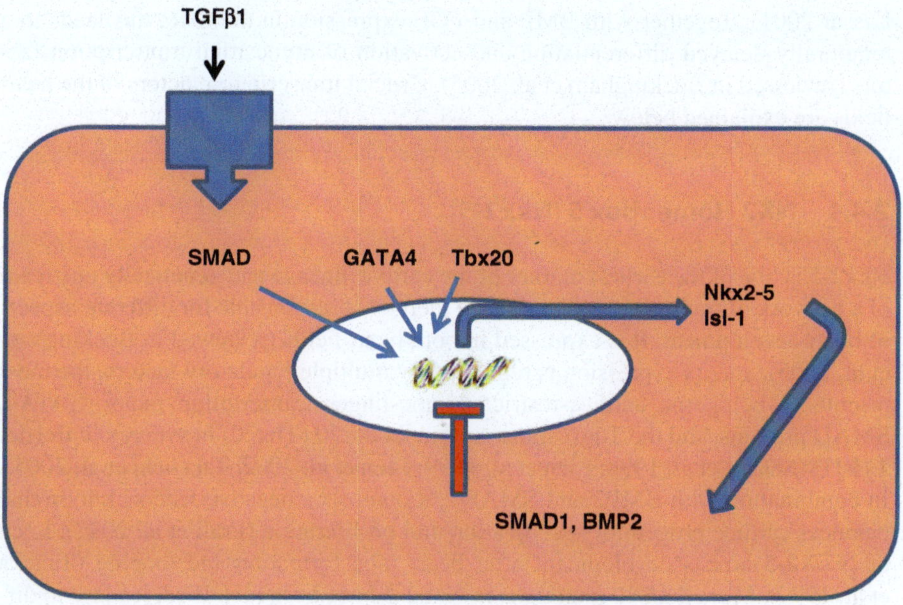

Fig. 2.2 Effect of TGFβ1 on the differentiation of cardiac progenitor cells into cardiomyocytes. Classical TGFβ pathways contribute to the induction of cardiac differentiation markers like Nkx and Isl-1

2.4 Marker Genes of the Two Heart Fields

At roughly week 2 of gestation in human, predetermined cardiac precursors spread out and join at the midline to form the cardiac crescent. In the cardiac crescent, myocardial differentiation is initiated and myocardial transcription factors are first detected. This region is often referred to as the first heart field (FHF) (reviewed in Buckingham et al. 2005). Cells from the FHF migrate and coalesce along the ventral midline to form the primitive heart tube. At this stage, approximately week 3 of gestation in human, the primitive heart tube is two layered, separated by extracellular matrix, and will exclusively give rise to the left ventricle. A second mesodermal heart field appears (SHF) having the same origin and being governed by similar cardiogenic signals. Progenitors from the SHF elongate both poles of the primitive heart tube and exclusively give rise to the outflow tract. Anatomically this means that the FHF contributes primarily to the left ventricle, while the SHF contributes primarily to the right ventricle and outflow tract. As a consequence, perturbations in the FHF will have a greater effect on the formation or function of the left ventricle, while perturbations in the SHF will primarily affect the outflow tract. In all other regions of the heart, each of the two heart fields may compensate for the other (reviewed in Buckingham et al. 2005). Interestingly, although the signals governing the heart fields are similar, the anatomical localization of the SHF is closer to the repressing signals such as the Wnt signal secreted from the neural tube (Tzahor and

Lassar 2001). Together with BMP and FGF expression in the SHF, this leads to a temporally delayed differentiation and activation of myocardial transcription factors (reviewed in Buckingham et al. 2005). Crucial transcription factors of the heart fields are explained below.

2.4.1 NK2 Homeobox 5 (Nkx2-5)

Nkx2-5 is one of the earliest marker of the cardiac lineage and genetically upstream of a network of transcription factors, and thereby is responsible for different aspects of heart development. It is expressed in both heart fields (reviewed in Buckingham et al. 2005), and its expression is regulated by multiple regulatory factors, the most prominent being the cardiac-restricted zinc-finger transcription factor GATA4, SMAD proteins, and the T-box transcription factor 20 (Tbx20) in synergy with Islet 1 (Isl1) (Searcy et al. 1998; Lien et al. 1999; Lien et al. 2002; Takeuchi et al. 2005). In combination with BMP2 and SMAD1, it controls a negative feedback loop that balances cardiac progenitor specification and proliferation (Prall et al. 2007). Lack of NKX2-5 arrests development after heart tube formation and looping (Tanaka et al. 1999). More than 40 mutations in Nkx2-5 have been identified, leading to different forms of CHD including defects in septations or the maturation of the conduction system with development of progressive cardiomyopathy or heart block (Schott et al. 1998; Pashmforoush et al. 2004; Reamon-Buettner and Borlak 2010).

2.4.2 GATA-Binding Protein 4 (GATA4)

GATA4 belongs to a family of six zinc-finger transcription factors containing a highly conserved DNA-binding domain and is one of the earliest transcription factors expressed in the cardiac progenitors. GATA4 expression is critical for cardiogenesis, differentiation, and morphogenesis and persists in the adult cardiomyocyte throughout life (reviewed in Molkentin 2000; Pikkarainen et al. 2004). Loss of GATA4 results in embryonic death as a result of severe developmental abnormalities such as a generalized disruption of the ventral body pattern and the complete lack of the primitive heart tube. Interestingly, the splanchnic mesoderm can nevertheless develop, and the progenitor cells differentiate further into primitive myocytes as judged by the expression of contractile proteins. Together, this data suggests that GATA4 is not essential for the specification of the cardiac cell lineages but instead plays a central role in the progenitor migration and/or the folding morphogenesis (Kuo et al. 1997; Molkentin et al. 1997). The effects of GATA4 on embryonic morphogenesis is dosage sensitive, such that lowered levels of the transcription factor cause a common atrioventricular canal, double-outlet right ventricle, and hypoplastic ventricular myocardium while surprisingly leaving the expression level of known target genes unaffected (Pu et al. 2004). GATA4 acts as co-regulator and forms multi-protein transcriptional complexes with numerous cardiac transcription factors, such as NKX2-5, TBX5, and the serum response factor (SRF), to control expression of a large set of downstream genes involved in differentiation and morphogenesis (Sepulveda et al. 1998; Searcy et al. 1998; Lien et al. 1999, 2002; Belaguli et al. 2000; Liberatore et al. 2002; Maitra et al. 2009). GATA4, together with Isl1, directly activates MEF2C transcription, jointly serving as the earliest transcriptional regulators controlling outflow tract and right ventricle development (Dodou et al. 2004). GATA4 also recruits MEF2 proteins to cooperatively activate target genes such as atrial natriuretic peptide (ANP), B-type natriuretic peptide, and to a lesser degree α-cardiac actin and α-myosin heavy chain (Dodou et al. 2004). Due to its versatile functions, mutations in GATA4 are a common cause in human CHD presenting as septal or endocardial cushion defects, right ventricular hypoplasia, and cardiomyopathy.

2.4.3 T-Box Transcription Factor 5 (Tbx5)

Tbx5 is a member of a conserved family of genes that share a common DNA-binding domain, the T-box. In vertebrate embryos, Tbx5 is expressed in the developing heart, forelimb, eye, and liver during morphogenesis and patterning (Liberatore et al. 2000). Although Tbx5 expression is uniform throughout the entire cardiac crescent, it seems that it favors an early onset of differentiation, which is a feature of the FHF lineage (Bruneau et al. 1999; Herrmann et al. 2011). With formation of the primitive heart tube, Tbx5 expression is gradually stronger in the posterior and weaker in the anterior segments and pronounced in the left ventricle but not in the right ventricle or outflow tract. The expression remains restricted to the atria and left ventricle, which is in line with its proposed function in the FHF (Bruneau et al. 1999). As transcription factor, TBX5 interacts with NKX2-5 and GATA4 transcription factors to synergistically regulate heart-specific genes in the nucleus and associates with NKX2-5 to promote cardiomyocyte differentiation (Hiroi et al. 2001; Bimber et al. 2007); it physically associates through its DNA-binding domains with MEF2C to synergistically activate the α-myosin heavy-chain promoter (Ghosh et al. 2009), and it acts in concert with TBX20 to control heart morphogenesis (Brown et al. 2005). Interestingly, while the nuclear function is well studied, TBX5 shuttles from nuclear to cytoplasmic sites where it forms a complex with the PDZ-LIM protein, LMP4, on the actin cytoskeleton. Although far from understood, the expression pattern suggests a specific role for TBX5 in epicardium formation (Bimber et al. 2007).

Based on this, it is not surprising that a nonsense mutation in Tbx5 causes a CHD, namely, the Holt-Oram syndrome, which presents with severe hypoplasia of posterior domains combined with defects in the atrial and occasionally ventricular septum and with impairments of the cardiac conduction system (Basson et al. 1997; Bruneau et al. 2001). In line with this is the finding that both ANP and connexin 40 (cx40) have been identified as TBX5 target genes, thereby offering a molecular explanation for the disease phenotype (Bruneau et al. 2001).

2.4.4 HAND Transcription Factors

The two related family members of the HAND proteins (heart and neural crest derivatives expressed transcripts 1 and 2) belong to the basic-helix-loop-helix (bHLH) family of transcription factors and are asymmetrically expressed in the developing heart. While both are present in cardiac progenitor cells, their expression is differentially regulated starting with the formation of the primitive heart tube and subsequent looping (reviewed in Srivastava 1999). In the primitive heart tube, HAND1 is primarily seen in regions forming the atria, the left ventricle, and the outflow tract. After the looping, it is more restricted to the outer curvature of the left ventricle and the outflow tract. In contrast, HAND2 expression is present throughout the primitive heart tube but restricted to the right ventricle and the outflow tract after looping. Nevertheless, the expression pattern of HAND1 and HAND2 is not completely exclusive, and experimental data are suggestive for a certain degree of functional redundancy (Firulli et al. 2010). For proper heart development, a sequential expression of both family members has been described as such that HAND1 expression is needed for the determination of epicardial precursors that are then dependent on HAND2 function (Barnes et al. 2011). The peculiar difference in its expression pattern is also reflected by developmental abnormalities caused by Hand mutations.

Lack of HAND1 is embryonically lethal at a very early stage and presents with abnormalities in placentation and cardiogenesis. The heart development is disturbed at the stage of looping and comes along with abnormal differentiation of the ventricular myocardium (Firulli et al. 1998, 2010; Riley et al. 1998). If HAND1 is present but its expression insufficient, heart morphogenesis takes place but the left ventricular myocardium is hypoplastic (Firulli et al. 2010). The molecular reason for that phenotype could be that HAND1 sits at a switch point between proliferation and differentiation. Indeed, genetic upregulation of HAND1 expression results in a significant extension of the heart tube and extraneous looping which has been attributed to an increase in proliferation of precursors, while lack of HAND1 shifts their phenotype toward differentiation. This suggests that HAND1 maintains an adequate pool of undifferentiated myocyte precursors to prevent cardiac hypoplasia (Risebro et al. 2006). In contrast to HAND1, HAND2 is also expressed in the SHF and crucial for embryonic viability. If genetically deleted, embryos die during early development and show signs of a hypoplastic right ventricle (Srivastava et al. 1997; Tsuchihashi et al. 2011). HAND2 is specifically required for proper patterning of the outflow tract, and loss of its expression in the neural crest results in misalignments and ventricular septum defects (Holler et al. 2010). If HAND2 expression is insufficient at later embryonic stages, it can cause a variety of CHDs (Tsuchihashi et al. 2011).

Like other transcription factors, HAND proteins can form multi-protein complexes to activate target genes. Unlike many others, HAND proteins form different homo- and heterodimer combinations. Moreover, HAND proteins revealed a broad dimerization profile with multiple other bHLH proteins reflecting the mechanisms by which they regulate transcription (Firulli et al. 2000). There is evidence that HAND2 and NKX2-5 act synergistically in development and target gene activation (Yamagishi et al. 2001; Thattaliyath et al. 2002a).

In the adult heart, HAND1 mutations and loss of function also cause hypoplastic walls and ventricular septum defects (Cheng et al. 2012), while supranormal expression levels lead to mild cardiac hypertrophy and a reduction in life expectancy. This is paralleled by a decrease of connexin 43 (cx43) protein in the cardiac intercalated discs which may increase the cardiac susceptibility to become arrhythmic (Breckenridge et al. 2009). A downregulation of Hand expression was observed in rodent hearts with signs of hypertrophy as well as in human cardiomyopathy. This might be related to the re-initiation of the so-called fetal gene program that allows the adult heart to cope with stressful situations (Thattaliyath et al. 2002b; Kubin et al. 2011). But despite the fact that Hand dysregulations are associated with disturbances in the patterning of the outflow tract and ventricular septum defects, HAND1 mutations are not related to the Tetralogy of Fallot (Esposito et al. 2011).

2.4.5 Myocyte-Specific Enhancer Factor 2 (MEF2C)

MEF2C is a family member of the MEF2 transcription factors also known as MADS box transcription enhancer factors (Molkentin et al. 1996). From *Drosophila* studies, it is known that it is crucial for mesoderm differentiation and subsequent fate determination (Nguyen et al. 1994). As one of the earliest markers of the cardiac muscle lineage, MEF2C is expressed in the progenitor cells of the SHF as a direct transcriptional target of ISL1 and GATA factors (Edmondson et al. 1994; Dodou et al. 2004). If MEF2C is lacking, the heart tube develops but does not loop and the outflow tract and right ventricle do not shape properly (Lin et al. 1997; Dodou et al. 2004). Furthermore, MEF2C null mutants develop severe vascular abnormalities, which are accompanied by a reduction of angiopoietin 1 and vascular endothelial growth factor (VEGF) expression in the myocardium (Lin et al. 1998; Bi et al. 1999). MEF2 transcription factors have two functional domains, the MEF2 domains which facilitate the binding of A/T-rich DNA sequence associated with muscle-specific genes and dimerization to form homo- or heterodimers, and the carboxyl terminus for transcriptional activation (Molkentin et al. 1996; Lin et al. 1997). A direct transcriptional target of MEF2C is Bop, which encodes a muscle-restricted protein and is essential for cardiac differentiation and morphogenesis. Phenotypic analyses suggest that MEF2C and BOP transcription factors may act synergistically to develop the SHF and its derivatives (Gottlieb et al. 2002; Phan et al. 2005). Dysregulation of MEF2C signaling (by mutations or transcriptional activation by stress-activated kinase p38) is involved in the development of CHD and cardiac hypertrophy. Surprisingly, MEF2C seems to be involved in both, in the essential physiological postnatal growth of cardiomyocytes as well as in the pathologic hypertrophy in response to stress (Kolodziejczyk et al. 1999; Munoz et al. 2009; Kodo et al. 2012) (Fig. 2.3).

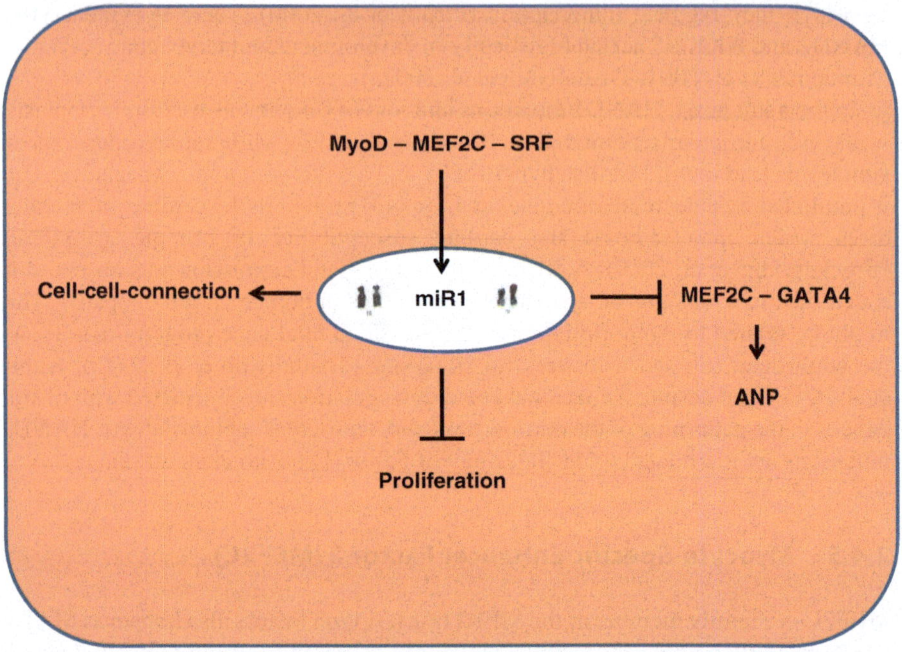

Fig. 2.3 Interaction of MEF2C as a cardiac differentiation factor on cardiogenesis via induction of miR1

2.4.6 T-Box Transcription Factor 1 (Tbx1)

Tbx1 is expressed in progenitor cells of the SHF and acts as a positive modulator of their proliferation while inhibiting their differentiation. As such it is a fundamental key regulator of the progenitor cell homeostasis. Indeed, loss- and gain-of-function experiments in the mouse showed premature or reduced progenitor cell differentiation, respectively, and hypoplasia of SHF-dependent segments, such as pharyngeal patterning and cardiovascular defects (Zhang et al. 2006; Chen et al. 2009, 2012). Quite interestingly, loss- and gain-of-function mutations in human can result in similar disease phenotypes (Zweier et al. 2007). The fibroblast growth factor (FGF) family members eight and ten are direct targets of TBX1 (Xu et al. 2004; Hu et al. 2004) as well as Wnt5a (Chen et al. 2012). Furthermore, TBX1 binds to and decreases the level of another master regulator, namely, SRF. With the appearance of differentiation markers, the Tbx1 expression is lost (Chen et al. 2009). Coherently, deletion of Tbx1 in myocytes or endothelial cells of the outflow tract does not cause a severe phenotype, while its deletion in NKX2-5 positive, early progenitor cells results in outflow tract abnormalities (Xu et al. 2004). Fate mapping revealed cell lineage heterogeneities within the SHF such that Tbx1 descendants populate most of the lateral and posterior outflow tract but not the anterior part. Descendants are found in the right ventricle flanking the ventricular septum, in the atria, and in the large vessels in discrete zones giving rise to cardiomyocytes, smooth muscle cells, and endothelial cells (Huynh et al. 2007; Chen et al. 2009). It is known that Tbx1 missense mutations or the frequent micro-deletions of chromosome 22q11 are responsible for inherited cardiovascular defects such as the DiGeorge or velo-cardio-facial syndrome (DGS/VCFS). This disease is seen with an incidence of 1 in 4000–5000 live births. Tbx1 null mice phenocopy to a great extent the DGS/VCF syndrome, which is indicative for TBX1 being central in the disease etiology (Merscher et al. 2001; Jerome and Papaioannou 2001; Lindsay et al. 2001; Yagi et al. 2003; Zweier et al. 2007).

2.4.7 Islet 1 (Isl1)

ISL1 was originally described as a LIM homeodomain transcription factor that binds to the enhancer region of the insulin gene (Tanizawa et al. 1994). Later its role in heart development became evident. It is transiently expressed in undifferentiated progenitor cells of the SHF prior to their differentiation (Cai et al. 2003). It was long thought that ISL1 represents a specific marker for the SHF but it is now known that its expression is also found in the neural crest (Engleka et al. 2012) and in progenitors residing in the postnatal heart (Laugwitz et al. 2005). Lack of Isl1 expression in the SHF causes disturbances in the formation of the outflow tract, the right ventricle, and parts of the atria (Cai et al. 2003). ISL1 presence is crucial for progenitor cell proliferation, survival, and migration. One of the first factors identified to directly regulate Isl1 expression was β-catenin (Lin et al. 2007). β-catenin signaling is responsible for the progenitor pool expansion and is balanced by Notch1 signaling, which is driving its differentiation (Kwon et al. 2009). Furthermore, the LIM protein Ajuba is restricting the SHF progenitor pool by binding ISL1, repressing its transcriptional activity, and is required for the transcriptional auto-repression of Isl1 (Witzel et al. 2012). Also, Isl1 is a direct transcriptional target of forkhead transcription factors (Kang et al. 2009). ISL1 directly regulates the transcriptional expression of FGF10 during heart formation (Golzio et al. 2012), and inversely, FGF and BMP growth factors are downregulated if ISL1 is lacking (Cai et al. 2003). Since ISL1 positive progenitors have the potential to differentiate into cardiomyocytes, smooth muscle cells, and possibly endothelial cells, they might gain importance for future approaches of genetic heart repair (Laugwitz et al. 2005; Bu et al. 2009; Kwon et al. 2009).

2.4.8 Forkhead Box H1 (Foxh1)

FOXH1 is a DNA-binding transcription factor in the TGF-SMAD signaling pathway that is required for successful gastrulation, head formation, and development of the SHF (Watanabe et al. 2002; Both et al. 2004). Its expression declines as development progresses (Weisberg et al. 1998). As expected for a SHF transcription factor, Foxh1 mutants form a primitive heart tube but lack the right ventricle and parts of the outflow tract (Both et al. 2004). The phenotype that Foxh1 mutants develop resembles to a great extent the phenotype of MEF2C mutants, which is a direct target gene of FOXH1 (Both et al. 2004). In response to activin/TGFβ signals, FOXH1 associates with SMAD proteins to mediate SMAD-dependent activation of TGF response element in target genes, and it physically and functionally interacts with NKX2-5. In the FOXH1/SMAD transcription complexes, FOXH1 mediates the DNA binding (Chen et al. 1997; Weisberg et al. 1998; Both et al. 2004).

2.4.9 T-Box Transcription Factor 20 (Tbx20)

TBX20 is a member of the T-box family of transcriptional regulators, which shows evolutionary conserved expression in the developing heart. At least four different Tbx20 isoforms exist, generated by alternative splicing, which can work as strong transactivators or transrepressors (Stennard et al. 2003). Tbx20 is expressed in the cardiac crescent as well as in the SHF and is required for the progression from the linear heart tube to a multi-chambered heart (Singh et al. 2005). Nevertheless, lack of TBX20 causes malformation of the SHF derived heart such as hypoplasia of the outflow tract and the right ventricle, which is paralleled by a decrease in expression of transcription factors that are specifically important for the formation of the SHF, i.e., Nkx2-5 and MEF2C (Takeuchi et al. 2005). Furthermore, TBX20 acts upstream of Wnt signaling and is central in endocardial cushion formation and valve remodeling during cardiogenesis (Cai et al. 2013).

TBX20 and other members of the T-box family are co-expressed in the developing heart and can act synergistically or repressive. TBX5 physically interacts and functions synergistically with TBX20. Consequently, the lack of either factor causes very similar defects in heart development (Brown et al. 2005). In contrast, the expression of Tbx2 is increased in Tbx20 null mice as it is expressed throughout the developing heart. Normally, the expression of Tbx2 is restricted to the outflow tract and atrioventricular canal. The lack of Tbx2 repression by TBX20 accounts for most of the cardiac phenotype, showing that coordination of gene expression by TBX20 is hierarchical and crucial for cardiogenesis (Harrelson et al. 2004; Cai et al. 2005; Stennard et al. 2005; Singh et al. 2005). TBX20 also interacts with other cardiac transcription factors such as NKX2-5, GATA4 and GATA5, and acts synergistically with ISL1 and GATA4 to activate both the MEF2C and Nkx2-5 enhancers (Stennard et al. 2003; Takeuchi et al. 2005) and functions as an enhancer for ISL1 for regulating the transcription of FGF10 (Golzio et al. 2012).

The function of TBX20 is dose dependent, such that a decrease in TBX20 level causes the development of a persistent truncus arteriosus, a hypoplastic right ventricle, and valve dysformation (Takeuchi et al. 2005; Cai et al. 2013). Indeed, Tbx20 gene polymorphism has been associated with the development of CHD (Chen et al. 2015). But also in the adult heart, TBX20 has key roles in maintenance and cardiac function. If TBX20 is lacking (or mutated) in cardiomyocytes, the heart rapidly develops contractile dysfunction and cardiomyopathies leading to premature death (Stennard et al. 2005; Sakabe et al. 2012).

2.4.10 Fibroblast Growth Factors 8 and 10 (FGF8 and FGF10)

Fgf8 and Fgf10 are expressed in the anterior part of the SHF. Cardiac progenitors from that region specifically contribute to the arterial pole (Kelly et al. 2001; Watanabe et al. 2010). Interestingly, lack of FGF8 and/or 10 gives rise to distinct morphological defects although their signaling functionally overlaps during development. FGF8/10 expression levels affect cardiogenesis in a dose-sensitive manner as shown for the formation of the right ventricle and outflow tract (Watanabe et al. 2010). Increasing levels of FGF8 raise cardiac cell number and inhibit forelimb formation (Sorrell and Waxman 2011). Furthermore, FGF8 has chemotactic effects on cardiac neural crest cells thereby directing their migration into the pharyngeal arches (Abu-Issa et al. 2002; Sato et al. 2011). FGF10 is also needed for limb development and lung formation (Min et al. 1998; Sekine et al. 1999) and, in conjugation with FGFR2b, is essential for cardiac fibroblast migration into the developing myocardium. Lack of FGF10 results in a decrease in cell immigration and myocardial proliferation and as a consequence a thinned cardiac wall (Vega-Hernández et al. 2011). Taken together, FGF signaling is primarily required for progenitor migration, proliferation, and normal cardiogenesis and not for early cell specification (Urness et al. 2011; Rochais et al. 2014).

Deficiencies in FGF signaling are responsible for the development of CHD. FGF8 deficiency phenocopies the human chromosome 22q11 micro-deletion, which is known as DiGeorge (DGS) and the velo-cardio-facial syndrome (VCFS) (Frank et al. 2002). Recently, it was described that Six1 and Eya1 genetically interact with its downstream effector FGF8 as well as TBX1 and together are crucial for the mammalian cardio-cranio-facial morphogenesis (Guo et al. 2011). Therapeutically it is interesting that FGF10 expression is sufficient to promote cardiomyocyte differentiation in embryonic or induced pluripotent stem cells and cell-cycle reentry in adult cardiomyocyte which gives it great potential for innovative therapies (Chan et al. 2010; Rochais et al. 2014).

2.5 The Proepicardium and Descendants

The developing heart tube recruits its cells from the FHF and SHF and consists of two layers, namely, the endocardium and the myocardium. The adult vertebrate heart, however, needs an outer lining that demarcates the organ and allows proper contraction, i.e., the epicardium, and it needs epicardium-derived cells (EPDCs) such as coronary vascular smooth muscle and endothelial cells, perivascular fibroblasts and myocardial fibroblasts. These essential components of the heart originate from a transient extra-cardiac embryonic tissue, the so-called proepicardium (Mikawa and Gourdie 1996; Dettman et al. 1998), which arises during the heart looping as an outgrowth from the right cardiac inflow segment at approximately day 21 of human gestation. The proepicardium disappears by the end of week five of human gestation. Interestingly, lower organisms which only have a tubular heart lack the epicardium (reviewed in Männer et al. 2001; Maya-Ramos et al. 2013).

Considering its origin, it is not surprising that progenitors expressing NKX2-5 and ISL1 contribute to the proepicardium. The actual fate of the common precursors toward epicardial or myocardial lineage, however, is determined by cross talk between FGF and BMP in which FGF signaling is dominant (van Wijk et al. 2009). In this, FGF signaling prevents apoptosis, maintains proliferation, and promotes villous outgrowth of the proepicardium without affecting the induction or maintenance of proepicardium-specific marker gene expression (Torlopp et al. 2010). FGF signaling arises from the proepicardium and stimulates differentiation into the epicardial lineage, while BMP signaling arises from the myocardium and stimulates differentiation into the cardiomyocyte lineage (Kruithof et al. 2006; van Wijk et al. 2009). Although debated, it seems possible that EPDCs can also differentiate into the cardiomyocytes lineage (Cai et al. 2008; Zhou et al. 2008a, Christoffels et al. 2009).

Initially, two independent lineage markers have been described for progenitors of the proepicardium, namely, Tbx18 and Wilms' tumor 1 (Wt1) (Kraus et al. 2001; Cai et al. 2008; Zhou et al. 2008b). TBX18 has been shown to have a role in the formation of the venous pole of the developing heart as well as the sinoatrial node. Interestingly, TBX18-expressing progenitors seem to descend from cells previously being negative for NKX2–5 (Christoffels et al. 2006; Wiese et al. 2009). WT1 is a zinc-finger transcription factor that is expressed in the proepicardium, the epicardium, and the subepicardial mesenchyme. It is required for the maintenance of progenitors in the undifferentiated state by repressing the epithelial phenotype in epicardial cells (Martínez-Estrada et al. 2010). Lack of WT1 causes disturbances in the epicardial layer, absence of the subepicardial mesenchyme, defects in cardiac vascularization and shows pericardial bleeding and embryonic death (Moore et al. 1999; Wagner et al. 2005). By identifying the two markers, Scleraxis and Semaphorin3D, a hitherto unrecognized subpopulation of progenitors from the proepicardium was described. Despite being morphologically homogeneous, the proepicardial progenitor pool seems to be heterogeneous. Scleraxis and Semaphorin3D specifically mark a distinct sub-compartment within the proepicardium that gives rise to coronary vascular endothelium and to the early sinus venosus and cardiac endocardium (Katz et al. 2012).

To ultimately form the epicardium, proepicardial cells proliferate and migrate onto the entire surface of the heart and form a single layer of cells attached directly to the myocard. The cell adhesion event is facilitated by the vascular cell adhesion molecule (VCAM-1) and α4 integrin. It has been shown that α4 integrin is expressed in the epicardium, whereas VCAM-1 is expressed in the myocardium. Lack of either factor results in the loss of epicardium formation (Kwee et al. 1995; Yang et al. 1995). The epicardium is involved in reciprocal signaling with the surrounding tissue (reviewed in Sucov et al. 2009; Pérez Pomares and de la Pompa 2011), and it forms mesenchymal cells through epithelial-to-mesenchymal transition (EMT). EMT is a critical step and by far not exclusive for cardiogenesis (reviewed in Lim and Thiery 2012). The transformed mesenchymal cells then contribute to the different lineages of the developing heart as discussed above. WT1 is essential for repression of the epithelial phenotype in epicardial cells through direct regulation of target genes, particularly components of the Wnt/β-catenin signaling pathway (Martínez-Estrada et al. 2010; Gise et al. 2011). Of note is that lack of β-catenin in epicardium-derived progenitor cells does not impair the formation of the proepicardium, the migration of proepicardial cells to the heart, or the formation of the primitive epicardium. Instead it disturbs subsequent processes such as the expansion of the subepicardial space, the invasion of the myocardium, and differentiation of epicardium-derived mesenchymal cells into coronary smooth muscle cells (Zamora et al. 2007). In epicardial cells, RALDH2, an enzyme involved in retinoic acid synthesis, is a direct transcriptional target of WT1. Lack of WT1 also regulates morphogen receptors involved in the proliferation, migration, and differentiation of epicardial and EPDCs (Guadix et al. 2011). Retinoic acid signaling stimulates myocardial expansion indirectly via induction of hepatic erythropoietin which in turn activates epicardial insulin-like growth factor (IGF2) (Brade et al. 2011). Epicardial IGF signaling is required for normal cardiogenesis by directing ventricular cardiomyocyte proliferation (Li et al. 2011). Extensive cross talks between retinoic acid signaling and other signaling cascades have been described. Studies on epicardial retinoic acid receptors revealed also a direct retinoid-dependent Wnt signaling pathway that is required for myocardial growth and coronary artery formation (Merki et al. 2005). Furthermore, retinoic acid signaling induces epicardial FGF signals that are essential for cardiomyoblast proliferation and coronary vascular development (Lavine et al. 2006). Another signaling cascade involved is the Notch pathway which modulates muscularization by affecting cell differentiation in the proepicardium and pericardial mesoderm as well as smooth muscle differentiation, which is important for the formation of a functional coronary system (del Monte et al. 2011; Grieskamp et al. 2011).

2.6 The Cardiac Neural Crest

2.6.1 Induction and Migration

Multipotent neural crest cells delineate from the dorsal neural tube. A subregion is called cardiac neural crest (CNC) arising between the mid-otic placode and the posterior border of the third somite. The CNC contributes progenitor cells (CNCCs), the third population needed for cardiogenesis. Functionally, the CNC is a blend between the cranial and trunk crest because like the cranial crest, it generates ectomesenchyme, while like the trunk crest, it lacks the ability to regenerate (reviewed in Hutson and Kirby 2007; Scholl and Kirby 2009). After EMT and delamination, CNCCs migrate to the caudal three pharyngeal arches and from there to the cardiac outflow tract where ectomesenchyme from the CNC forms the aorticopulmonary septum dividing the arterial outlet of the developing heart (Kirby et al. 1983). CNCCs add to the atrioventricular cushion that forms the heart valves, provides the developing heart with parasympathetic innervation, and plays a role in the maturation of the cardiac conduction system (Poelmann and Gittenberger-de Groot 1999; Waldo et al. 1999; Nakamura et al. 2006; Gurjarpadhye et al. 2007). If CNCCs are lacking, cardiogenesis is severely disturbed showing abnormal patterning of the great arteries, disturbances in the formation of the cardiac outflow tract and septation, and myocardial dysfunction (reviewed in Hutson and Kirby 2007; Scholl and Kirby 2009). CNCCs require an extensive number of signals for induction, migration, proliferation, differentiation, and survival including Wnt/β-catenin, FGFs, BMPs, and retinoic acid (reviewed in Aybar and Mayor 2002; Meulemans and Bronner-Fraser 2004; Scholl and Kirby 2009). Cell migration and the multitude of secretory signals make it difficult to study the regulation and role of CNCCs since relative levels of synergistic or opposing signals in the particular location define the progenitor's response. An important role in the induction process has the canonical Wnt signaling, while the noncanonical Wnt signaling directs cell migration (de Calisto et al. 2005). Critical transcription factors such as SOX9 and SLUG/SNAIL trigger neural crest cells to undergo EMT and to delaminate, while FOXD3 regulates the expression of cell-cell adhesion molecules that are important for the progenitor's cell migration (Cheung et al. 2005). Slug expression is critically regulated by BMP signaling (Sakai et al. 2005) of which gradients are built up by the expression of the BMP inhibitor, noggin (Sela-Donenfeld and Kalcheim 1999; Tribulo et al. 2003). For the early event of delamination, the cell status is of great importance as neural crest cells delaminate synchronously only in the S phase of the cell cycle (Burstyn-Cohen and Kalcheim 2002).

For their targeted migration, CNCCs follow a definite pattern controlled by numerous signals. Important for the understanding of their fate is the finding that neural crest cells are not irreversibly committed before delamination; they rather have a developmental plasticity (Trainor et al. 2002). Conserved in different species, neural crest cells express a combination of Eph receptors and ephrins which are important mediators of cell-contact-dependent interactions. By complementary expression of EphA4/EphB1 receptors and ephrin-B2 intermingling of different arches, neural crest is restricted and the migration of CNCCs is directed (Smith et al. 1997; Mellott and Burke 2008). On their journey, CNCCs are exposed to different signals that determine their fate in a subtle spatiotemporal manner (Trainor et al. 2002). FGF signaling is an important modulator for migration and is itself modulated by CNCCs such that lack of FGF8 signaling causes cell death in the neural crest but ablation of the neural crest causes an increase in the FGF8 signal. One function for FGF8 seems to be the induction of FGF2 which itself is chemotactic for mesencephalic neural crest cell migration (Kubota and Ito 2000; Hutson et al. 2006). Important modulators of CNCC migration into the cardiac outflow tract are semaphorins (Sema) which are secreted or transmembrane glycoproteins that act together with their receptors, plexins, in many processes (reviewed in Kruger et al. 2005). Correct navigation of CNCCs is regulated by a network of repulsive and attractive signals. For recognition migrating CNCCs express PLEXIN-A2, PLEXIN-D1, and neuropilin. PLEXIN-A2 recognizes the membrane-bound ligands, SEMA6A and SEMA6B, which are expressed in the dorsal neural tube and the lateral pharyngeal arch mesenchyme. Since they repel CNCCs, this region is recognized as a migrating route, while SEMA3C is a CNCC attractant expressed in the target region, the cardiac outflow tract. If semaphorin signaling is disrupted, proper migration is impaired causing cardiovascular malformations such as a persistent truncus arteriosus and aortic arch interruption (Eickholt et al. 1999; Brown et al. 2001; Feiner et al. 2001; Toyofuku et al. 2008).

2.6.2 Signal Conversion in the Target Region

Naturally, the signaling cascades during migration and in the targeting region are partially overlapping. Nevertheless, some signaling cascades predominantly affect the post-migratory CNCC development. This has been shown for endothelin (ET) signaling which explicitly affects the post-migratory and cell-autonomous process of aortic arch patterning but not migration of CNCCs. Endothelin 1 (ET1) is activated through cleavage by ECE-1 and mediates its effects via its receptors ET_A and ET_B. The ET_A receptor is expressed in ectomesenchymal cells of the pharyngeal arches and cardiac outflow tract, whereas its ligand ET1 is expressed by arch epithelium, paraxial mesoderm-derived arch core, and the arch vessel endothelium. It was shown by pharmacological and genetic means that disruption of this signaling disturbs the morphogenesis of the face and heart leading to a clinical phenocopy of the DiGeorge syndrome with defects in cardiac outflow and great vessel structures (Kempf et al. 1998; Clouthier et al. 1998, 2003; Yanagisawa et al. 1998).

A very similar observation was made for the role of platelet-derived growth factor receptors (PDGFR) of which two are expressed on CNCCs, namely, PDGFRα and PDGFRβ. If not properly expressed in CNCCs, either receptor causes a distinct phenotype. Lack of PDGFRα results in craniofacial and aortic arch development defects, while lack of PDGFRβ causes a ventricular septal defect. Obviously, the receptors are functionally redundant since lack of both results in a more severe phenotype with higher penetrance such as CNCC-related heart abnormalities including persistent truncus arteriosus and a ventricular septal defect. As for the ET signaling, CNCC delamination, migration, proliferation, and differentiation seem not to be affected (Richarte et al. 2007).

The transforming growth factor β (TGFβ) superfamily of secreted, hormonally active, polypeptides controls cellular functions from embryonic development to tissue homeostasis and has been implicated in NCC development. Ligands of that family, such as TGFβs, BMPs, growth and differentiation factors (GDFs), activins, and nodal, bind to a type II receptor dimer, which recruits a type I receptor in a sequential or cooperative manner, typically forming a hetero-tetrameric complex. Upon complexing, the type I receptor is phosphorylated and, thus, activated which then phosphorylates intracellular SMAD proteins (R-SMAD). R-SMAD together with a Co-SMAD (SMAD4) translocates to the nucleus where it modulates the gene expression of specific target genes (Massagué 1998; Derynck et al. 1998; Massagué and Chen 2000). Defects in development have been described for all levels of the complex TGFβ signaling network. Interestingly, despite the signal redundancy, different TGFβ signals can act synergistically or highly specific during cardiogenesis (Sanford et al. 1997; Kim et al. 2001). BMP signaling is needed for cushion formation and septation in the developing heart which are the precursors of valves and septa. It has been shown that BMP6 and BMP7 expression is overlapping in adjacent sites such as the cardiac cushions. Expression of neither BMP seems to be required for cardiogenesis, while lack of both causes a marked delay in the formation of endocardial cushions in the outflow tract and subsequently defects in the valve morphogenesis as well as chamber septation, altogether leading to embryonic death due to heart failure (Kim et al. 2001). At the level of type II receptors, it has been described that a lack of TGFBR2 in the neural crest causes two distinct cardiovascular malformations, which are persistent truncus arteriosus and interrupted aortic arch. Interestingly, pharyngeal organ defects as described for the DiGeorge syndrome are not present, which is indicative for a perturbation of CNCCs at later, post-migratory stages (Choudhary et al. 2006). The TGFβ type I receptor activin receptor-like kinase 5 (ALK5) is involved in GDF and TGFβ signaling. Although not needed for CNCC migration, failure of ALK5 signaling causes severe cardiovascular and pharyngeal defects, such as inappropriate remodeling of pharyngeal arch arteries, abnormal aortic sac development, failure in pharyngeal organ migration, and persistent truncus arteriosus (Wang et al. 2006). The type I receptors, ALK2 and ALK3, are important for BMP signaling. Failure in signaling of either receptor causes cardiovascular defects such as persistent truncus arteriosus and abnormal maturation of the aortic arch if lacking ALK2, or shortened cardiac outflow tract and defective septation if lacking ALK3. However, while ALK3 is not needed for many aspects of early neural crest biology, ALK2 is required for CNCC migration to the outflow tract (Stottmann et al. 2004; Kaartinen et al. 2004). Taken together, TGFβ/SMAD signaling has multiple cell-autonomous and non-cell-autonomous effects on the survival of CNCCs and on coordinating the movement of multiple cell lineages in the positioning and the remodeling of the outflow tract (Jia et al. 2007).

2.7 The Need for Polarities and Asymmetries in Shaping the Heart

During development at the stage of the primitive heart tube, a dorsal-ventral polarity becomes necessary. This is needed for the heart tube to rotate rightward, making the ventral surface of the heart tube the ballooning outer curvature, while the dorsal surface of the heart tube becomes the inner curvature. The outer curvature is the side of actual growth because it shapes in a segmented fashion the individual ventricles of the future four-chambered heart. The inner curvature is a place of active remodeling, migration, alignment, and separation of the two heart poles to ultimately split the two blood circuits (reviewed in Christoffels et al. 2004; Srivastava 2006). A cocktail of transcription factors regulate the expression and repression of factors that define the unique identity of outer or inner curvature cells. Prominent examples are FOXH1 and the TBX proteins, TBX20 and TBX2. If FOXH1 is lacking, the expression of typical outer curvature markers such as ANP is disturbed, and heart development phenocopies the lack of MEF2C which is a direct target of FOXH1 (Both et al. 2004). The inner curvature is marked by TBX2 expression which is a repressor of genes that are typically found on the outer curvature (Harrelson et al. 2004). As pointed out before, TBX20 is an upstream repressor of TBX2 thereby allowing cardiac chamber formation in the outer curvature (Cai et al. 2005; Stennard et al. 2005; Singh et al. 2005). Another important asymmetry signal arises early during development at the node and is transduced to the Pitx2 gene. PITX2 is a bicoid-related homeodomain transcription factor that is critical for the determination of the left-right asymmetry of the developing heart, thereby being involved in patterning of the aortic arches, outflow tract, and atrioventricular valves and cushions (Liu et al. 2002; Franco and Campione 2003; Ai et al. 2006). This PITX2-regulated asymmetric pathway is independent of cardiac looping and is active in the myocardium, the CNCCs (target of canonical Wnt signaling) (Kioussi et al. 2002), and the SHF (target of TBX1) (Nowotschin et al. 2006). Disturbance of PITX2 expression correlates with the development of CHD. Ectopically expression in the developing myocardium causes the formation of a double-outlet right ventricle. In contrast, lack of PITX2 causes a number of severe cardiovascular defects which include atrial isomerism, double inlet left ventricle, transposition of the great arteries, persistent truncus arteriosus, and abnormal aortic arch remodeling (Franco and Campione 2003).

2.8 Cardiogenic Control by Noncoding RNAs

The regulatory control of gene expression as described above relies not only on the expression and interplay of transcription factors. It is also driven by the stability and fate of messenger RNAs (mRNAs) as well as the expression of epigenetic regulators, which add a new complexity to all developmental stages. This is facilitated by micro-RNAs (miRNAs) and long noncoding RNAs (lncRNAs), and its impact on embryogenesis, development, and aging is currently discussed (Boettger and Braun 2012; Devaux et al. 2015).

2.8.1 miRNAs and Cell Fate Decisions

The noncoding miRNAs are genomically encoded by 21–24 nucleotides that target mRNAs for translational inhibition or degradation. A single miRNA can target a number of mRNAs and, vice versa, individual mRNAs can also be targeted by different miRNAs. In the genome, miRNAs can occur as intergenic or intronic genes encoded as either singular miRNAs or in clusters. Most miRNAs are transcribed as pre-miRNAs by RNA polymerase II, often as a polycistronic primary transcript. miRNAs are 5′ capped and 3′ polyadenylated, and as for other genes, expression of miRNAs is tightly controlled by various transcriptional and epigenetic regulators. Maturation of pre-miRNAs needs sequential processing steps, a nuclear export, and miRNA integration into a RNA-induced silencing complex (RISC). The first processing step occurs in the nucleus in a complex containing factors such as Drosha and DGCR8. The processed pre-miRNA is then, with the help of RanGTP and

Exportin5, exported to the cytosol where miRNA hairpin structure is processed in a complex containing RNase III, Dicer. Only the mature miRNA is integrated into the RISC where it binds a complementary RNA strand to facilitate gene silencing. Regulators are active at every level of miRNA maturation, which makes it a complex network of signals to fine-tune the dosage of numerous key regulatory proteins (reviewed in Boettger and Braun 2012).

The expression of miRNAs is required for proper development. Lack of components of the maturation machinery, such as Dicer that processes most of the miRNAs, causes premature embryonic death. Studies using models with spatiotemporal control of Dicer expression gave a hint on the general function of miRNAs during different developmental stages. If Dicer is lacking in the developing heart (NKX2-5-positive cells), cardiac outflow tract alignment and chamber septation are disturbed. Interestingly, the development of the phenotype is preceded by upregulation of some critical factors (PITX2c and SEMA3C), while apoptosis of mesenchymal cells is decreased (Saxena and Tabin 2010). If Dicer is lost in NCC, multiple craniofacial and cardiovascular structures are malformed. Dicer proved to be important for the viability of cranial NCCs but not of the intermediate subset of CNCCs. Lack of Dicer impairs migration and patterning of CNCCs resulting in cardiovascular abnormalities such as aortic arch interruption, double-outlet right ventricle, and ventricular septal defects (Huang et al. 2010). Also in the adult heart, miRNAs play a crucial role in the development and progression of disease. Lack of Dicer in the adult cardiomyocyte causes cardiac remodeling, misexpression of cardiac contractile proteins, and profound sarcomere disarray ultimately leading to heart failure. Interestingly, the failing heart shows a decreased Dicer expression which can renormalize if the heart is supported by ventricle assist devices (Chen et al. 2008; da Costa Martins et al. 2008).

2.8.2 miR-1/miR133a

Studies using models with a global depletion of miRNAs clearly showed that miRNAs are major players in orchestrating cardiac development. However, these approaches could not identify the actual miRNA that is responsible for observed malformations. Only a spatiotemporal depletion of individual miRNAs can give a better picture of how miRNAs interfere with the known signaling cascades. The miRNAs 1 (miR-1) and 133a (miR-133a) were among the first to be described and best to be studied in cardiogenesis and heart disease. In the mammalian genome, miR-1 and miR-133a are encoded in duplicates as hetero-clusters on different chromosomes. The miR-1-1/133a-2 cluster is intergenically encoded on human chromosome 20, while the miR-1-2/1331-1 cluster is encoded in an antisense direction within an intron of the Mib1 gene on human chromosome 18. The expression of the miR-1-2/1331-1 cluster is, however, completely independent from the transcriptional activity of the Mib1 gene. Both clusters are highly conserved and expressed in myogenic progenitors (reviewed in Srivastava 2006; Boettger and Braun 2012). The transcriptional activity of miR-1-1 and miR-1-2 in cardiac and skeletal muscle precursor cells is under the control of differentiation regulators such as SRF (heart muscle) and MYOD and MEF2 (skeletal muscle) (Zhao et al. 2005; Liu et al. 2007). The Mef2 enhancer is located intragenically between the miR-1-2/1331-1 coding regions and is activated at the stage of the primitive heart tube. It then controls transcription throughout the atrial and ventricular chambers of the heart (Liu et al. 2007). Cardiogenesis depends on the expression of miR-1 in particular for the regulation of cardiac morphogenesis, the electrical conduction system, and the cell-cycle control as concluded from the differential gene expression when lacking miR-1-2 (Zhao et al. 2007). Also, HAND2 and the Notch ligand Delta were predicted to be miR-1-targeted genes (Zhao et al. 2005; Kwon et al. 2005). Later it was empirically shown that miR-133a also binds HAND2 simultaneously with miR-1 (Vo et al. 2010). Not fully understood are the effects of miR-1/133a expression on the cellular fate. Depending on the conditions, and maybe expression of cofactors, miR-1 and miR-133a can have opposing effects on cardiac and skeletal muscle formation such that miR-1 promotes myogenesis while miR-133a inhibits differentiation and promotes proliferation (Chen et al. 2006; Liu et al. 2008). In contrast, an increase of miR-1 is also critical since it specifically leads to a decrease of proliferation in ventricular cardiomyocytes.

It is now accepted that the miR-1/133a cluster has functions beyond cardiogenesis, also playing an important role in the adult heart. In cardiac hypertrophy, miR-1/133a is downregulated. Overexpressed in vitro miR-1 or 133a can inhibit cardiac hypertrophy, while inhibition of miR-133a causes cardiac hypertrophy. This phenotype was explained by identified target genes for miR-133a such as RhoA (a GDP-GTP exchange protein), Cdc42 (a signal transduction kinase), and Nelf-A/WHSC2 (a nuclear factor). All of these factors have been discussed in the context of cardiac hypertrophy or cardiogenesis (Carè et al. 2007). Another study showed that miR-1 controls the expression of calmodulin, thereby downregulating calcium-calmodulin signaling through calcineurin to the nuclear factor of activated T-cell (NFAT) transcription factor. Furthermore miR-1 negatively regulates the expression of MEF2C and GATA4. As a consequence miR-1 decreases the hypertrophic response of cardiomyocytes in vitro and the adult heart in vivo (Ikeda et al. 2009).

2.8.3 Other Cardiogenic miRNAs

Myosins are muscle-expressed proteins that convert chemical energy in the form of ATP into contractile force. It was shown in the mouse that three different myosins encode related miRNAs in their introns. Specifically, the fast myosin Myh6 co-expresses miR-208a which acts as a regulator of the slow myosins, Myh7/miR-208b and Myh7b/miR-499. The intronic miRNAs miR-208b and miR-499 have redundant roles in activating and repressing myofiber gene programs of slow and fast fibers, respectively (van Rooij et al. 2009). This is of interest because under cardiac stress conditions, such as pressure overload or if overexpressed, miR-208a can activate a hypertrophic response via activation of Myh7 expression (van Rooij et al. 2007; Callis et al. 2009). Furthermore, miR-208a is required for proper cardiac conduction and expression of the cardiac transcription factors, homeodomain-only protein and GATA4, and the gap junction protein cx40 (Callis et al. 2009). In a sum, miR-208a may be one regulator of cardiomyocyte dedifferentiation as observed in the stressed heart (Kubin et al. 2011). miR-499 is transcriptionally downregulated by p53 and may also be a regulator of cardiac stress response. In the postischemic heart, miR499 targets catalytic subunits of calcineurin and may through its suppression of calcineurin-mediated dephosphorylation of dynamin-related protein-1 (Drp1) inhibit apoptosis of cardiomyocytes and activate the mitochondrial fission program (Wang et al. 2011).

Also the miRNA cluster miR17-92 was shown to be critical in cardiogenesis. Lack of miR17-92 results in lung hypoplasia and cardiac malformations such as ventricular septum defects and causes premature death shortly after birth (Ventura et al. 2008). BMP signaling was shown to transcriptionally regulate miR17-92, which then targets Isl1 and Tbx1. This makes it likely that parts of the malformations seen when BMP signaling is impaired are based on a miRNA-mediated mechanism (Wang et al. 2010).

The miR-15 miRNA family includes miR-15a, miR-15b, miR-16-1, miR-16-2, miR-195, and miR-497, of which some have been described to be essential during cardiogenesis and in cardiac stress response. miR-195 is postnatally upregulated at a time when cardiomyocytes mitotically arrest, terminally differentiate, and adapt their metabolism. Since miR-195 targets a number of cell-cycle genes such as checkpoint kinase 1, it was speculated that miR-195 may be the central regulator. Indeed, overexpression of miR-195 in the embryonic heart causes cardiac malformations including hypoplastic ventricle and ventricular septum defects (Porrello et al. 2011). miR-15 family members are also involved in stress response after ischemic injury. If silenced in vitro cardiomyocytes are rendered resistant to hypoxia-induced cell death, while silencing of miR-15 after infarct in vivo reduces the infarct size and enhances cardiac function (Hullinger et al. 2012). One mechanism that might be responsible for this outcome could be that miR-15b specifically targets ADP-ribosylation factor-like 2, which localizes to the mitochondrial adenine nucleotide transporter 1, the exchanger of ADP/ATP. This affects mitochondrial integrity and ATP production (Nishi et al. 2010). How this correlates to the finding of the postnatal metabolic adaptation is not clear. However, it shows that miRNAs can enforce their signals not only via the regulation of transcription factors but might also be responsible for actively inducing a metabolic shift.

2.8.4 Long Noncoding RNAs (lncRNAs) in Cardiomyocyte Differentiation

Another area of regulatory species that recently bloomed is the field of lncRNAs. By definition lncRNAs are noncoding RNA sequences that are not processed into small RNAs such as miRNAs and are not members of classical housekeeping RNAs such as tRNAs or rRNAs, which is why they were originally considered as "junk RNA." lncRNAs are transcribed by RNA polymerase II, 5′ capped, spliced, and 3′

polyadenylated altogether resembling the processing of regular mRNA. Interestingly, the number of coding sequences did not rise significantly during the development from worm to mankind, whereas the number of lncRNAs did. It is believed that the number of lncRNAs exceeds the number of coding sequences. The classification of lncRNAs follows their genomic integration pattern as intergenic (between two coding sequences), intronic (within introns of a coding sequence), bidirectional (within one kb of promoters in antisense direction), enhancer (within enhancer region of promoters), sense (on the sense strand of coding sequence, intron/exon overlaps possible), or antisense (on the antisense strand of coding sequence, intron/exon overlaps possible). According to their mode of action, lncRNAs are sorted into five groups: (1) signal lncRNAs regulate transcription in spatiotemporal manner in response to various stimuli, (2) decoy lncRNA sequester regulatory factors leading to transcriptional repression, (3) guide lncRNAs complex with regulators and guide the complex to specific target genes, (4) scaffold lncRNAs complex simultaneously with several transcriptional regulators, and (5) enhancer lncRNAs make steric contact between enhancers and promoters by maintaining chromosomal loops. Although individual lncRNAs are not highly conserved across species, their modes of action are similar and, thus, the expression of lncRNAs dramatically increases the complexity of signaling (reviewed in Batista and Chang 2013; Devaux et al. 2015).

Few examples have been described that illustrate the importance of lncRNAs as regulators. For instance, Fendrr is an lncRNA that has been shown to be essential in cardiogenesis and body wall development. In mouse, its expression is restricted to the nascent lateral plate mesoderm where it is expressed in cardiac progenitors. Loss of expression is embryonically lethal but shows upregulation of some transcription factors controlling lateral plate or cardiac mesoderm differentiation, namely, Gata6 and Nkx2-5. This was attributed to epigenetic modification. Indeed, Fendrr binds to the polycomb-repressive complex 2 (PRC2) and TrxG/MLL histone-modifying complexes, which regulate the activation state of certain genes that are involved in the control of pluripotency, lineage fate, and differentiation. Furthermore, it was shown that Fendrr can bind to the promoters of Foxf1 and Pitx2. Lack of Fendrr is accompanied by a decreased PRC2 occupancy and altered methylation state at target gene promoters thereby changing the chromatin signatures. This is of importance since changes in the epigenetic signatures of regulatory elements can persist throughout several stages of differentiation (Grote et al. 2013). Another example for lncRNA during mouse cardiogenesis is Braveheart (Bvht), which in embryonic stem cells was shown to be important for lineage commitment toward cardiac fate. Bvht acts upstream of MesP1, thereby directing the gene expression of the core cardiovascular gene network. Most likely this is accomplished via interacting with the PRC2 component SUZ12, making Bvht as well an epigenetic regulator of gene expression and cell-fate decision (Klattenhoff et al. 2013).

In the adult heart, a cluster of lncRNA transcripts from Myh7 loci was described to be needed to keep the differentiated state of cardiomyocytes. These so-called myosin heavy-chain-associated RNA transcripts (Myheart or Mhrt) are abundantly expressed in the unstressed heart. Pathological stress activates the BRG1/HDAC/PARP chromatin repressor complex that inhibits Mhrt transcription. While in the fetal heart BRG1 promotes myocyte proliferation by maintaining BMP10 and p57kip2 repression, its activation in the adult heart precedes cardiac remodeling and development of cardiomyopathies displayed as hypertrophy and heart failure. Interestingly, restoring Mhrt transcript levels in stressed hearts can prevent cardiomyopathy by binding of Mhrt transcripts to BRG1. This interaction interferes with its genomic DNA binding and action of the repressor complex (Hang et al. 2010; Han et al. 2014). lncRNAs have a great potential as biomarkers or therapeutic targets; however, its biology is far from being understood or exhausted in practice (Matkovich et al. 2014; Ounzain et al. 2014).

2.9 Metabolism at the Crossroads of Differentiation

The heart is the first organ to start working and the last to resign. Throughout life cardiomyocytes need sufficient supply of ATP to contract and have to adapt the ever-changing workload to the actual metabolism without risking accumulation of metabolites. However, the source of ATP changes during development. While the fetal heart is made of proliferative cardiomyocytes and mainly glycolytic, the adult heart is made of terminally differentiated cardiomyocytes depending on mitochondrial oxidative phosphorylation with fatty acid β-oxidation becoming the major fuel. The adjustment from glycolysis to oxidative phosphorylation is essential for survival and occurs rapidly during the first few days after birth. Interestingly, the stressed or failing adult heart can return to fetal gene expression and a more glycolytic metabolism at any age during life (reviewed in Lopaschuk and Jaswal 2010; Doenst et al. 2013; Szibor et al. 2014).

There is an intimate relationship between mitochondrial morphology and well-being and cellular viability. Despite the fact that fetal cardiomyocytes are metabolically glycolytic, mitochondria are central players at all stages of cardiomyocyte differentiation. In fact it was shown that although mitochondrial function inversely correlates with cellular stemness, mitochondrial competence and maturation are a prerequisite for cardiomyocyte specification, differentiation, and excitation-contraction coupling (Chung et al. 2007; Mandal et al. 2011; Vega-Naredo et al. 2014). In other words, not the increase in contractility generates a need for more mitochondria but instead the gain of mitochondrial competence allows differentiation and, with it, proper contraction. The underlying mechanisms are now fully understood, but it was shown that the closure of the mitochondrial permeability pore is crucial for mitochondrial maturation leading to a more polarized mitochondrial membrane potential, less production of reactive oxygen species, and more cristae (Kasahara et al. 2013). Especially mitochondrial fusion is required for proper cardiomyocyte development. If lacking mitofusins 1 and 2, or optic atrophy 1 in the embryonic mouse heart or in mouse embryonic stem cells, cardiogenesis is severely impaired. This is paralleled by decreased levels of critical differentiation and transcription factors such as TGFβ, BMPs, SRF, GATA4, and MEF2 (Kasahara et al. 2013).

2.10 Transcriptional Regulators of Mitochondrial Metabolism

The complex interplay between cell-fate specification and metabolic adaptation from fetal to postnatal life needs the integration of a number of signaling networks. Some signaling networks that govern the metabolic preference of cells are described in more detail below.

2.10.1 Hypoxia-Inducible Factor 1 Alpha (HIF1α)

HIF1α forms together with the aryl hydrocarbon receptor nuclear translocator (ARNT) the heterodimeric transcription factor hypoxia-inducible factor-1 (HIF1). While ARNT is constitutively expressed, HIF1α levels are tightly controlled in an oxygen-sensitive manner. So-called HIF prolyl-hydroxylases hydroxylate HIF1α at a conserved proline residue, which allows its recognition and ubiquitination, followed by a rapid degradation in the proteasome. HIF1α is a key regulator for the adaptation of cells and tissues to hypoxia during all stages of life starting from development to aging in health and disease. HIF1α has been extensively studied in the fields of cancer biology and vascularization as well as energy metabolism, but has also been implicated in playing a central role during development and in cardiogenesis (Iyer et al. 1998; Ryan et al. 1998) (reviewed in Bishop and Ratcliffe 2015). The developing heart is extraordinary vulnerable to oxygen deprivation. Therefore, it is not surprising that lack of HIF1α causes extensive placental, vascular, and heart defects, resulting in embryonic lethality. Loss of HIF1α in specific cell-types involved in cardiogenesis causes cardiac malformations such as septation and conotruncal heart defects (Kenchegowda et al. 2014; O'Reilly et al. 2014). If HIF1α is lacking only in fetal cardiomyocytes, development is aborted at the stage of cardiac looping, and levels of core transcription factors such as MEF2C and TBX5 are decreased. Both MEF2C and Tbx5 are direct transcriptional target genes of HIF1α (Krishnan et al. 2008). Furthermore, HIF signaling regulates the degree of epicardial EMT and migration of epicardial derived cells into the myocardium and is thereby critical in the patterning of the coronary vasculogenesis (Tao et al. 2013). Importantly, HIF signaling not only activates a network of hypoxic response genes and shifts the metabolism to glycolysis; in parallel it actively represses mitochondrial oxidative phosphorylation by inducing pyruvate dehydrogenase kinase 1 (PDK1). PDK1 inhibits by phosphorylation the activity of the pyruvate dehydrogenase (PDH), which converts pyruvate to acetyl CoA that is fed into the Krebs cycle (Kim et al. 2006; Papandreou et al. 2006). While the fetal heart is primarily glycolytic, the postnatal heart switches quickly to lipid oxidation with mitochondrial oxidative phosphorylation as primary source for ATP. As the availability of oxygen rises with birth, the levels of HIF1α drop quickly. This adaptation of the heart to postnatal metabolism is partly facilitated by HAND1, which as well is under direct transcriptional control of HIF1α (Nau et al. 2002; Breckenridge et al. 2013).

2.10.2 The Peroxisome Proliferator-Activated Receptor (PPAR)-γ Coactivator-1 α (PGC1α)

PGC1α is a member of a family of transcription coactivators that are central in energy metabolism and mitochondrial biogenesis. Its signaling shifts the metabolism of muscle cells from glycolytic to more oxidative (reviewed in Huss and Kelly 2004; Liang and Ward 2006). Activation of PGC1α is especially vital for the postnatal maturation of cardiomyocytes, and this signal is, depending on circumstances, mediated by different PPARs (PPAR α, β, δ but not γ) that form heterodimers with other nuclear receptors such as thyroid hormone receptor α1 and retinoid-activated receptor α (Gilde et al. 2003; Burkart et al. 2007; Buroker et al. 2008). As a coactivator of PPARα, PGC1α controls the transcription of genes involved in mitochondrial fatty acid oxidation (Vega et al. 2000). Specifically, PPAR signaling controls the cardiac fatty acid oxidative flux by expression of the muscle-type carnitine palmitoyl-transferase I (M-CPT I). CPT I catalyzes the first reaction in the mitochondrial import of long-chain fatty acids. Furthermore, lack of PPAR signaling impairs the rates of palmitate oxidation in part due to the transcriptional control of malonyl-CoA decarboxylase. Together this demonstrates that PPAR is a central regulator of fatty acid import and oxidation (Mascaró et al. 1998; Brandt et al. 1998; Campbell et al. 2002). Although impairment of PPAR signaling is partially compensated for by increases in the rates of glucose oxidation and glycolysis, it is now clear that also in the adult cardiomyocyte, the activity of PPAR regulation is necessary for adaptational processes, while impairments or disturbances of this regulation lead to disease (Cheng et al. 2004) (reviewed in Finck and Kelly 2002). For instance, in the adult heart, hypoxic conditions stabilize HIF1α, which in turn decreases PPAR-regulated gene expression and shifts the metabolism back to glycolysis (Razeghi et al. 2001). Interestingly, it was shown that HIF signaling can oppose PGC1α effects not only by activating opposing pathways but also by decreasing the DNA-binding activity of PPARs (Belanger et al. 2007).

2.11 Concluding Remarks

Shaping the heart and differentiating cardiomyocytes is a complex process that needs the accurate interplay of a myriad of different cells of various sources. This process follows an evolutionary conserved choreography in a very delicate spatiotemporal manner, which is orchestrated on many different levels ranging from transcription factors and epigenetic regulators to noncoding RNAs. Each cell can be a signal transmitter and recipient and ultimately responds to the sum of signals by different means such as cell-fate determination, migration, differentiation, or even controlled cell death. A profound understanding of this complex development is needed to better understand mechanisms underlying CHD and in order to prevent or treat it before its onset. It is also important to approach heart diseases with novel regenerative therapies. Induced dedifferentiation and redifferentiation might add an important plasticity to the diseased heart to cope with stress (reviewed in Szibor et al. 2014). Finally, non-cardiomyocytes can be reprogrammed with little effort to cardiomyocyte-like cells with respect to marker gene expression and contractility (Ieda et al. 2010; Qian et al. 2012). It is the hope of many patients suffering from CHD or heart failure of other etiology that recapitulating the process of heart development in vitro could give rise to radically novel cell replacement therapies (Kattman et al. 2006; Jaenisch and Young 2008; Murry and Keller 2008). But despite all achievements and enthusiasm in the field, the way toward that goal will still be long and challenging.

References

Abu-Issa R, Smyth G, Smoak I, Yamamura K-I, Meyers EN (2002) Fgf8 is required for pharyngeal arch and cardiovascular development in the mouse. Development 129:4613–4625

Ai D, Liu W, Ma L, Dong F, Lu M-F, Wang D, Verzi MP, Cai C, Gage PJ, Evans S, Black BL, Brown NA, Martin JF (2006) Pitx2 regulates cardiac left-right asymmetry by patterning second cardiac lineage-derived myocardium. Dev Biol 296:437–449

Alsan BH, Schultheiss TM (2002) Regulation of avian cardiogenesis by Fgf8 signaling. Development 129:1935–1943

Andrée B, Duprez D, Vorbusch B, Arnold HH, Brand T (1998) BMP-2 induces ectopic expression of cardiac lineage markers and interferes with somite formation in chicken embryos. Mech Dev 70:119–131

Arai A, Yamamoto K, Toyama J (1997) Murine cardiac progenitor cells require visceral embryonic endoderm and primitive streak for terminal differentiation. Dev Dyn 210:344–353

Ávila P, Mercier L-A, Dore A, Marcotte F, Mongeon F-P, Ibrahim R, Asgar A, Miro J, Andelfinger G, Mondésert B, de Guise P, Poirier N, Khairy P (2014) Adult congenital heart disease: a growing epidemic. Can J Cardiol 30:S410–S419

Aybar MJ, Mayor R (2002) Early induction of neural crest cells: lessons learned from frog, fish and chick. Curr Opin Genet Dev 12:452–458

Barnes RM, Firulli BA, VanDusen NJ, Morikawa Y, Conway SJ, Cserjesi P, Vincentz JW, Firulli AB (2011) Hand2 loss-of-function in Hand1-expressing cells reveals distinct roles in epicardial and coronary vessel development. Circ Res 108:940–949

Basson CT, Bachinsky DR, Lin RC, Levi T, Elkins JA, Soults J, Grayzel D, Kroumpouzou E, Traill TA, Leblanc-Straceski J, Renault B, Kucherlapati R, Seidman JG, Seidman CE (1997) Mutations in human TBX5 [corrected] cause limb and cardiac malformation in Holt-Oram syndrome. Nat Genet 15:30–35

Batista PJ, Chang HY (2013) Long noncoding RNAs: cellular address codes in development and disease. Cell 152:1298–1307

Belaguli NS, Sepulveda JL, Nigam V, Charron F, Nemer M, Schwartz RJ (2000) Cardiac tissue enriched factors serum response factor and GATA-4 are mutual coregulators. Mol Cell Biol 20:7550–7558

Belanger AJ, Luo Z, Vincent KA, Akita GY, Cheng SH, Gregory RJ, Jiang C (2007) Hypoxia-inducible factor 1 mediates hypoxia-induced cardiomyocyte lipid accumulation by reducing the DNA binding activity of peroxisome proliferator-activated receptor alpha/retinoid X receptor. Biochem Biophys Res Commun 364:567–572

Bi W, Drake CJ, Schwarz JJ (1999) The transcription factor MEF2C-null mouse exhibits complex vascular malformations and reduced cardiac expression of angiopoietin 1 and VEGF. Dev Biol 211:255–267

Bimber B, Dettman RW, Simon H-G (2007) Differential regulation of Tbx5 protein expression and sub-cellular localization during heart development. Dev Biol 302:230–242

Bishop T, Ratcliffe PJ (2015) HIF hydroxylase pathways in cardiovascular physiology and medicine. Circ Res 117:65–79

Boettger T, Braun T (2012) A new level of complexity: the role of microRNAs in cardiovascular development. Circ Res 110:1000–1013

Bondue A, Lapouge G, Paulissen C, Semeraro C, Iacovino M, Kyba M, Blanpain C (2008) Mesp1 acts as a master regulator of multipotent cardiovascular progenitor specification. Cell Stem Cell 3:69–84

Brade T, Kumar S, Cunningham TJ, Chatzi C, Zhao X, Cavallero S, Li P, Sucov HM, Ruiz-Lozano P, Duester G (2011) Retinoic acid stimulates myocardial expansion by induction of hepatic erythropoietin which activates epicardial Igf2. Development 138:139–148

Brandt JM, Djouadi F, Kelly DP (1998) Fatty acids activate transcription of the muscle carnitine palmitoyltransferase I gene in cardiac myocytes via the peroxisome proliferator-activated receptor alpha. J Biolo Chem 273:23786–23792

Breckenridge RA, Zuberi Z, Gomes J, Orford R, Dupays L, Felkin LE, Clark JE, Magee AI, Ehler E, Birks EJ, Barton PJR, Tinker A, Mohun TJ (2009) Overexpression of the transcription factor Hand1 causes predisposition towards arrhythmia in mice. J Mol Cell Cardiol 47:133–141

Breckenridge RA, Piotrowska I, Ng K-E, Ragan TJ, West JA, Kotecha S, Towers N, Bennett M, Kienesberger PC, Smolenski RT, Siddall HK, Offer JL, Mocanu MM, Yelon DM, Dyck JRB, Griffin JL, Abramov AY, Gould AP, Mohun TJ (2013) Hypoxic regulation of hand1 controls the fetal-neonatal switch in cardiac metabolism. PLoS Biol 11:e1001666

Brown CB, Feiner L, Lu MM, Li J, Ma X, Webber AL, Jia L, Raper JA, Epstein JA (2001) PlexinA2 and semaphorin signaling during cardiac neural crest development. Development 128:3071–3080

Brown DD, Martz SN, Binder O, Goetz SC, Price BMJ, Smith JC, Conlon FL (2005) Tbx5 and Tbx20 act synergistically to control vertebrate heart morphogenesis. Development 132:553–563

Bruneau BG, Logan M, Davis N, Levi T, Tabin CJ, Seidman JG, Seidman CE (1999) Chamber-specific cardiac expression of Tbx5 and heart defects in Holt-Oram syndrome. Dev Biol 211:100–108

Bruneau BG, Nemer G, Schmitt JP, Charron F, Robitaille L, Caron S, Conner DA, Gessler M, Nemer M, Seidman CE, Seidman JG (2001) A murine model of Holt-Oram syndrome defines roles of the T-box transcription factor Tbx5 in cardiogenesis and disease. Cell 106:709–721

Bu L, Jiang X, Martin-Puig S, Caron L, Zhu S, Shao Y, Roberts DJ, Huang PL, Domian IJ, Chien KR (2009) Human ISL1 heart progenitors generate diverse multipotent cardiovascular cell lineages. Nature 460:113–117

Buckingham M, Meilhac S, Zaffran S (2005) Building the mammalian heart from two sources of myocardial cells. Nat Rev Genet 6:826–835

Burkart EM, Sambandam N, Han X, Gross RW, Courtois M, Gierasch CM, Shoghi K, Welch MJ, Kelly DP (2007) Nuclear receptors PPARbeta/delta and PPARalpha direct distinct metabolic regulatory programs in the mouse heart. J Clini Invest 117:3930–3939

Buroker NE, Ning X-H, Portman M (2008) Cardiac PPARalpha protein expression is constant as alternate nuclear receptors and PGC-1 coordinately increase during the postnatal metabolic transition. PPAR Res 2008:279531

Burstyn-Cohen T, Kalcheim C (2002) Association between the cell cycle and neural crest delamination through specific regulation of G1/S transition. Dev Cell 3:383–395

Cai C-L, Liang X, Shi Y, Chu P-H, Pfaff SL, Chen J, Evans S (2003) Isl1 identifies a cardiac progenitor population that proliferates prior to differentiation and contributes a majority of cells to the heart. Dev Cell 5:877–889

Cai C-L, Zhou W, Yang L, Bu L, Qyang Y, Zhang X, Li X, Rosenfeld MG, Chen J, Evans S (2005) T-box genes coordinate regional rates of proliferation and regional specification during cardiogenesis. Development 132:2475–2487

Cai C-L, Martin JC, Sun Y, Cui L, Wang L, Ouyang K, Yang L, Bu L, Liang X, Zhang X, Stallcup WB, Denton CP, McCulloch A, Chen J, Evans SM (2008) A myocardial lineage derives from Tbx18 epicardial cells. Nature 454:104–108

Cai X, Zhang W, Hu J, Zhang L, Sultana N, Wu B, Cai W, Zhou B, Cai C-L (2013) Tbx20 acts upstream of Wnt signaling to regulate endocardial cushion formation and valve remodeling during mouse cardiogenesis. Development 140:3176–3187

Callis TE, Pandya K, Seok HY, Tang R-H, Tatsuguchi M, Huang Z-P, Chen J-F, Deng Z, Gunn B, Shumate J, Willis MS, Selzman CH, Wang D-Z (2009) MicroRNA-208a is a regulator of cardiac hypertrophy and conduction in mice. J Clin Invest 119:2772–2786

Campbell FM, Kozak R, Wagner A, Altarejos JY, Dyck JRB, Belke DD, Severson DL, Kelly DP, Lopaschuk GD (2002) A role for peroxisome proliferator-activated receptor alpha (PPARalpha) in the control of cardiac malonyl-CoA levels: reduced fatty acid oxidation rates and increased glucose oxidation rates in the hearts of mice lacking PPARalpha are associated with higher concentrations of malonyl-CoA and reduced expression of malonyl-CoA decarboxylase. J Biolo Chem 277:4098–4103

Carè A, Catalucci D, Felicetti F, Bonci D, Addario A, Gallo P, Bang M-L, Segnalini P, Gu Y, Dalton ND, Elia L, Latronico MVG, Høydal M, Autore C, Russo MA, Dorn GW, Ellingsen O, Ruiz-Lozano P, Peterson KL, Croce CM, Peschle C, Condorelli G (2007) MicroRNA-133 controls cardiac hypertrophy. Nat Med 13:613–618

Chan SS-K, Li H-J, Hsueh Y-C, Lee DS, Chen J-H, Hwang S-M, Chen C-Y, Shih E, Hsieh PCH (2010) Fibroblast growth factor-10 promotes cardiomyocyte differentiation from embryonic and induced pluripotent stem cells. PLoS One 5:e14414

Chen X, Weisberg E, Fridmacher V, Watanabe M, Naco G, Whitman M (1997) Smad4 and FAST-1 in the assembly of activin-responsive factor. Nature 389:85–89

Chen J-F, Mandel EM, Thomson JM, Wu Q, Callis TE, Hammond SM, Conlon FL, Wang D-Z (2006) The role of microRNA-1 and microRNA-133 in skeletal muscle proliferation and differentiation. Nat Genet 38:228–233

Chen J-F, Murchison EP, Tang R, Callis TE, Tatsuguchi M, Deng Z, Rojas M, Hammond SM, Schneider MD, Selzman CH, Meissner G, Patterson C, Hannon GJ, Wang D-Z (2008) Targeted deletion of Dicer in the heart leads to dilated cardiomyopathy and heart failure. Proc Natl Acad Sci U S A 105:2111–2116

Chen L, Fulcoli FG, Tang S, Baldini A (2009) Tbx1 regulates proliferation and differentiation of multipotent heart progenitors. Circ Res 105:842–851

Chen L, Fulcoli FG, Ferrentino R, Martucciello S, Illingworth EA, Baldini A (2012) Transcriptional control in cardiac progenitors: Tbx1 interacts with the BAF chromatin remodeling complex and regulates Wnt5a. PLoS Genet 8:e1002571

Chen J, Sun F, Fu J, Zhang H (2015) Association of TBX20 gene polymorphism with congenital heart disease in Han Chinese neonates. Pediatr Cardiol 36:737–742

Cheng L, Ding G, Qin Q, Huang Y, Lewis W, He N, Evans RM, Schneider MD, Brako FA, Xiao Y, Chen YE, Yang Q (2004) Cardiomyocyte-restricted peroxisome proliferator-activated receptor-delta deletion perturbs myocardial fatty acid oxidation and leads to cardiomyopathy. Nat Med 10:1245–1250

Cheng Z, Lib L, Li Z, Liu M, Yan J, Wang B, Ma X (2012) Two novel HAND1 mutations in Chinese patients with ventricular septal defect. Clin Chim Acta 413:675–677

Cheung M, Chaboissier M-C, Mynett A, Hirst E, Schedl A, Briscoe J (2005) The transcriptional control of trunk neural crest induction, survival, and delamination. Dev Cell 8:179–192

Choudhary B, Ito Y, Makita T, Sasaki T, Chai Y, Sucov HM (2006) Cardiovascular malformations with normal smooth muscle differentiation in neural crest-specific type II TGFbeta receptor (Tgfbr2) mutant mice. Dev Biol 289:420–429

Christoffels VM, Burch JBE, Moorman AFM (2004) Architectural plan for the heart: early patterning and delineation of the chambers and the nodes. Trends Cardiovasc Med 14:301–307

Christoffels VM, Mommersteeg MTM, Trowe M-O, Prall OWJ, de Gier-de Vries C, Soufan AT, Bussen M, Schuster-Gossler K, Harvey RP, Moorman AFM, Kispert A (2006) Formation of the venous pole of the heart from an Nkx2-5-negative precursor population requires Tbx18. Circ Res 98:1555–1563

Christoffels VM, Grieskamp T, Norden J, Mommersteeg MTM, Rudat C, Kispert A (2009) Tbx18 and the fate of epicardial progenitors. Nature 458:E8–E9, – discussion E9–10

Chung S, Dzeja PP, Faustino RS, Perez-Terzic C, Behfar A, Terzic A (2007) Mitochondrial oxidative metabolism is required for the cardiac differentiation of stem cells. Nat Clin Pract Cardiovasc Med 4(Suppl 1):S60–S67
Clevers H, Nusse R (2012) Wnt/β-catenin signaling and disease. Cell 149:1192–1205
Clouthier DE, Hosoda K, Richardson JA, Williams SC, Yanagisawa H, Kuwaki T, Kumada M, Hammer RE, Yanagisawa M (1998) Cranial and cardiac neural crest defects in endothelin-A receptor-deficient mice. Development 125:813–824
Clouthier DE, Williams SC, Hammer RE, Richardson JA, Yanagisawa M (2003) Cell-autonomous and nonautonomous actions of endothelin-A receptor signaling in craniofacial and cardiovascular development. Dev Biol 261:506–519
Costello I, Pimeisl I-M, Dräger S, Bikoff EK, Robertson EJ, Arnold SJ (2011) The T-box transcription factor Eomesodermin acts upstream of Mesp1 to specify cardiac mesoderm during mouse gastrulation. Nat Cell Biol 13:1084–1091
da Costa Martins PA, Bourajjaj M, Gladka M, Kortland M, van Oort RJ, Pinto YM, Molkentin JD, De Windt LJ (2008) Conditional dicer gene deletion in the postnatal myocardium provokes spontaneous cardiac remodeling. Circulation 118:1567–1576
David R, Brenner C, Stieber J, Schwarz F, Brunner S, Vollmer M, Mentele E, Müller-Höcker J, Kitajima S, Lickert H, Rupp R, Franz WM (2008) MesP1 drives vertebrate cardiovascular differentiation through Dkk-1-mediated blockade of Wnt-signalling. Nat Cell Biol 10:338–345
De Calisto J, Araya C, Marchant L, Riaz CF, Mayor R (2005) Essential role of non-canonical Wnt signalling in neural crest migration. Development 132:2587–2597
del Monte G, Casanova JC, Guadix JA, MacGrogan D, Burch JBE, Pérez Pomares JM, de la Pompa JL (2011) Differential Notch signaling in the epicardium is required for cardiac inflow development and coronary vessel morphogenesis. Circ Res 108:824–836
Derynck R, Zhang Y, Feng XH (1998) Smads: transcriptional activators of TGF-beta responses. Cell 95:737–740
Dettman RW, Denetclaw W, Ordahl CP, Bristow J (1998) Common epicardial origin of coronary vascular smooth muscle, perivascular fibroblasts, and intermyocardial fibroblasts in the avian heart. Dev Biol 193:169–181
Devaux Y, Zangrando J, Schroen B, Creemers EE, Pedrazzini T, Chang C-P, Dorn GW, Thum T, Heymans S, Cardiolinc network (2015) Long noncoding RNAs in cardiac development and ageing. Nat Rev Cardiol 12:415–425
Dodou E, Verzi MP, Anderson JP, Xu S-M, Black BL (2004) Mef2c is a direct transcriptional target of ISL1 and GATA factors in the anterior heart field during mouse embryonic development. Development 131:3931–3942
Doenst T, Nguyen TD, Abel ED (2013) Cardiac metabolism in heart failure: implications beyond ATP production. Circ Res 113:709–724
Dyer MA, Farrington SM, Mohn D, Munday JR, Baron MH (2001) Indian hedgehog activates hematopoiesis and vasculogenesis and can respecify prospective neurectodermal cell fate in the mouse embryo. Development 128:1717–1730
Edmondson DG, Lyons GE, Martin JF, Olson EN (1994) Mef2 gene expression marks the cardiac and skeletal muscle lineages during mouse embryogenesis. Development 120:1251–1263
Eickholt BJ, Mackenzie SL, Graham A, Walsh FS, Doherty P (1999) Evidence for collapsin-1 functioning in the control of neural crest migration in both trunk and hindbrain regions. Development 126:2181–2189
Engleka KA, Manderfield LJ, Brust RD, Li L, Cohen A, Dymecki SM, Epstein JA (2012) Islet1 derivatives in the heart are of both neural crest and second heart field origin. Circ Res 110:922–926
Esposito G, Butler TL, Blue GM, Cole AD, Sholler GF, Kirk EP, Grossfeld P, Perryman BM, Harvey RP, Winlaw DS (2011) Somatic mutations in NKX2–5, GATA4, and HAND1 are not a common cause of tetralogy of Fallot or hypoplastic left heart. Am J Med Genet A 155A:2416–2421
Feiner L, Webber AL, Brown CB, Lu MM, Jia L, Feinstein P, Mombaerts P, Epstein JA, Raper JA (2001) Targeted disruption of semaphorin 3C leads to persistent truncus arteriosus and aortic arch interruption. Development 128:3061–3070

Finck BN, Kelly DP (2002) Peroxisome proliferator-activated receptor alpha (PPARalpha) signaling in the gene regulatory control of energy metabolism in the normal and diseased heart. J Mol Cell Cardiol 34:1249–1257

Firulli AB, McFadden DG, Lin Q, Srivastava D, Olson EN (1998) Heart and extra-embryonic mesodermal defects in mouse embryos lacking the bHLH transcription factor Hand1. Nat Genet 18:266–270

Firulli BA, Hadzic DB, McDaid JR, Firulli AB (2000) The basic helix-loop-helix transcription factors dHAND and eHAND exhibit dimerization characteristics that suggest complex regulation of function. J Biol Chem 275:33567–33573

Firulli BA, McConville DP, Byers JS, Vincentz JW, Barnes RM, Firulli AB (2010) Analysis of a Hand1 hypomorphic allele reveals a critical threshold for embryonic viability. Dev Dyn 239:2748–2760

Franco D, Campione M (2003) The role of Pitx2 during cardiac development. Linking left-right signaling and congenital heart diseases. Trends Cardiovasc Med 13:157–163

Frank DU, Fotheringham LK, Brewer JA, Muglia LJ, Tristani-Firouzi M, Capecchi MR, Moon AM (2002) An Fgf8 mouse mutant phenocopies human 22q11 deletion syndrome. Development 129:4591–4603

Ghosh TK, Song FF, Packham EA, Buxton S, Robinson TE, Ronksley J, Self T, Bonser AJ, Brook JD (2009) Physical interaction between TBX5 and MEF2C is required for early heart development. Mol Cell Biol 29:2205–2218

Gilde AJ, van der Lee KAJM, Willemsen PHM, Chinetti G, van der Leij FR, van der Vusse GJ, Staels B, van Bilsen M (2003) Peroxisome proliferator-activated receptor (PPAR) alpha and PPARbeta/delta, but not PPARgamma, modulate the expression of genes involved in cardiac lipid metabolism. Circ Res 92:518–524

Golzio C, Havis E, Daubas P, Nuel G, Babarit C, Munnich A, Vekemans M, Zaffran S, Lyonnet S, Etchevers HC (2012) ISL1 directly regulates FGF10 transcription during human cardiac outflow formation. PLoS One 7, e30677

Gottlieb PD, Pierce SA, Sims RJ, Yamagishi H, Weihe EK, Harriss JV, Maika SD, Kuziel WA, King HL, Olson EN, Nakagawa O, Srivastava D (2002) Bop encodes a muscle-restricted protein containing MYND and SET domains and is essential for cardiac differentiation and morphogenesis. Nat Genet 31:25–32

Grieskamp T, Rudat C, Lüdtke TH-W, Norden J, Kispert A (2011) Notch signaling regulates smooth muscle differentiation of epicardium-derived cells. Circ Res 108:813–823

Grote P, Wittler L, Hendrix D, Koch F, Währisch S, Beisaw A, Macura K, Bläss G, Kellis M, Werber M, Herrmann BG (2013) The tissue-specific lncRNA Fendrr is an essential regulator of heart and body wall development in the mouse. Dev Cell 24:206–214

Guadix JA, Ruiz-Villalba A, Lettice L, Velecela V, Muñoz-Chapuli R, Hastie ND, Pérez Pomares JM, Martínez-Estrada OM (2011) Wt1 controls retinoic acid signalling in embryonic epicardium through transcriptional activation of Raldh2. Development 138:1093–1097

Guo C, Sun Y, Zhou B, Adam RM, Li X, Pu WT, Morrow BE, Moon A, Li X (2011) A Tbx1-Six1/Eya1-Fgf8 genetic pathway controls mammalian cardiovascular and craniofacial morphogenesis. J Clin Invest 121:1585–1595

Gurjarpadhye A, Hewett KW, Justus C, Wen X, Stadt H, Kirby ML, Sedmera D, Gourdie RG (2007) Cardiac neural crest ablation inhibits compaction and electrical function of conduction system bundles. Am J Physiol Heart Circ Physiol 292:H1291–H1300

Han P, Li W, Lin C-H, Yang J, Shang C, Nurnberg ST, Jin KK, Xu W, Lin C-Y, Lin C-J, Xiong Y, Chien H-C, Zhou B, Ashley E, Bernstein D, Chen P-S, Chen H-SV, Quertermous T, Chang C-P (2014) A long noncoding RNA protects the heart from pathological hypertrophy. Nature 514:102–106

Hang CT, Yang J, Han P, Cheng H-L, Shang C, Ashley E, Zhou B, Chang C-P (2010) Chromatin regulation by Brg1 underlies heart muscle development and disease. Nature 466:62–67

Harrelson Z, Kelly RG, Goldin SN, Gibson-Brown JJ, Bollag RJ, Silver LM, Papaioannou VE (2004) Tbx2 is essential for patterning the atrioventricular canal and for morphogenesis of the outflow tract during heart development. Development 131:5041–5052

Harvey RP (2002) Patterning the vertebrate heart. Nat Rev Genet 3:544–556
Herrmann F, Bundschu K, Kühl SJ, Kühl M (2011) Tbx5 overexpression favors a first heart field lineage in murine embryonic stem cells and in Xenopus laevis embryos. Dev Dyn 240:2634–2645
Hiroi Y, Kudoh S, Monzen K, Ikeda Y, Yazaki Y, Nagai R, Komuro I (2001) Tbx5 associates with Nkx2-5 and synergistically promotes cardiomyocyte differentiation. Nat Genet 28:276–280
Holler KL, Hendershot TJ, Troy SE, Vincentz JW, Firulli AB, Howard MJ (2010) Targeted deletion of Hand2 in cardiac neural crest-derived cells influences cardiac gene expression and outflow tract development. Dev Biol 341:291–304
Hu T, Yamagishi H, Maeda J, McAnally J, Yamagishi C, Srivastava D (2004) Tbx1 regulates fibroblast growth factors in the anterior heart field through a reinforcing autoregulatory loop involving forkhead transcription factors. Development 131:5491–5502
Huang Z-P, Chen J-F, Regan JN, Maguire CT, Tang R-H, Dong XR, Majesky MW, Wang D-Z (2010) Loss of microRNAs in neural crest leads to cardiovascular syndromes resembling human congenital heart defects. Arterioscler Thromb Vasc Biol 30:2575–2586
Hullinger TG, Montgomery RL, Seto AG, Dickinson BA, Semus HM, Lynch JM, Dalby CM, Robinson K, Stack C, Latimer PA, Hare JM, Olson EN, van Rooij E (2012) Inhibition of miR-15 protects against cardiac ischemic injury. Circ Res 110:71–81
Huss JM, Kelly DP (2004) Nuclear receptor signaling and cardiac energetics. Circ Res 95:568–578
Hutson MR, Kirby ML (2007) Model systems for the study of heart development and disease. Cardiac neural crest and conotruncal malformations. Semin Cell Dev Biol 18:101–110
Hutson MR, Zhang P, Stadt HA, Sato AK, Li Y-X, Burch J, Creazzo TL, Kirby ML (2006) Cardiac arterial pole alignment is sensitive to FGF8 signaling in the pharynx. Dev Biol 295:486–497
Huynh T, Chen L, Terrell P, Baldini A (2007) A fate map of Tbx1 expressing cells reveals heterogeneity in the second cardiac field. Genesis 45:470–475
Ieda M, Fu J-D, Delgado-Olguin P, Vedantham V, Hayashi Y, Bruneau BG, Srivastava D (2010) Direct reprogramming of fibroblasts into functional cardiomyocytes by defined factors. Cell 142:375–386
Ikeda S, He A, Kong SW, Lu J, Bejar R, Bodyak N, Lee K-H, Ma Q, Kang PM, Golub TR, Pu WT (2009) MicroRNA-1 negatively regulates expression of the hypertrophy-associated calmodulin and Mef2a genes. Mol Cell Biol 29:2193–2204
Iyer NV, Kotch LE, Agani F, Leung SW, Laughner E, Wenger RH, Gassmann M, Gearhart JD, Lawler AM, Yu AY, Semenza GL (1998) Cellular and developmental control of O2 homeostasis by hypoxia-inducible factor 1 alpha. Genes Dev 12:149–162
Jaenisch R, Young R (2008) Stem cells, the molecular circuitry of pluripotency and nuclear reprogramming. Cell 132:567–582
Jensen B, Wang T, Christoffels VM, Moorman AFM (2013) Evolution and development of the building plan of the vertebrate heart. Biochim Biophys Acta 1833:783–794
Jerome LA, Papaioannou VE (2001) DiGeorge syndrome phenotype in mice mutant for the T-box gene, Tbx1. Nat Genet 27:286–291
Jia Q, McDill BW, Li S-Z, Deng C, Chang C-P, Chen F (2007) Smad signaling in the neural crest regulates cardiac outflow tract remodeling through cell autonomous and non-cell autonomous effects. Dev Biol 311:172–184
Kaartinen V, Dudas M, Nagy A, Sridurongrit S, Lu MM, Epstein JA (2004) Cardiac outflow tract defects in mice lacking ALK2 in neural crest cells. Development 131:3481–3490
Kang J, Nathan E, Xu S-M, Tzahor E, Black BL (2009) Isl1 is a direct transcriptional target of Forkhead transcription factors in second-heart-field-derived mesoderm. Dev Biol 334:513–522
Kasahara A, Cipolat S, Chen Y, Dorn GW, Scorrano L (2013) Mitochondrial fusion directs cardiomyocyte differentiation via calcineurin and Notch signaling. Science 342:734–737
Kattman SJ, Huber TL, Keller GM (2006) Multipotent flk-1+ cardiovascular progenitor cells give rise to the cardiomyocyte, endothelial, and vascular smooth muscle lineages. Dev Cell 11:723–732

Katz TC, Singh MK, Degenhardt K, Rivera-Feliciano J, Johnson RL, Epstein JA, Tabin CJ (2012) Distinct compartments of the proepicardial organ give rise to coronary vascular endothelial cells. Dev Cell 22:639–650

Kelly RG, Brown NA, Buckingham ME (2001) The arterial pole of the mouse heart forms from Fgf10-expressing cells in pharyngeal mesoderm. Dev Cell 1:435–440

Kempf H, Linares C, Corvol P, Gasc JM (1998) Pharmacological inactivation of the endothelin type A receptor in the early chick embryo: a model of mispatterning of the branchial arch derivatives. Development 125:4931–4941

Kenchegowda D, Liu H, Thompson K, Luo L, Martin SS, Fisher SA (2014) Vulnerability of the developing heart to oxygen deprivation as a cause of congenital heart defects. J Am Heart Assoc 3:e000841

Kim RY, Robertson EJ, Solloway MJ (2001) Bmp6 and Bmp7 are required for cushion formation and septation in the developing mouse heart. Dev Biol 235:449–466

Kim J-W, Tchernyshyov I, Semenza GL, Dang CV (2006) HIF-1-mediated expression of pyruvate dehydrogenase kinase: a metabolic switch required for cellular adaptation to hypoxia. Cell Metab 3:177–185

Kioussi C, Briata P, Baek SH, Rose DW, Hamblet NS, Herman T, Ohgi KA, Lin C, Gleiberman A, Wang J, Brault V, Ruiz-Lozano P, Nguyen HD, Kemler R, Glass CK, Wynshaw-Boris A, Rosenfeld MG (2002) Identification of a Wnt/Dvl/beta-Catenin –> Pitx2 pathway mediating cell-type-specific proliferation during development. Cell 111:673–685

Kirby ML, Gale TF, Stewart DE (1983) Neural crest cells contribute to normal aorticopulmonary septation. Science 220:1059–1061

Kitajima S, Takagi A, Inoue T, Saga Y (2000) MesP1 and MesP2 are essential for the development of cardiac mesoderm. Development 127:3215–3226

Klattenhoff CA, Scheuermann JC, Surface LE, Bradley RK, Fields PA, Steinhauser ML, Ding H, Butty VL, Torrey L, Haas S, Abo R, Tabebordbar M, Lee RT, Burge CB, Boyer LA (2013) Braveheart, a long noncoding RNA required for cardiovascular lineage commitment. Cell 152:570–583

Kodo K, Nishizawa T, Furutani M, Arai S, Ishihara K, Oda M, Makino S, Fukuda K, Takahashi T, Matsuoka R, Nakanishi T, Yamagishi H (2012) Genetic analysis of essential cardiac transcription factors in 256 patients with non-syndromic congenital heart defects. Circ J 76:1703–1711

Kolodziejczyk SM, Wang L, Balazsi K, DeRepentigny Y, Kothary R, Megeney LA (1999) MEF2 is upregulated during cardiac hypertrophy and is required for normal post-natal growth of the myocardium. Curr Biol 9:1203–1206

Kraus F, Haenig B, Kispert A (2001) Cloning and expression analysis of the mouse T-box gene Tbx18. Mech Dev 100:83–86

Krishnan J, Ahuja P, Bodenmann S, Knapik D, Perriard E, Krek W, Perriard J-C (2008) Essential role of developmentally activated hypoxia-inducible factor 1alpha for cardiac morphogenesis and function. Circ Res 103:1139–1146

Kruger RP, Aurandt J, Guan K-L (2005) Semaphorins command cells to move. Nat Rev Mol Cell Biol 6:789–800

Kruithof BPT, van Wijk B, Somi S, Kruithof-de Julio M, Pérez Pomares JM, Weesie F, Wessels A, Moorman AFM, van den Hoff MJB (2006) BMP and FGF regulate the differentiation of multipotential pericardial mesoderm into the myocardial or epicardial lineage. Dev Biol 295:507–522

Kubin T, Pöling J, Kostin S, Gajawada P, Hein S, Rees W, Wietelmann A, Tanaka M, Lörchner H, Schimanski S, Szibor M, Warnecke H, Braun T (2011) Oncostatin m is a major mediator of cardiomyocyte dedifferentiation and remodeling. Cell Stem Cell 9:420–432

Kubota Y, Ito K (2000) Chemotactic migration of mesencephalic neural crest cells in the mouse. Dev Dyn 217:170–179

Kuo CT, Morrisey EE, Anandappa R, Sigrist K, Lu MM, Parmacek MS, Soudais C, Leiden JM (1997) GATA4 transcription factor is required for ventral morphogenesis and heart tube formation. Genes Dev 11:1048–1060

Kwee L, Baldwin HS, Shen HM, Stewart CL, Buck C, Buck CA, Labow MA (1995) Defective development of the embryonic and extraembryonic circulatory systems in vascular cell adhesion molecule (VCAM-1) deficient mice. Development 121:489–503

Kwon C, Han Z, Olson EN, Srivastava D (2005) MicroRNA1 influences cardiac differentiation in Drosophila and regulates Notch signaling. Proc Natl Acad Sci U S A 102:18986–18991

Kwon C, Qian L, Cheng P, Nigam V, Arnold J, Srivastava D (2009) A regulatory pathway involving Notch1/beta-catenin/Isl1 determines cardiac progenitor cell fate. Nat Cell Biol 11:951–957

Lartillot N, Lespinet O, Vervoort M, Adoutte A (2002) Expression pattern of Brachyury in the mollusc Patella vulgata suggests a conserved role in the establishment of the AP axis in Bilateria. Development 129:1411–1421

Laugwitz K-L, Moretti A, Lam J, Gruber P, Chen Y, Woodard S, Lin L-Z, Cai C-L, Lu MM, Reth M, Platoshyn O, Yuan JX-J, Evans S, Chien KR (2005) Postnatal isl1+ cardioblasts enter fully differentiated cardiomyocyte lineages. Nature 433:647–653

Lavine KJ, White AC, Park C, Smith CS, Choi K, Long F, Hui C-C, Ornitz DM (2006) Fibroblast growth factor signals regulate a wave of Hedgehog activation that is essential for coronary vascular development. Genes Dev 20:1651–1666

Li P, Cavallero S, Gu Y, Chen THP, Hughes J, Hassan AB, Brüning JC, Pashmforoush M, Sucov HM (2011) IGF signaling directs ventricular cardiomyocyte proliferation during embryonic heart development. Development 138:1795–1805

Liang H, Ward WF (2006) PGC-1alpha: a key regulator of energy metabolism. Adv Physiol Educ 30:145–151

Liberatore CM, Searcy-Schrick RD, Yutzey KE (2000) Ventricular expression of tbx5 inhibits normal heart chamber development. Dev Biol 223:169–180

Liberatore CM, Searcy-Schrick RD, Vincent EB, Yutzey KE (2002) Nkx-2.5 gene induction in mice is mediated by a Smad consensus regulatory region. Dev Biol 244:243–256

Lien CL, Wu C, Mercer B, Webb R, Richardson JA, Olson EN (1999) Control of early cardiac-specific transcription of Nkx2-5 by a GATA-dependent enhancer. Development 126:75–84

Lien C-L, McAnally J, Richardson JA, Olson EN (2002) Cardiac-specific activity of an Nkx2-5 enhancer requires an evolutionarily conserved Smad binding site. Dev Biol 244:257–266

Lim J, Thiery JP (2012) Epithelial-mesenchymal transitions: insights from development. Development 139:3471–3486

Lin Q, Schwarz J, Bucana C, Olson EN (1997) Control of mouse cardiac morphogenesis and myogenesis by transcription factor MEF2C. Science 276:1404–1407

Lin Q, Lu J, Yanagisawa H, Webb R, Lyons GE, Richardson JA, Olson EN (1998) Requirement of the MADS-box transcription factor MEF2C for vascular development. Development 125:4565–4574

Lin L, Cui L, Zhou W, Dufort D, Zhang X, Cai C-L, Bu L, Yang L, Martin J, Kemler R, Rosenfeld MG, Chen J, Evans SM (2007) Beta-catenin directly regulates Islet1 expression in cardiovascular progenitors and is required for multiple aspects of cardiogenesis. Proc Natl Acad Sci U S A 104:9313–9318

Lindsay EA, Vitelli F, Su H, Morishima M, Huynh T, Pramparo T, Jurecic V, Ogunrinu G, Sutherland HF, Scambler PJ, Bradley A, Baldini A (2001) Tbx1 haploinsufficieny in the DiGeorge syndrome region causes aortic arch defects in mice. Nature 410:97–101

Lindsley RC, Gill JG, Murphy TL, Langer EM, Cai M, Mashayekhi M, Wang W, Niwa N, Nerbonne JM, Kyba M, Murphy KM (2008) Mesp1 coordinately regulates cardiovascular fate restriction and epithelial-mesenchymal transition in differentiating ESCs. Cell Stem Cell 3:55–68

Liu C, Liu W, Palie J, Lu M-F, Brown NA, Martin JF (2002) Pitx2c patterns anterior myocardium and aortic arch vessels and is required for local cell movement into atrioventricular cushions. Development 129:5081–5091

Liu N, Williams AH, Kim Y, McAnally J, Bezprozvannaya S, Sutherland LB, Richardson JA, Bassel-Duby R, Olson EN (2007) An intragenic MEF2-dependent enhancer directs muscle-specific expression of microRNAs 1 and 133. Proc Natl Acad Sci U S A 104:20844–20849

Liu N, Bezprozvannaya S, Williams AH, Qi X, Richardson JA, Bassel-Duby R, Olson EN (2008) microRNA-133a regulates cardiomyocyte proliferation and suppresses smooth muscle gene expression in the heart. Genes Dev 22:3242–3254

Lopaschuk GD, Jaswal JS (2010) Energy metabolic phenotype of the cardiomyocyte during development, differentiation, and postnatal maturation. J Cardiovasc Pharmacol 56:130–140

Maitra M, Schluterman MK, Nichols HA, Richardson JA, Lo CW, Srivastava D, Garg V (2009) Interaction of Gata4 and Gata6 with Tbx5 is critical for normal cardiac development. Dev Biol 326:368–377

Mandal S, Lindgren AG, Srivastava AS, Clark AT, Banerjee U (2011) Mitochondrial function controls proliferation and early differentiation potential of embryonic stem cells. Stem Cells 29:486–495

Männer J, Pérez-Pomares JM, Macías D, Muñoz-Chápuli R (2001) The origin, formation and developmental significance of the epicardium: a review. Cells Tissues Organs (Print) 169:89–103

Marcellini S, Technau U, Smith JC, Lemaire P (2003) Evolution of Brachyury proteins: identification of a novel regulatory domain conserved within Bilateria. Dev Biol 260:352–361

Martínez-Estrada OM, Lettice LA, Essafi A, Guadix JA, Slight J, Velecela V, Hall E, Reichmann J, Devenney PS, Hohenstein P, Hosen N, Hill RE, Muñoz-Chapuli R, Hastie ND (2010) Wt1 is required for cardiovascular progenitor cell formation through transcriptional control of Snail and E-cadherin. Nat Genet 42:89–93

Marvin MJ, Di Rocco G, Gardiner A, Bush SM, Lassar AB (2001) Inhibition of Wnt activity induces heart formation from posterior mesoderm. Genes Dev 15:316–327

Mascaró C, Acosta E, Ortiz JA, Marrero PF, Hegardt FG, Haro D (1998) Control of human muscle-type carnitine palmitoyltransferase I gene transcription by peroxisome proliferator-activated receptor. J Biolo Chem 273:8560–8563

Massagué J (1998) TGF-beta signal transduction. Annu Rev Biochem 67:753–791

Massagué J, Chen YG (2000) Controlling TGF-beta signaling. Genes Dev 14:627–644

Matkovich SJ, Edwards JR, Grossenheider TC, de Guzman SC, Dorn GW (2014) Epigenetic coordination of embryonic heart transcription by dynamically regulated long noncoding RNAs. Proc Natl Acad Sci U S A 111:12264–12269

Maya-Ramos L, Cleland J, Bressan M, Mikawa T (2013) Induction of the proepicardium. J Dev Biol 1:82–91

Mellott DO, Burke RD (2008) Divergent roles for Eph and ephrin in avian cranial neural crest. BMC Dev Biol 8:56

Merki E, Zamora M, Raya A, Kawakami Y, Wang J, Zhang X, Burch J, Kubalak SW, Kaliman P, Izpisúa Belmonte JC, Chien KR, Ruiz-Lozano P (2005) Epicardial retinoid X receptor alpha is required for myocardial growth and coronary artery formation. Proc Natl Acad Sci U S A 102:18455–18460

Merscher S, Funke B, Epstein JA, Heyer J, Puech A, Lu MM, Xavier RJ, Demay MB, Russell RG, Factor S, Tokooya K, Jore BS, Lopez M, Pandita RK, Lia M, Carrion D, Xu H, Schorle H, Kobler JB, Scambler P, Wynshaw-Boris A, Skoultchi AI, Morrow BE, Kucherlapati R (2001) TBX1 is responsible for cardiovascular defects in velo-cardio-facial/DiGeorge syndrome. Cell 104:619–629

Meulemans D, Bronner-Fraser M (2004) Gene-regulatory interactions in neural crest evolution and development. Dev Cell 7:291–299

Mikawa T, Gourdie RG (1996) Pericardial mesoderm generates a population of coronary smooth muscle cells migrating into the heart along with ingrowth of the epicardial organ. Dev Biol 174:221–232

Min H, Danilenko DM, Scully SA, Bolon B, Ring BD, Tarpley JE, DeRose M, Simonet WS (1998) Fgf-10 is required for both limb and lung development and exhibits striking functional similarity to Drosophila branchless. Genes Dev 12:3156–3161

Molkentin JD (2000) The zinc finger-containing transcription factors GATA-4, –5, and –6. Ubiquitously expressed regulators of tissue-specific gene expression. J Biol Chem 275:38949–38952

Molkentin JD, Black BL, Martin JF, Olson EN (1996) Mutational analysis of the DNA binding, dimerization, and transcriptional activation domains of MEF2C. Mol Cell Biol 16:2627–2636

Molkentin JD, Lin Q, Duncan SA, Olson EN (1997) Requirement of the transcription factor GATA4 for heart tube formation and ventral morphogenesis. Genes Dev 11:1061–1072

Moore AW, McInnes L, Kreidberg J, Hastie ND, Schedl A (1999) YAC complementation shows a requirement for Wt1 in the development of epicardium, adrenal gland and throughout nephrogenesis. Development 126:1845–1857

Moorman AFM, Christoffels VM (2003) Cardiac chamber formation: development, genes, and evolution. Physiol Rev 83:1223–1267

Munoz JP, Collao A, Chiong M, Maldonado C, Adasme T, Carrasco L, Ocaranza P, Bravo R, Gonzalez L, Díaz-Araya G, Hidalgo C, Lavandero S (2009) The transcription factor MEF2C mediates cardiomyocyte hypertrophy induced by IGF-1 signaling. Biochem Biophys Res Commun 388:155–160

Murry CE, Keller G (2008) Differentiation of embryonic stem cells to clinically relevant populations: lessons from embryonic development. Cell 132:661–680

Nakamura T, Colbert MC, Robbins J (2006) Neural crest cells retain multipotential characteristics in the developing valves and label the cardiac conduction system. Circ Res 98:1547–1554

Nau PN, Van Natta T, Ralphe JC, Teneyck CJ, Bedell KA, Caldarone CA, Segar JL, Scholz TD (2002) Metabolic adaptation of the fetal and postnatal ovine heart: regulatory role of hypoxia-inducible factors and nuclear respiratory factor-1. Pediatr Res 52:269–278

Nemer M (2008) Genetic insights into normal and abnormal heart development. Cardiovasc Pathol 17:48–54

Nguyen HT, Bodmer R, Abmayr SM, McDermott JC, Spoerel NA (1994) D-mef2: a Drosophila mesoderm-specific MADS box-containing gene with a biphasic expression profile during embryogenesis. Proc Natl Acad Sci U S A 91:7520–7524

Nishi H, Ono K, Iwanaga Y, Horie T, Nagao K, Takemura G, Kinoshita M, Kuwabara Y, Mori RT, Hasegawa K, Kita T, Kimura T (2010) MicroRNA-15b modulates cellular ATP levels and degenerates mitochondria via Arl2 in neonatal rat cardiac myocytes. J Biol Chem 285:4920–4930

Nowotschin S, Liao J, Gage PJ, Epstein JA, Campione M, Morrow BE (2006) Tbx1 affects asymmetric cardiac morphogenesis by regulating Pitx2 in the secondary heart field. Development 133:1565–1573

O'Reilly VC, Lopes Floro K, Shi H, Chapman BE, Preis JI, James AC, Chapman G, Harvey RP, Johnson RS, Grieve SM, Sparrow DB, Dunwoodie SL (2014) Gene-environment interaction demonstrates the vulnerability of the embryonic heart. Dev Biol 391:99–110

Ounzain S, Pezzuto I, Micheletti R, Burdet F, Sheta R, Nemir M, Gonzales C, Sarre A, Alexanian M, Blow MJ, May D, Johnson R, Dauvillier J, Pennacchio LA, Pedrazzini T (2014) Functional importance of cardiac enhancer-associated noncoding RNAs in heart development and disease. J Mol Cell Cardiol 76:55–70

Paige SL, Plonowska K, Xu A, Wu SM (2015) Molecular regulation of cardiomyocyte differentiation. Circ Res 116:341–353

Pandur P, Läsche M, Eisenberg LM, Kühl M (2002) Wnt-11 activation of a non-canonical Wnt signalling pathway is required for cardiogenesis. Nature 418:636–641

Papandreou I, Cairns RA, Fontana L, Lim AL, Denko NC (2006) HIF-1 mediates adaptation to hypoxia by actively downregulating mitochondrial oxygen consumption. Cell Metab 3:187–197

Pashmforoush M, Lu JT, Chen H, Amand TS, Kondo R, Pradervand S, Evans SM, Clark B, Feramisco JR, Giles W, Ho SY, Benson DW, Silberbach M, Shou W, Chien KR (2004) Nkx2-5 pathways and congenital heart disease; loss of ventricular myocyte lineage specification leads to progressive cardiomyopathy and complete heart block. Cell 117:373–386

Pérez Pomares JM, de la Pompa JL (2011) Signaling during epicardium and coronary vessel development. Circ Res 109:1429–1442

Phan D, Rasmussen TL, Nakagawa O, McAnally J, Gottlieb PD, Tucker PW, Richardson JA, Bassel-Duby R, Olson EN (2005) BOP, a regulator of right ventricular heart development, is a direct transcriptional target of MEF2C in the developing heart. Development 132:2669–2678

Pikkarainen S, Tokola H, Kerkelä R, Ruskoaho H (2004) GATA transcription factors in the developing and adult heart. 63:196–207

Poelmann RE, Gittenberger-de Groot AC (1999) A subpopulation of apoptosis-prone cardiac neural crest cells targets to the venous pole: multiple functions in heart development? Dev Biol 207:271–286

Porrello ER, Johnson BA, Aurora AB, Simpson E, Nam Y-J, Matkovich SJ, Dorn GW, van Rooij E, Olson EN (2011) MiR-15 family regulates postnatal mitotic arrest of cardiomyocytes. Circ Res 109:670–679

Prall OWJ, Menon MK, Solloway MJ, Watanabe Y, Zaffran S, Bajolle F, Biben C, McBride JJ, Robertson BR, Chaulet H, Stennard FA, Wise N, Schaft D, Wolstein O, Furtado MB, Shiratori H, Chien KR, Hamada H, Black BL, Saga Y, Robertson EJ, Buckingham ME, Harvey RP (2007) An Nkx2-5/Bmp2/Smad1 negative feedback loop controls heart progenitor specification and proliferation. Cell 128:947–959

Pu WT, Ishiwata T, Juraszek AL, Ma Q, Izumo S (2004) GATA4 is a dosage-sensitive regulator of cardiac morphogenesis. Dev Biol 275:235–244

Qian L, Huang Y, Spencer CI, Foley A, Vedantham V, Liu L, Conway SJ, Fu J-D, Srivastava D (2012) In vivo reprogramming of murine cardiac fibroblasts into induced cardiomyocytes. Nature 485:593–598

Rana MS, Christoffels VM, Moorman AFM (2013) A molecular and genetic outline of cardiac morphogenesis. Acta Physiol (Oxf) 207:588–615

Razeghi P, Young ME, Abbasi S, Taegtmeyer H (2001) Hypoxia in vivo decreases peroxisome proliferator-activated receptor alpha-regulated gene expression in rat heart. Biochem Biophys Res Commun 287:5–10

Reamon-Buettner SM, Borlak J (2010) NKX2-5: an update on this hypermutable homeodomain protein and its role in human congenital heart disease (CHD). Hum Mutat 31:1185–1194

Richarte AM, Mead HB, Tallquist MD (2007) Cooperation between the PDGF receptors in cardiac neural crest cell migration. Dev Biol 306:785–796

Riley P, Anson-Cartwright L, Cross JC (1998) The Hand1 bHLH transcription factor is essential for placentation and cardiac morphogenesis. Nat Genet 18:271–275

Risebro CA, Smart N, Dupays L, Breckenridge R, Mohun TJ, Riley PR (2006) Hand1 regulates cardiomyocyte proliferation versus differentiation in the developing heart. Development 133:4595–4606

Rochais F, Sturny R, Chao C-M, Mesbah K, Bennett M, Mohun TJ, Bellusci S, Kelly RG (2014) FGF10 promotes regional foetal cardiomyocyte proliferation and adult cardiomyocyte cell-cycle re-entry. 104:432–442

Ryan HE, Lo J, Johnson RS (1998) HIF-1 alpha is required for solid tumor formation and embryonic vascularization. EMBO J 17:3005–3015

Sakabe NJ, Aneas I, Shen T, Shokri L, Park S-Y, Bulyk ML, Evans SM, Nobrega MA (2012) Dual transcriptional activator and repressor roles of TBX20 regulate adult cardiac structure and function. Hum Mol Genet 21:2194–2204

Sakai D, Tanaka Y, Endo Y, Osumi N, Okamoto H, Wakamatsu Y (2005) Regulation of Slug transcription in embryonic ectoderm by beta-catenin-Lef/Tcf and BMP-Smad signaling. Dev Growth Differ 47:471–482

Sanford LP, Ormsby I, Gittenberger-de Groot AC, Sariola H, Friedman R, Boivin GP, Cardell EL, Doetschman T (1997) TGFbeta2 knockout mice have multiple developmental defects that are non-overlapping with other TGFbeta knockout phenotypes. Development 124: 2659–2670

Sato A, Scholl AM, Kuhn EN, Kuhn EB, Stadt HA, Decker JR, Pegram K, Hutson MR, Kirby ML (2011) FGF8 signaling is chemotactic for cardiac neural crest cells. Dev Biol 354:18–30

Saxena A, Tabin CJ (2010) miRNA-processing enzyme Dicer is necessary for cardiac outflow tract alignment and chamber septation. Proc Natl Acad Sci U S A 107:87–91

Schneider VA, Mercola M (2001) Wnt antagonism initiates cardiogenesis in Xenopus laevis. Genes Dev 15:304–315

Scholl AM, Kirby ML (2009) Signals controlling neural crest contributions to the heart. Wiley Interdiscip Rev Syst Biol Med 1:220–227

Schott JJ, Benson DW, Basson CT, Pease W, Silberbach GM, Moak JP, Maron BJ, Seidman CE, Seidman JG (1998) Congenital heart disease caused by mutations in the transcription factor NKX2-5. Science 281:108–111

Schultheiss TM, Burch JB, Lassar AB (1997) A role for bone morphogenetic proteins in the induction of cardiac myogenesis. Genes Dev 11:451–462

Searcy RD, Vincent EB, Liberatore CM, Yutzey KE (1998) A GATA-dependent nkx-2.5 regulatory element activates early cardiac gene expression in transgenic mice. Development 125:4461–4470

Sekine K, Ohuchi H, Fujiwara M, Yamasaki M, Yoshizawa T, Sato T, Yagishita N, Matsui D, Koga Y, Itoh N, Kato S (1999) Fgf10 is essential for limb and lung formation. Nat Genet 21:138–141

Sela-Donenfeld D, Kalcheim C (1999) Regulation of the onset of neural crest migration by coordinated activity of BMP4 and Noggin in the dorsal neural tube. Development 126:4749–4762

Sepulveda JL, Belaguli N, Nigam V, Chen CY, Nemer M, Schwartz RJ (1998) GATA-4 and Nkx-2.5 coactivate Nkx-2 DNA binding targets: role for regulating early cardiac gene expression. Mol Cell Biol 18:3405–3415

Singh MK, Christoffels VM, Dias JM, Trowe M-O, Petry M, Schuster-Gossler K, Bürger A, Ericson J, Kispert A (2005) Tbx20 is essential for cardiac chamber differentiation and repression of Tbx2. Development 132:2697–2707

Smith A, Robinson V, Patel K, Wilkinson DG (1997) The EphA4 and EphB1 receptor tyrosine kinases and ephrin-B2 ligand regulate targeted migration of branchial neural crest cells. Curr Biol 7:561–570

Sorrell MRJ, Waxman JS (2011) Restraint of Fgf8 signaling by retinoic acid signaling is required for proper heart and forelimb formation. Dev Biol 358:44–55

Srivastava D (1999) HAND proteins: molecular mediators of cardiac development and congenital heart disease. Trends Cardiovasc Med 9:11–18

Srivastava D (2006) Making or breaking the heart: from lineage determination to morphogenesis. Cell 126:1037–1048

Srivastava D, Thomas T, Lin Q, Kirby ML, Brown D, Olson EN (1997) Regulation of cardiac mesodermal and neural crest development by the bHLH transcription factor, dHAND. Nat Genet 16:154–160

Stennard FA, Costa MW, Elliott DA, Rankin S, Haast SJP, Lai D, McDonald LPA, Niederreither K, Dolle P, Bruneau BG, Zorn AM, Harvey RP (2003) Cardiac T-box factor Tbx20 directly interacts with Nkx2-5, GATA4, and GATA5 in regulation of gene expression in the developing heart. Dev Biol 262:206–224

Stennard FA, Costa MW, Lai D, Biben C, Furtado MB, Solloway MJ, McCulley DJ, Leimena C, Preis JI, Dunwoodie SL, Elliott DE, Prall OWJ, Black BL, Fatkin D, Harvey RP (2005) Murine T-box transcription factor Tbx20 acts as a repressor during heart development, and is essential for adult heart integrity, function and adaptation. Development 132:2451–2462

Stottmann RW, Choi M, Mishina Y, Meyers EN, Klingensmith J (2004) BMP receptor IA is required in mammalian neural crest cells for development of the cardiac outflow tract and ventricular myocardium. Development 131:2205–2218

Sucov HM, Gu Y, Thomas S, Li P, Pashmforoush M (2009) Epicardial control of myocardial proliferation and morphogenesis. Pediatr Cardiol 30:617–625

Szibor M, Pöling J, Warnecke H, Kubin T, Braun T (2014) Remodeling and dedifferentiation of adult cardiomyocytes during disease and regeneration. Cell Mol Life Sci 71:1907–1916

Takeuchi JK, Mileikovskaia M, Koshiba-Takeuchi K, Heidt AB, Mori AD, Arruda EP, Gertsenstein M, Georges R, Davidson L, Mo R, Hui C-C, Henkelman RM, Nemer M, Black BL, Nagy A, Bruneau BG (2005) Tbx20 dose-dependently regulates transcription factor networks required for mouse heart and motoneuron development. Development 132:2463–2474

Tanaka M, Chen Z, Bartunkova S, Yamasaki N, Izumo S (1999) The cardiac homeobox gene Csx/Nkx2.5 lies genetically upstream of multiple genes essential for heart development. Development 126:1269–1280

Tanizawa Y, Riggs AC, Dagogo-Jack S, Vaxillaire M, Froguel P, Liu L, Donis-Keller H, Permutt MA (1994) Isolation of the human LIM/homeodomain gene islet-1 and identification of a simple sequence repeat polymorphism [corrected]. Diabetes 43:935–941

Tao J, Doughman Y, Yang K, Ramirez-Bergeron D, Watanabe M (2013) Epicardial HIF signaling regulates vascular precursor cell invasion into the myocardium. Dev Biol 376:136–149

Thattaliyath BD, Firulli BA, Firulli AB (2002a) The basic-helix-loop-helix transcription factor HAND2 directly regulates transcription of the atrial naturetic peptide gene. J Mol Cell Cardiol 34:1335–1344

Thattaliyath BD, Livi CB, Steinhelper ME, Toney GM, Firulli AB (2002b) HAND1 and HAND2 are expressed in the adult-rodent heart and are modulated during cardiac hypertrophy. Biochem Biophys Res Commun 297:870–875

Torlopp A, Schlueter J, Brand T (2010) Role of fibroblast growth factor signaling during proepicardium formation in the chick embryo. Dev Dyn 239:2393–2403

Toyofuku T, Yoshida J, Sugimoto T, Yamamoto M, Makino N, Takamatsu H, Takegahara N, Suto F, Hori M, Fujisawa H, Kumanogoh A, Kikutani H (2008) Repulsive and attractive semaphorins cooperate to direct the navigation of cardiac neural crest cells. Dev Biol 321:251–262

Trainor PA, Ariza-McNaughton L, Krumlauf R (2002) Role of the isthmus and FGFs in resolving the paradox of neural crest plasticity and prepatterning. Science 295:1288–1291

Tribulo C, Aybar MJ, Nguyen VH, Mullins MC, Mayor R (2003) Regulation of Msx genes by a Bmp gradient is essential for neural crest specification. Development 130:6441–6452

Tsuchihashi T, Maeda J, Shin CH, Ivey KN, Black BL, Olson EN, Yamagishi H, Srivastava D (2011) Hand2 function in second heart field progenitors is essential for cardiogenesis. Dev Biol 351:62–69

Tzahor E (2007) Wnt/beta-catenin signaling and cardiogenesis: timing does matter. Dev Cell 13:10–13

Tzahor E, Lassar AB (2001) Wnt signals from the neural tube block ectopic cardiogenesis. Genes Dev 15:255–260

Urness LD, Bleyl SB, Wright TJ, Moon AM, Mansour SL (2011) Redundant and dosage sensitive requirements for Fgf3 and Fgf10 in cardiovascular development. Dev Biol 356:383–397

van der Linde D, Konings EEM, Slager MA, Witsenburg M, Helbing WA, Takkenberg JJM, Roos-Hesselink JW (2011) Birth prevalence of congenital heart disease worldwide: a systematic review and meta-analysis. J Am Coll Cardiol 58:2241–2247

van Rooij E, Sutherland LB, Qi X, Richardson JA, Hill J, Olson EN (2007) Control of stress-dependent cardiac growth and gene expression by a microRNA. Science 316:575–579

van Rooij E, Quiat D, Johnson BA, Sutherland LB, Qi X, Richardson JA, Kelm RJ, Olson EN (2009) A family of microRNAs encoded by myosin genes governs myosin expression and muscle performance. Dev Cell 17:662–673

van Wijk B, van den Berg G, Abu-Issa R, Barnett P, van der Velden S, Schmidt M, Ruijter JM, Kirby ML, Moorman AFM, van den Hoff MJB (2009) Epicardium and myocardium separate from a common precursor pool by crosstalk between bone morphogenetic protein- and fibroblast growth factor-signaling pathways. Circ Res 105:431–441

Vega RB, Huss JM, Kelly DP (2000) The coactivator PGC-1 cooperates with peroxisome proliferator-activated receptor alpha in transcriptional control of nuclear genes encoding mitochondrial fatty acid oxidation enzymes. Mol Cell Biol 20:1868–1876

Vega-Hernández M, Kovacs A, De Langhe S, Ornitz DM (2011) FGF10/FGFR2b signaling is essential for cardiac fibroblast development and growth of the myocardium. Development 138:3331–3340

Vega-Naredo I, Loureiro R, Mesquita KA, Barbosa IA, Tavares LC, Branco AF, Erickson JR, Holy J, Perkins EL, Carvalho RA, Oliveira PJ (2014) Mitochondrial metabolism directs stemness and differentiation in P19 embryonal carcinoma stem cells. Cell Death Differ 21:1560–1574

Ventura A, Young AG, Winslow MM, Lintault L, Meissner A, Erkeland SJ, Newman J, Bronson RT, Crowley D, Stone JR, Jaenisch R, Sharp PA, Jacks T (2008) Targeted deletion reveals essential and overlapping functions of the miR-17 through 92 family of miRNA clusters. Cell 132:875–886

Vo NK, Dalton RP, Liu N, Olson EN, Goodman RH (2010) Affinity purification of microRNA-133a with the cardiac transcription factor, Hand2. Proc Natl Acad Sci U S A 107:19231–19236

von Both I, Silvestri C, Erdemir T, Lickert H, Walls JR, Henkelman RM, Rossant J, Harvey RP, Attisano L, Wrana JL (2004) Foxh1 is essential for development of the anterior heart field. Dev Cell 7:331–345

von Gise A, Zhou B, Honor LB, Ma Q, Petryk A, Pu WT (2011) WT1 regulates epicardial epithelial to mesenchymal transition through β-catenin and retinoic acid signaling pathways. Dev Biol 356:421–431

Wagner N, Wagner K-D, Theres H, Englert C, Schedl A, Scholz H (2005) Coronary vessel development requires activation of the TrkB neurotrophin receptor by the Wilms' tumor transcription factor Wt1. Genes Dev 19:2631–2642

Waldo KL, Lo CW, Kirby ML (1999) Connexin 43 expression reflects neural crest patterns during cardiovascular development. Dev Biol 208:307–323

Wang J, Nagy A, Larsson J, Dudas M, Sucov HM, Kaartinen V (2006) Defective ALK5 signaling in the neural crest leads to increased postmigratory neural crest cell apoptosis and severe outflow tract defects. BMC Dev Biol 6:51

Wang J, Greene SB, Bonilla-Claudio M, Tao Y, Zhang J, Bai Y, Huang Z, Black BL, Wang F, Martin JF (2010) Bmp signaling regulates myocardial differentiation from cardiac progenitors through a MicroRNA-mediated mechanism. Dev Cell 19:903–912

Wang J-X, Jiao J-Q, Li Q, Long B, Wang K, Liu J-P, Li Y-R, Li P-F (2011) miR-499 regulates mitochondrial dynamics by targeting calcineurin and dynamin-related protein-1. Nat Med 17:71–78

Watanabe M, Rebbert ML, Andreazzoli M, Takahashi N, Toyama R, Zimmerman S, Whitman M, Dawid IB (2002) Regulation of the Lim-1 gene is mediated through conserved FAST-1/FoxH1 sites in the first intron. Dev Dyn 225:448–456

Watanabe Y, Miyagawa-Tomita S, Vincent SD, Kelly RG, Moon AM, Buckingham ME (2010) Role of mesodermal FGF8 and FGF10 overlaps in the development of the arterial pole of the heart and pharyngeal arch arteries. Circ Res 106:495–503

Weisberg E, Winnier GE, Chen X, Farnsworth CL, Hogan BL, Whitman M (1998) A mouse homologue of FAST-1 transduces TGF beta superfamily signals and is expressed during early embryogenesis. Mech Dev 79:17–27

Wiese C, Grieskamp T, Airik R, Mommersteeg MTM, Gardiwal A, Vries de Gier-de C, Schuster-Gossler K, Moorman AFM, Kispert A, Christoffels VM (2009) Formation of the sinus node head and differentiation of sinus node myocardium are independently regulated by Tbx18 and Tbx3. Circ Res 104:388–397

Willert K, Nusse R (2012) Wnt proteins. Cold Spring Harb Perspect Biol 4:a007864

Witzel HR, Jungblut B, Choe CP, Crump JG, Braun T, Dobreva G (2012) The LIM protein Ajuba restricts the second heart field progenitor pool by regulating Isl1 activity. Dev Cell 23:58–70

Xu H, Morishima M, Wylie JN, Schwartz RJ, Bruneau BG, Lindsay EA, Baldini A (2004) Tbx1 has a dual role in the morphogenesis of the cardiac outflow tract. Development 131: 3217–3227

Yagi H, Furutani Y, Hamada H, Sasaki T, Asakawa S, Minoshima S, Ichida F, Joo K, Kimura M, Imamura S-I, Kamatani N, Momma K, Takao A, Nakazawa M, Shimizu N, Matsuoka R (2003) Role of TBX1 in human del22q11.2 syndrome. Lancet 362:1366–1373

Yamagishi H, Yamagishi C, Nakagawa O, Harvey RP, Olson EN, Srivastava D (2001) The combinatorial activities of Nkx2.5 and dHAND are essential for cardiac ventricle formation. Dev Biol 239:190–203

Yanagisawa H, Hammer RE, Richardson JA, Williams SC, Clouthier DE, Yanagisawa M (1998) Role of Endothelin-1/Endothelin-A receptor-mediated signaling pathway in the aortic arch patterning in mice. J Clini Invest 102:22–33

Yang JT, Rayburn H, Hynes RO (1995) Cell adhesion events mediated by alpha 4 integrins are essential in placental and cardiac development. Development 121:549–560

Zamora M, Männer J, Ruiz-Lozano P (2007) Epicardium-derived progenitor cells require beta-catenin for coronary artery formation. Proc Natl Acad Sci U S A 104:18109–18114

Zhang XM, Ramalho-Santos M, McMahon AP (2001) Smoothened mutants reveal redundant roles for Shh and Ihh signaling including regulation of L/R symmetry by the mouse node. Cell 106:781–792

Zhang Z, Huynh T, Baldini A (2006) Mesodermal expression of Tbx1 is necessary and sufficient for pharyngeal arch and cardiac outflow tract development. Development 133:3587–3595

Zhao Y, Samal E, Srivastava D (2005) Serum response factor regulates a muscle-specific microRNA that targets Hand2 during cardiogenesis. Nature 436:214–220

Zhao Y, Ransom JF, Li A, Vedantham V, von Drehle M, Muth AN, Tsuchihashi T, Mcmanus MT, Schwartz RJ, Srivastava D (2007) Dysregulation of cardiogenesis, cardiac conduction, and cell cycle in mice lacking miRNA-1-2. Cell 129:303–317

Zhou B, von Gise A, Ma Q, Rivera-Feliciano J, Pu WT (2008a) Nkx2-5- and Isl1-expressing cardiac progenitors contribute to proepicardium. Biochem Biophys Res Commun 375:450–453

Zhou B, Ma Q, Rajagopal S, Wu SM, Domian I, Rivera-Feliciano J, Jiang D, von Gise A, Ikeda S, Chien KR, Pu WT (2008b) Epicardial progenitors contribute to the cardiomyocyte lineage in the developing heart. Nature 454:109–113

Zweier C, Sticht H, Aydin-Yaylagül I, Campbell CE, Rauch A (2007) Human TBX1 missense mutations cause gain of function resulting in the same phenotype as 22q11.2 deletions. Am J Hum Genet 80:510–517

Excitation–Contraction Coupling of Cardiomyocytes

3

Jens Kockskämper

Abstract

Cardiomyocytes are electrically excitable cells built to contract. The cellular processes linking electrical excitation, i.e. the sarcolemmal action potential, with contraction are referred to as excitation–contraction coupling. An increase in intracellular Ca^{2+} concentration ($[Ca^{2+}]_i$) is the key mediator of excitation–contraction coupling. Cardiac action potentials are characterised by a long plateau phase carried by Ca^{2+} influx through L-type Ca^{2+} channels. L-type Ca^{2+} influx triggers Ca^{2+} release from the sarcoplasmic reticulum (SR) through Ca^{2+} release channels or ryanodine receptors. This Ca^{2+}-induced Ca^{2+} release causes a large increase in $[Ca^{2+}]_i$ from ≈ 100 nM in diastole to ≈ 1 μM in systole. Ca^{2+} binding to the myofilaments causes contraction. Ca^{2+} removal from the cytosol by the SR-Ca^{2+}-ATPase (SERCA) and the sarcolemmal Na^+–Ca^{2+} exchanger (NCX) mediates relaxation. The amplitude of the $[Ca^{2+}]_i$ transient (CaT) decides about the strength of contraction. Increases in L-type Ca^{2+} current, SERCA activity, SR Ca^{2+} load and fractional release, IP_3 signalling and $[Na^+]_i$ all increase CaT amplitude. Under conditions of Ca^{2+} overload, SR Ca^{2+} release also occurs spontaneously, i.e. in the absence of an action potential, and may elicit life-threatening arrhythmias via activation of electrogenic NCX and subsequent membrane depolarisation. Atrial and ventricular myocytes share these basic principles of excitation–contraction coupling and $[Ca^{2+}]_i$ regulation. However, there are important differences between these types of cardiomyocytes regarding action potential configuration, sarcolemmal structure (transverse tubules) and subcellular Ca^{2+} regulation. Remodelling of excitation–contraction coupling occurs in cardiac disease such as heart failure and atrial fibrillation and represents a potential therapeutic target.

J. Kockskämper
Institut für Pharmakologie und Klinische Pharmazie, Philipps-Universität Marburg, Karl-von-Frisch-Str. 1, Marburg 35032, Germany
e-mail: jens.kockskaemper@staff.uni-marburg.de

K.-D. Schlüter (ed.), *Cardiomyocytes – Active Players in Cardiac Disease*,
DOI 10.1007/978-3-319-31251-4_3

3.1 Introduction

A beating heart – for many of us – is the embodiment of life. Excitation–contraction (EC) coupling refers to the cellular mechanisms that make the heart beat. EC coupling, therefore, is a central theme in cardiac physiology and pathology, and the last three decades have seen enormous progress in understanding the cellular, subcellular and sometimes even molecular processes involved in EC coupling in unprecedented detail. Still, there is much more to learn and the topic keeps fascinating scientists around the world. It is obvious, therefore, that here I cannot deal with this topic in its entirety. Rather I have tried to provide a framework of the basic principles of EC coupling in cardiomyocytes. At times, I have simplified things or omitted details for the sake of clarity and conciseness. Particular focus has been given to some more recent aspects of EC coupling in cardiomyocytes, e.g. important ultrastructural and functional differences between different types of cardiomyocytes. Readers, who are interested to learn more on cardiac EC coupling, or some aspects thereof, that can be provided here, are referred to some excellent recent and "classical" textbooks dealing with (parts of) this topic (Bers 2001).

3.2 Excitation–Contraction Coupling in Cardiomyocytes: Basic Principles

Excitation–contraction (EC) coupling refers to the cellular processes linking electrical excitation of the cardiomyocyte, i.e. the sarcolemmal action potential (AP), with contraction, i.e. activation of the myofilaments. The essential link between excitation and contraction is an increase in intracellular Ca^{2+} concentration ($[Ca^{2+}]_i$). The $[Ca^{2+}]_i$ increase is brought about by sarcolemmal Ca^{2+} influx during the AP triggering ensuing Ca^{2+} release from the sarcoplasmic reticulum (SR), the major intracellular Ca^{2+} store. Sarcolemmal Ca^{2+} influx occurs predominantly through L-type Ca^{2+} channels (LTCC) and SR Ca^{2+} release mostly through ryanodine receptors (RyR). This so-called Ca^{2+}-induced Ca^{2+} release (CICR) is characteristic for cardiomyocytes (Fabiato and Fabiato 1975; Niggli and Lederer 1990). It distinguishes them from skeletal muscle, in which SR Ca^{2+} release is triggered by voltage (depolarisation) rather than Ca^{2+} influx (Armstrong et al. 1972). Following termination of SR Ca^{2+} release, $[Ca^{2+}]_i$ starts to decline. $[Ca^{2+}]_i$ decline is mediated by several mechanisms, the two most important ones being Ca^{2+} reuptake into the SR via the SR Ca^{2+}-ATPase (SERCA), which is regulated by phospholamban (PLB), and Ca^{2+} extrusion from the cell via the sarcolemmal Na^+–Ca^{2+} exchanger (NCX). The $[Ca^{2+}]_i$ increase induced by CICR elicits contraction by binding of Ca^{2+} to troponin C, which initiates actin–myosin interaction. $[Ca^{2+}]_i$ decline, in turn, causes unbinding of Ca^{2+} from troponin C and the ensuing relaxation.

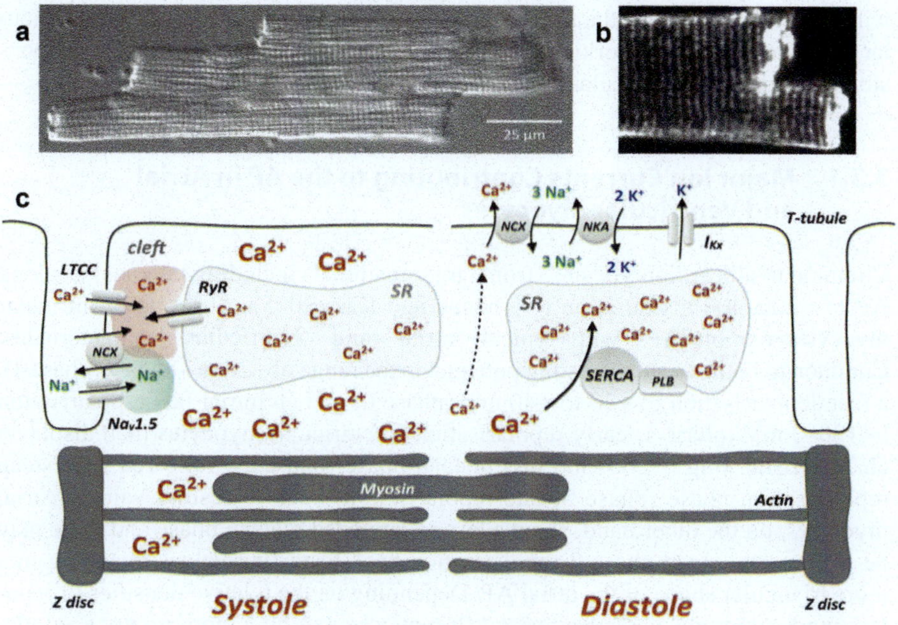

Fig. 3.1 Excitation–contraction (EC) coupling in ventricular cardiomyocytes. (**a**). Light microscopic (differential interference contrast) image of a rat ventricular myocyte (same cell as in Fig. 3.2a) revealing the typical cross striations of cardiomyocytes. (**b**). Confocal image showing part of a rat ventricular myocyte stained with the membrane dye di-8-ANEPPS (same cell as in Fig. 3.2b). Image illustrates the regular T-tubular membrane system typical for ventricular myocytes. (**c**). Schematic presentation of EC coupling in a ventricular myocyte. *Left side* shows the mechanisms underlying the $[Ca^{2+}]_i$ increase in systole, *right side* the mechanisms responsible for $[Ca^{2+}]_i$ decline in diastole. Abbreviations as in the text. *IKx* refers to various K^+ currents

As discussed below, there are important differences in EC coupling between the various types of cardiomyocytes, between species or in cardiac disease. The basic principles of EC coupling outlined here, however, are valid for all types of mammalian cardiomyocytes. They are illustrated schematically in Fig. 3.1 for a ventricular cardiomyocyte.

3.3 The Action Potential: Where It All Begins

Cardiac excitation originates in the sinoatrial (SA) node, where autonomously active pacemaker cells generate an AP. This AP propagates via the atria to the atrioventricular (AV) node and, further, via the bundle of His, the right and left bundle branches and the Purkinje fibres to the ventricular myocardium. Pacemaker cells of the SA node and the cardiac conduction system display some unique properties that will be discussed later. In the following, I will focus on the AP in atrial and ventricular myocytes from the "working" myocardium, the cardiomyocytes that are responsible for atrial and ventricular contraction.

3.3.1 Major Ion Currents Contributing to the AP in Atrial and Ventricular Myocytes

Common to all cardiomyocytes (from larger mammals including humans) is a long AP of ≈ 200–400 ms duration (Nerbonne and Kass 2005). Atrial and ventricular myocytes exhibit APs with typical "atrial" and "ventricular" characteristics. Common to both is a stable resting potential in the range of −75 to −85 mV (phase 4), a fast depolarisation to +20 to +40 mV (phase 0) and an initial fast repolarisation to +20–0 mV (phase 1, early repolarisation). Ventricular myocytes then display a characteristic long (≈200–300 ms) plateau phase (phase 2), followed by a steep repolarisation phase (phase 3, late repolarisation) back to resting values. Atrial myocytes, on the other hand, show a less pronounced plateau phase and there may be slower, less steep phase 3 repolarisation back to resting values, resulting in a more triangular shape of the atrial AP. Depending on the relative densities of repolarizing K^+ currents, however, atrial APs may also exhibit a more pronounced plateau phase and steeper phase 3 repolarisation (Wang et al. 1993). In general, atrial APs are shorter than ventricular APs. The differences in shape and duration of atrial and ventricular APs are caused by differences in the underlying ion currents. The negative resting potential is caused mainly (but not exclusively) by the inward rectifier K^+ current, I_{K1}. The density of I_{K1} is lower in atrial myocytes (Varro et al. 1993). Hence, the resting potential in atrial myocytes is usually a few millivolts less negative than in ventricular myocytes. In both atrial and ventricular myocytes, phase 0 depolarisation is carried by inward Na^+ current, I_{Na}, and phase 2 depolarisation (the plateau phase) by inward L-type Ca^{2+} current, $I_{Ca,L}$, which is also the main trigger for SR Ca^{2+} release. Repolarisation differs substantially between atrial and ventricular myocytes (Nerbonne and Kass 2005). Phase 1 repolarisation is caused by inactivation of I_{Na} and activation of transient outward K^+ currents, I_{to}. In ventricular myocytes, I_{to} is made up of a fast ($I_{to,fast}$) and a slow ($I_{to,slow}$) component, whereas atrial myocytes exhibit only the fast component, $I_{to,fast}$. Phase 3 repolarisation in ventricular myocytes is caused by activation of the delayed rectifier K^+ currents, I_{Kr} ("rapid") and I_{Ks} ("slow"). Furthermore, there is a large contribution of I_{K1} to ventricular phase 3 repolarisation. In atrial myocytes, I_{Kr}, I_{Ks} and I_{K1} also contribute to repolarisation, but their current densities are considerably lower than in ventricular myocytes (Nerbonne and Kass 2005; Hibino et al. 2010). In addition, atrial myocytes express another rapidly activating ("ultrarapid"), voltage-dependent K^+ current, I_{Kur}. I_{Kur} contributes to both early and late repolarisation in atrial myocytes. Because I_{Kur} is specifically expressed in atrial myocytes and virtually absent in ventricular myocytes, it is a promising target for treatment of atrial arrhythmias and efforts are underway to develop selective I_{Kur} blockers (Dobrev et al. 2012).

Table 3.1 summarises the major ion currents, including those generated by electrogenic ion transporters, contributing to the AP in human atrial and ventricular myocytes as well as the genes and proteins underlying these currents.

3.3.2 Na^+–Ca^{2+} Exchanger and Na^+–K^+ ATPase Contribute to the AP in Atrial and Ventricular Myocytes

There is an important contribution of electrogenic ion transporters to both the AP and the resting potential. The two most important electrogenic ion transporters in cardiomyocytes are NCX and the Na^+–K^+ pump or Na^+–K^+ ATPase (NKA) (Table 3.1). Both are expressed at high density in atrial and ventricular myocardium (Gaborit et al. 2007). NCX1, the cardiac isoform, is made up of a single polypeptide encoded by the *SLC8A1* gene. It exchanges 3 Na^+ for 1 Ca^{2+} (Reeves and Hale 1984) and, thus, generates a current, I_{NCX}. The reversal potential for I_{NCX} is given by $E_{NCX}=3\ E_{Na}-2\ E_{Ca}$. Thus, depending on membrane potential (E_m) and transsarcolemmal [Na^+] and [Ca^{2+}] gradients, NCX can run either in Na^+ influx–Ca^{2+} efflux (forward) mode (when $E_m < E_{NCX}$) or in Na^+ efflux–Ca^{2+} influx (reverse) mode (when $E_m > E_{NCX}$). At the very beginning of the AP, when Na_v channels open to depolarise E_m and Na^+ rushes in to elevate local [Na^+], NCX runs in reverse mode, bringing Ca^{2+} into the cell and generating a repolarizing outward current (Bers 2002). During most of the AP, however, when $[Ca^{2+}]_i$ is elevated, and during the resting potential, NCX runs in forward mode to remove Ca^{2+} from the cell and, thus, contributes a depolarizing inward current (Bers 2002). NKA is a heteromer composed of an α subunit, a β subunit and phospholemman. There are three different α subunits in cardiomyocytes, of which α1 is most abundant. α1 NKA serves as the housekeeping subunit regulating bulk cytoplasmic [Na^+], whereas α2 and α3 NKA may be involved in local modulation of cardiomyocyte function including EC coupling (Shattock et al. 2015). For each ATP hydrolyzed, NKA exports 3 Na^+ and imports 2 K^+, thereby generating an outward current, I_{NKA}. The main function of NKA is to maintain the physiological transsarcolemmal [Na^+] and [K^+] gradients. Because of its electrogenic nature, however, NKA also contributes to the membrane potential, a fact that is often ignored. I_{NKA} makes the resting potential more negative, reduces the plateau phase and contributes to phase 3 repolarisation. In ventricular myocytes, inhibition of I_{NKA} depolarises the resting potential by a few millivolts and prolongs AP duration by as much as 21 % (Glitsch 2001). Similar effects are observed in atrial myocytes, where inhibition of I_{NKA} depolarises the resting potential by 7 mV and prolongs AP duration by 13 % (Workman et al. 2003). The density of I_{NKA} is lower in atrial than in ventricular myocytes (0.7–1.1 pA/pF versus 1.6–2.2 pA/pF), which may explain, in part, the smaller effect of I_{NKA} inhibition on AP prolongation and the less negative resting potential in atrial myocytes (Kockskamper and Glitsch 1997; Glitsch 2001; Workman et al. 2003).

Table 3.1 Major currents contributing to the action potential in human atrial and ventricular myocytes

Current	Gene	Protein	Functions	Phase of AP	Expression/current density
I_{Na}	*SCN5A*	$Na_v1.5$	Activation: fast initial depolarisation	Phase 0	Atria, ventricles
			Inactivation: early repolarisation	Phase 1	
$I_{to,fast}$	*KCND3*	$K_v4.3$	Early repolarisation	Phase 1	Atria >> ventricles
$I_{to,slow}$	*KCNA4*	$K_v1.4$	Early repolarisation	Phase 1	Ventricles
$I_{Ca,L}$	*CACNA1C*	$Ca_v1.2$	Late depolarisation (plateau phase)	Phase 2	Atria, ventricles
I_{Kr}	*KCNH2*	HERG	Late repolarisation	Phase 3	Atria, ventricles
I_{Ks}	*KCNQ1*	K_vLQT1	Late repolarisation	Phase 3	Atria, ventricles
I_{Kur}	*KCNA5*	$K_v1.5$	Early and late repolarisation	Phases 1, 3	Atria
I_{K1}	*KCNJ2*	Kir2.1	Late repolarisation, resting potential	Phases 3, 4	Ventricles > atria
	KCNJ12	Kir2.2			
	KCNJ4	Kir2.3			
I_{NCX}	*SLC8A1*	NCX1	Outward current and Ca^{2+} influx during phases 0 and 1	Phases 0–4	Ventricles > atria
			Inward current and Ca^{2+} efflux during phases 2–4		
I_{NKA}	*ATP1A1*	α1 NKA	Na^+ and K^+ homeostasis	Phases 0–4	Ventricles > atria
	ATP1A2	α2 NKA	Contribution of outward current during phases 0–4		
	ATP1A3	α3 NKA			

Genes denote the pore-forming α subunits of ion channels or, in case of NKA, the catalytic and ion-transporting α subunits

Modulatory subunits of channels and transporters are not shown

3.3.3 Additional Currents Modulating the Membrane Potential in Atrial and Ventricular Myocytes

A variety of additional currents may modulate membrane potential in atrial and ventricular cardiomyocytes. Some of these currents are considered "background" currents, whereas others become activated under certain physiological or pathophysiological conditions.

Parasympathetic (vagal) stimulation, via its transmitter acetylcholine (ACh), activates a K^+ current, $I_{K(ACh)}$ (Sakmann et al. 1983). ACh binds to M_2 muscarinic receptors coupled to the inhibitory G-protein, G_i. The βγ subunit of G_i then causes fast activation of $I_{K(ACh)}$ (Logothetis et al. 1987). Activation of $I_{K(ACh)}$ hyperpolarises the resting potential (by shifting towards E_K) and reduces AP duration (Belardinelli and Isenberg 1983). Furthermore, it also reduces membrane resistance, thus weakening the effect of depolarizing currents. $I_{K(ACh)}$ is most prominently expressed in the SA and AV node and in atrial myocardium (Hibino et al. 2010), where it mediates the negative chronotropic and dromotropic effects of vagal stimulation.

Ischaemia causes a number of metabolic changes including a fall in intracellular [ATP] and a rise in intracellular ADP, AMP and adenosine concentrations. The fall in [ATP] and the concomitant rise in [ADP] will activate the ATP-dependent K^+ current, $I_{K(ATP)}$ (Noma 1983). $I_{K(ATP)}$ is found both in atrial and ventricular myocytes. Extracellular adenosine acts via G_i-coupled A_1 adenosine receptors to activate a K^+ current, $I_{K(Ado)}$. Thus, adenosine and acetylcholine share the same effects by activation of the same population of K^+ channels (Belardinelli and Isenberg 1983; Kurachi et al. 1986). The combined activation of $I_{K(ATP)}$ and $I_{K(Ado)}$ will contribute to changes in cardiac function in ischaemia including the slowing of AV nodal conduction and AP shortening in ventricular myocardium.

$I_{K(ACh)}/I_{K(Ado)}$ and $I_{K(ATP)}$ – like I_{K1} – belong to the family of inwardly rectifying K^+ channels (Kir) (Hibino et al. 2010). Kir channel α subunits contain two membrane-spanning domains rather than the six membrane-spanning domains found in K_v channel α subunits (and in Na_v and Ca_v channel α subunits, where this motif occurs four times). More recently, yet another family of K^+ channels was identified with the unique structure of two-pore (2P) domain each with the motif of two membrane-spanning domains. They were termed K2P channels. Some K2P channels are also found in cardiomyocytes (e.g. TREK-1, TASK-1) where they might contribute to background current. TASK-1 is an acid-sensitive K2P channel α subunit that is expressed in human atrial and in rat ventricular myocytes (Putzke et al. 2007; Limberg et al. 2011). I_{TASK} contributes to AP repolarisation (Putzke et al. 2007; Limberg et al. 2011). It is expressed specifically in human atria and upregulated in chronic atrial fibrillation, thus constituting a potential new drug target (Schmidt et al. 2015).

Stretch activates or modulates several ion currents in the heart. In mammalian cardiomyocytes, stretch may activate a non-selective cation current, $I_{SAC,NS,}$ and a K^+ current, $I_{SAC,K}$. $I_{SAC,NS}$ may be formed by TRPC channel subunits, whereas one possible candidate for $I_{SAC,K}$ is TREK-1, a K2P channel α subunit (Maroto et al. 2005; Terrenoire et al. 2001). Activation of stretch-activated currents has important consequences for cardiac function. For example, activation of $I_{SAC,NS}$ has been shown to accelerate diastolic depolarisation in pacemaker myocytes, thus increasing heart rate, and to alter AP configuration in ventricular myocytes with potential induction of arrhythmias (Cooper et al. 2000; Kohl et al. 2001). Selective inhibition of these channels by GsMtx-4, a peptide isolated from spider venom, is able to prevent stretch-induced atrial fibrillation (Bode et al. 2001).

3.3.4 Regional Differences in AP Morphology Within the Heart

AP morphology differs between different regions and cell types within the heart. These differences arise because of the differential expression of ion channels. In addition to the differences between atrial and ventricular APs described above, there are notable differences between pacemaker myocytes and those of the working myocardium and between various regions within the ventricular myocardium (Nerbonne and Kass 2005).

Pacemaker myocytes (e.g. from the SA and AV node) lack a stable resting potential but rather exhibit diastolic (phase 4) depolarisation, which is carried, in part, by the "funny" current, I_f (DiFrancesco 1986, 2010). I_f is a non-selective cation current mainly conducting Na^+ and K^+. It is activated by hyperpolarisation and cyclic nucleotides (cAMP) and made up of HCN4 subunits (DiFrancesco 2010). Other mechanisms contribute to diastolic depolarisation in pacemaker myocytes including T-type Ca^{2+} current, $I_{Ca,T}$, and a "Ca^{2+} clock" recruiting depolarizing I_{NCX} via SR Ca^{2+} release (Huser et al. 2000; Lakatta et al. 2010).

Ventricular myocytes isolated from various regions within the ventricles exhibit considerable differences in AP duration. For example, subendocardial myocytes show longer APs than subepicardial myocytes, and these differences have been attributed largely to differential expression of voltage-dependent K^+ currents (Nerbonne and Kass 2005). The transmural gradient in AP duration and repolarisation times has been suggested to underlie the T-wave in the electrocardiogram. More recent (and older) studies, however, have challenged this notion (Janse et al. 2012). In the intact human heart, there is only a small transmural gradient in repolarisation times, and it is likely that rather apico-basal differences in repolarisation times determine the T-wave in the electrocardiogram (Janse et al. 2012).

3.4 The $[Ca^{2+}]_i$ Transient: The Essential Link Between Excitation and Contraction

The AP is the first step in the EC coupling process. It is followed by an increase in $[Ca^{2+}]_i$ which ultimately induces contraction via binding to the myofilaments. Figure 3.1c depicts the major processes occurring during EC coupling in a ventricular myocyte: sarcolemmal Ca^{2+} influx during the AP, largely through LTCC, triggers Ca^{2+} release from the SR, largely through RyR. This systolic $[Ca^{2+}]_i$ increase from baseline values of ≈ 100 nM to peak values of ≈ 1 μM $[Ca^{2+}]_i$ initiates contraction. Removal of Ca^{2+} from the cytoplasm, mainly by SERCA and NCX, causes diastolic $[Ca^{2+}]_i$ decrease to initiate relaxation. The transient increase and decrease of $[Ca^{2+}]_i$ during a twitch is termed the $[Ca^{2+}]_i$ transient (CaT). It is the essential link between cardiac excitation and contraction.

3.4.1 L-type Ca^{2+} Influx Triggers SR Ca^{2+} Release Through Ryanodine Receptors

During the AP, LTCC are activated and the ensuing Ca^{2+} influx triggers Ca^{2+} release from the SR via RyR (Fig. 3.1c, left). The major channel-forming α1 subunit of LTCC in both atrial and ventricular myocytes is α1C ($Ca_v1.2$, encoded by the *CACNA1C* gene). Atrial myocytes may also express some α1D ($Ca_v1.3$, encoded by the *CACNA1D* gene). LTCC are the target of an important drug class, the dihydropyridines (prototype: nifedipine), which is why they are sometimes called dihydropyridine receptor or DHPR. Ryanodine receptors (RyR) are named after ryanodine, a plant alkaloid that binds specifically to the channel protein. The cardiac isoform of RyR is type-2 (RyR2, encoded by the *RYR2* gene) in both atrial and ventricular myocytes. The RyR2 channel is a homotetramer and one of the largest proteins in cardiomyocytes with a molecular weight of >2 MDa.

LTCC have been studied extensively for more than three decades now, and this work has been summarised in excellent reviews (e.g. McDonald et al. 1994; Benitah et al. 2010). Ca^{2+} influx via LTCC serves two main functions in cardiomyocytes: (1) it provides trigger Ca^{2+} for CICR and (2) it provides Ca^{2+} for loading of the SR Ca^{2+} store. Key properties of LTCC include large conductance with long openings (hence "L"-type), high-voltage activation (at potentials less negative than −40 mV) and voltage- and Ca^{2+}-dependent inactivation. The latter is mediated by calmodulin bound to the channel at the C-terminus (Peterson et al. 1999) and occurs when local $[Ca^{2+}]$ is elevated through Ca^{2+} permeating the channel and through CICR from nearby RyR. Ca^{2+}-dependent inactivation provides a negative feedback mechanism limiting sarcolemmal Ca^{2+} entry and avoiding Ca^{2+} overload of cardiomyocytes (Bers 2008).

LTCC are regulated by phosphorylation through various protein kinases. Phosphorylation by protein kinase A (PKA) underlies the increase in L-type Ca^{2+} current mediated by β-adrenergic stimulation. How this is achieved, however, has not been clarified yet. One of the sites phosphorylated is S1928 of the α1C subunit, but recent data indicate that this (alone) cannot explain the functional effects of PKA stimulation on LTCC (Ganesan et al. 2006; Lemke et al. 2008). Other sites must contribute, and these may include S478 and S479 on the β2 subunit (Bunemann et al. 1999) or sites on the associated cytoskeletal protein *ahnak* (Benitah et al. 2010). Protein kinase C (PKC) also phosphorylates LTCC on the α1C and β2 subunit, and both stimulation and inhibition of the current have been reported (Benitah et al. 2010). Moreover, Ca^{2+}-/calmodulin-dependent protein kinase II (CaMKII) regulates LTCC. Stimulation of CaMKII causes facilitation of the L-type Ca^{2+} current (Anderson et al. 1994), and this may involve phosphorylation of the α1C subunit at S1512 and S1570 (Lee et al. 2006) and of the β2 subunit at T498 (Koval et al. 2010).

RyR2 represent the major SR Ca^{2+} release channel in cardiomyocytes. Opening of RyR2 decides about the amount of Ca^{2+} released from the SR during EC coupling and, hence, about the strength of contraction. Opening (and closing) of RyR2, therefore, is highly regulated (Fill and Copello 2002). Ca^{2+} is a major regulator of RyR2. Open probability (P_o) of RyR2 is increased by an increase of $[Ca^{2+}]$ on the cytosolic face of the channel and by an increase of $[Ca^{2+}]$ within the SR, i.e. luminal $[Ca^{2+}]$. The former occurs during Ca^{2+} influx through LTCC and is the major means of triggering RyR2 opening during EC coupling. The latter may occur as a result of β-adrenergic stimulation or cardiac glycoside action and accounts for the increase in fractional SR Ca^{2+} release at elevated SR Ca^{2+} load (see below). Other regulators of RyR2 include Mg^{2+} (inhibitory), ATP (stimulatory) and glycolytic intermediates (Fill and Copello 2002; Kockskamper et al. 2005; Zima et al. 2006). Experimentally, caffeine represents an important tool to study RyR2 function. Caffeine increases

RyR2 P_o by sensitising the channel to cytosolic Ca^{2+} (Rousseau and Meissner 1989). High concentrations of caffeine (10–20 mM) are used to assess SR Ca^{2+} load (see below). Other experimentally useful tools include ruthenium red and tetracaine, both decreasing RyR2 P_o. Recently, "stabilisers" of RyR2 (K201 (aka JTV-519) and S107) have been developed as potential new antiarrhythmic drugs (Wehrens et al. 2004a; Wehrens and Marks 2004; Lehnart et al. 2008).

RyR2 exist as macromolecular complexes with various adapter and regulatory proteins (Marx et al. 2000; Fill and Copello 2002; Farrell et al. 2003; Wehrens and Marks 2004; Lehnart et al. 2005). These include FK506-binding protein (FKBP) 12.6 (aka calstabin2, stabilising the closed state of the channel), calmodulin, S100A1 and sorcin (each inhibiting channel activity), triadin, junctin and calsequestrin (acting in concert to "sense" luminal [Ca^{2+}]), CaMKII, PKA anchored via mAKAP, protein phosphatase 1 (PP1) anchored via spinophilin, protein phosphatase 2A (PP2A) anchored via PR130 and cAMP-specific phosphodiesterase 4D. Thus, RyR2 activity is regulated in a very complex and local manner in the immediate vicinity of the channel both from the cytosolic and luminal side. Phosphorylation of RyR2 appears to be an important regulatory mechanism as suggested by the presence of two kinases (PKA and CaMKII) and two phosphatases (PP1 and PP2A) within the macromolecular complex. RyR2 can be phosphorylated by both PKA and CaMKII. It contains numerous phosphorylation sites (Rodriguez et al. 2003; Huttlin et al. 2010; Scriven et al. 2013). So far, only three of these have been identified and characterised in more detail: (1) S2809 is phosphorylated by both PKA and CaMKII in vitro and is assumed to be phosphorylated primarily by PKA in vivo (Marx et al. 2000; Rodriguez et al. 2003; Wehrens et al. 2004b), (2) S2815 is phosphorylated by CaMKII (Wehrens et al. 2004b) and (3) S2030 is phosphorylated by PKA (Xiao et al. 2005). Phosphorylation of S2809 and S2815 is associated with increased RyR2 channel activity, increased diastolic Ca^{2+} leak and arrhythmias (Wehrens et al. 2004b; Wehrens and Marks 2004). The role of S2809 phosphorylation is controversially debated (Bers 2012).

LTCC and RyR2 are found in the cardiac dyad (Fig. 3.1c, left), where sarcolemma and junctional SR come close to each other. The dyadic cleft is $\approx$ 12–15 nm wide and comprises a small volume in which accumulation of ions, particularly Ca^{2+}, can readily occur (Scriven et al. 2013). The functional unit within a dyad is called a couplon made up of a cluster of $\approx$ 10 LTCC and a juxtaposed cluster of $\approx$ 75–100 RyR2 (Scriven et al. 2010). More recent data suggest even smaller RyR2 cluster sizes with an average of $\approx$ 14 RyR2 per cluster (Baddeley et al. 2009; Kohl et al. 2013; Scriven et al. 2013). A ventricular myocyte contains $\approx$ 20–30,000 couplons (Bers 2008; Jayasinghe et al. 2009). The special architecture of the couplon ensures that Ca^{2+} entering through LTCC can increase sufficiently in the cleft to trigger SR Ca^{2+} release through RyR2 despite their rather low Ca^{2+} sensitivity (Bers 2008).

Before I discuss further details of Ca^{2+} regulation in cardiomyocytes, I will introduce some important structural and functional differences between atrial and ventricular myocytes with pivotal implications for subcellular Ca^{2+} signalling in EC coupling.

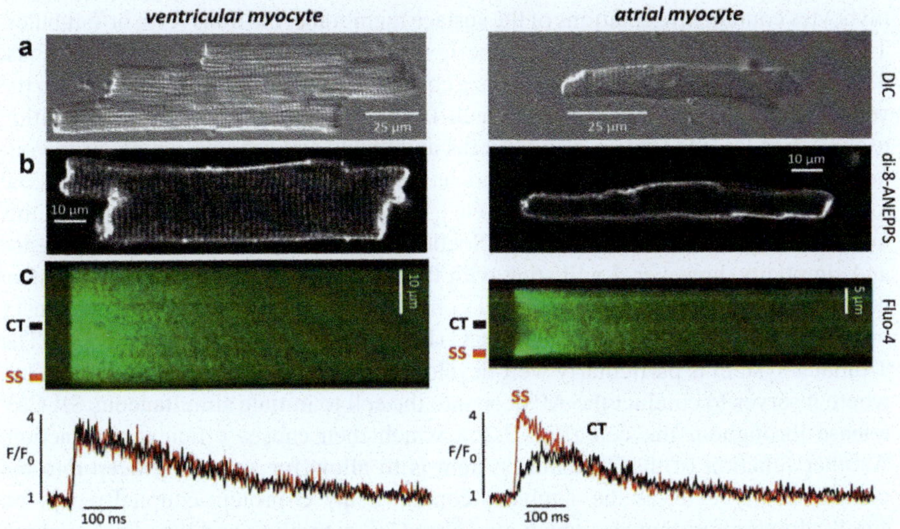

Fig. 3.2 Comparison of morphology and subcellularly resolved CaTs between atrial and ventricular cardiomyocytes. Rat ventricular myocytes are shown on the *left*, rat atrial myocytes on the *right*. (**a**). Light microscopic (differential interference contrast, DIC) images revealing typical cross striations in both types of cardiomyocytes. Images recorded on a Leica DMI3000B microscope equipped with a 63×/1.4 NA oil immersion objective lens. (**b**). Confocal images of di-8-ANEPPS-stained myocytes revealing a regular T-tubular system in the ventricular myocyte and the absence of such a system in the atrial myocyte. Cells were stained with di-8-ANEPPS (10 µM; 7 min; excitation, 488 nm; emission, >585 nm) and bathed in normal Tyrode's solution. (**c**). Linescan images of fluo-4 fluorescence (*top*) and normalised fluorescence traces (*bottom*), F/F_0, where F_0 denotes minimum fluorescence in diastole, of subcellular regions as indicated on the *left hand side* of the linescans. The line was positioned perpendicular to the longitudinal axis of the cells. A single CaT is shown (obtained during steady-state stimulation at 1 Hz). Note that Ca^{2+} (fluorescence) increases simultaneously throughout the cell (the entire *scan line*) in the ventricular myocyte, whereas there are substantial spatio-temporal differences in the atrial myocyte. Linescans were acquired at 1.15 ms per line from myocytes bathed in Tyrode's solution (1.5 mM $[Ca^{2+}]_o$) at room temperature and loaded with fluo-4/AM (7.5 µM; 20 min loading; 20 min wash; excitation, 488 nm; emission, >505 nm). Images in **b** and **c** were recorded on a Zeiss LSM510 confocal microscope equipped with a 63×/1.3 NA oil immersion objective lens. Thickness of the confocal plane was 1–1.5 µm. Images were processed with ImageJ (Recordings in **b** and **c** kindly provided by T. Kiess)

3.4.2 T-tubule or Not T-tubule: EC Coupling Differs Between Atrial and Ventricular Myocytes

There are important differences in EC coupling between atrial and ventricular myocytes with regard to ultrastructure and Ca^{2+} signalling (Fig. 3.2). Both atrial and ventricular myocytes are cross striated (Fig. 3.2a). Ventricular myocytes are very large, rod- or brick-shaped cells (length $\geq$ 100 µm, width $\geq$ 20 µm) with a volume of $\approx$ 30–35 pl (Satoh et al. 1996). By contrast, atrial myocytes – at least from smaller mammals – are much smaller, thin, more spindle-shaped cells (length $\leq$ 100 µm, width $\approx$ 10 µm) with a volume of $\approx$ 3–5 pl (Walden et al. 2009). Thus, atrial myocyte volume is $\approx$ six- to tenfold lower than ventricular myocyte volume. Ventricular myocytes contain invaginations of the surface membrane, the transverse or T-tubules. T-tubules occur regularly with sarcomeric spacings of a little less than 2 μm (Figs. 3.1b and 3.2b). The T-tubular system is a complex network penetrating deep into the ventricular myocyte (Soeller and Cannell 1999; Jayasinghe et al. 2009). In addition to the T-tubules, the system also contains axial parts and segments running in various directions. Therefore, some authors have suggested the terms "transverse-axial tubular system" or "sarcolemmal Z rete", which describe the morphology of this membrane system more accurately (Soeller and Cannell 1999). For convenience and simplicity, however, I will stick with the terms T-tubules and T-tubular system here. The T-tubules contain important components of the EC coupling machinery including Na_v channels, NCX and LTCC (Fig. 3.1c, left) (Scriven et al. 2000). The T-tubular system is particularly well-developed in ventricular myocytes (Fig. 3.2b), where it serves to conduct the AP deep into the cell to initiate simultaneous SR Ca^{2+} release throughout the cell (Fig. 3.2c), which then causes efficient contraction. Another function of the T-tubular system is to allow for signalling deep into the ventricular myocyte, as the T-tubules contain many G-protein-coupled receptors, e.g. β-adrenergic receptors, which may alter Ca^{2+} signalling and thus EC coupling. In contrast to ventricular myocytes, atrial myocytes from many mammalian species, including mouse, rat, rabbit and cat, contain no or only few T-tubules (Fig. 3.2b) (Huser et al. 1996; Dibb et al. 2013; Greiser et al. 2014). The absence of a regular T-tubular system in these atrial myocytes has profound functional consequences with regard to subcellular Ca^{2+} signalling (Fig. 3.2c). The systolic $[Ca^{2+}]_i$ increase in atrial myocytes does not occur synchronously, as in ventricular myocytes, but rather in two steps: first, Ca^{2+} influx through LTCC triggers Ca^{2+} release from the subsarcolemmal SR only and, second, the subsarcolemmal $[Ca^{2+}]_i$ increase propagates towards the cell centre via CICR from central SR to increase central $[Ca^{2+}]_i$ and initiate contraction. These subcellular inhomogeneities in the atrial CaT can be visualised by means of fast confocal Ca^{2+} imaging (Fig. 3.2c). In transverse linescan images, the atrial CaT appears U-shaped with central (CT) areas lagging behind the increase in subsarcolemmal (SS) $[Ca^{2+}]$ (Fig. 3.2c, right) (Berlin 1995; Huser et al. 1996). Furthermore, the subsarcolemmal $[Ca^{2+}]$ transient is usually larger in amplitude than the central $[Ca^{2+}]$ transient (Fig. 3.2c, right). This contrasts with the ventricular CaT, which appears homogenous starting simultaneously at a single line throughout the cell with no appreciable differences in either time course or amplitude between subsarcolemmal and central areas (Fig. 3.2c, left). In fast two-dimensional imaging, the atrial CaT appears as a peripheral ring of elevated subsarcolemmal $[Ca^{2+}]$ followed by a central increase in $[Ca^{2+}]$ (Kockskamper et al. 2001; Mackenzie et al. 2001; Blatter et al. 2003), whereas the ventricular CaT shows a simultaneous $[Ca^{2+}]$ increase throughout the myocyte.

Interestingly, the ventricular CaT can be transformed into an atrial-like CaT. Using formamide to detubulate ventricular myocytes, the ventricular CaT no longer appears homogenous but rather looks U-shaped like an atrial CaT (Brette et al. 2002, 2005). This finding provides convincing evidence that indeed the T-tubular system in ventricular myocytes is responsible for the subcellularly homogenous $[Ca^{2+}]_i$ increases during the electrically stimulated CaT and a lack thereof results in an atrial-like CaT with $[Ca^{2+}]_i$ increases starting in the periphery before propagating towards the cell centre.

Recent studies have revealed that atrial myocytes from larger mammals, including dog, sheep, cow and horse, contain a more regular T-tubular system (Dolber et al. 1994; Dibb et al. 2009; Lenaerts et al. 2009; Richards et al. 2011; Dibb et al. 2013; Trafford et al. 2013). An obvious and important question is whether human atrial myocytes possess a T-tubular system. While earlier observations suggested this not to be the case, recent data point towards the existence of a more regular T-tubular system human atrial myocytes, although only a fraction of cells exhibited some T-tubules (Legato 1973; Richards et al. 2011). Possible explanations for these disparate findings are developmental, growth- or size-dependent, regional and disease-dependent alterations in the T-tubular system in atrial myocytes from humans (and other large mammalian species). For example, in sheep, cow and horse atrial myocytes, T-tubular abundance increases with cell width, and cardiac disease reduces T-tubular density in sheep atrial myocytes (Richards et al. 2011; Dibb et al. 2009; Lenaerts et al. 2009). Thus, the issue of (changes of) a T-tubular system in human atrial myocytes will have to be clarified in future studies as it bears important implications for human atrial physiology and pathology.

3.4.3 Sparkling Stars: Ca^{2+} Sparks Are the Building Blocks of the CaT

Ca^{2+} sparks are elementary Ca^{2+} release events in cardiomyocytes (and other muscle cell types). They were first described in 1993 in rat ventricular myocytes (Cheng et al. 1993) when laser-scanning confocal microscopes and fluorescent Ca^{2+}-sensitive dyes (fluo-3) had become available to study intracellular Ca^{2+} signalling with high spatial and temporal resolution (Ca^{2+} imaging). Ca^{2+} sparks are locally confined, short-lived Ca^{2+} release events of small amplitude (Cheng et al. 1996; Niggli 1999). Figure 3.3 shows examples of Ca^{2+} sparks (marked by the asterisks) recorded from rat ventricular myocytes by means of confocal linescan imaging of fluo-4 fluorescence. The beginning of the recording (Fig. 3.3a, arrow) shows an electrically stimulated CaT. Afterwards, stimulation was switched off. The remaining events represent spontaneously occurring Ca^{2+} waves (hash marks) and Ca^{2+} sparks (asterisks). Comparison with the electrically stimulated CaT reveals the tiny nature of a Ca^{2+} spark, which explains why the detection of Ca^{2+} sparks is not possible by means of epifluorescence microscopy but rather requires confocal Ca^{2+} imaging methods (note that the Ca^{2+} spark in Fig. 3.3a (red asterisk) is not detectable in the black trace, which represents the fluorescence averaged over the entire line (whole cell (wc))). Ca^{2+} sparks have a spatial spread of $\approx$ 2–3 μm, they last for $\approx$ 30–50 ms and their amplitude amounts to $\approx$ 200–300 nM (Cheng et al. 1996; Niggli 1999; Bers 2002). In ventricular myocytes, Ca^{2+} sparks occur predominantly at the T-tubules (Cheng et al. 1996). In atrial myocytes that lack T-tubules, Ca^{2+} sparks with similar properties are observed both in the subsarcolemmal and central area of the cell (Sheehan et al. 2006).

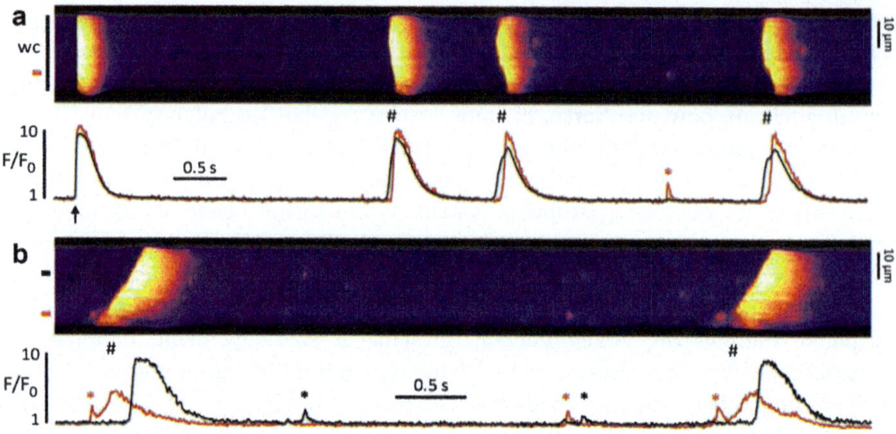

Fig. 3.3 CaT, Ca^{2+} waves and Ca^{2+} sparks in rat ventricular myocytes as revealed by confocal linescan imaging. (**a**). Linescan image (*top*) and normalised fluo-4 fluorescence traces (F/F_0, *bottom*) of subcellular regions from a rat ventricular myocyte as indicated on the *left*. The line was positioned perpendicular to the longitudinal axis of the cell. *Black trace* is averaged over the entire line (whole cell (wc)), *red trace* from the region where the Ca^{2+} spark (*asterisk*) was observed. The first event (marked by the *arrow*) is an electrically stimulated CaT. Afterwards, stimulation was switched off and the remaining events occur spontaneously. Ca^{2+} waves are indicated by the *hash marks* and the Ca^{2+} spark by the *asterisk*. (**b**). Similar recording from a different ventricular myocyte in the absence of electrical stimulation. Two Ca^{2+} waves (*hash marks*) and several Ca^{2+} sparks (*asterisks*) are visible. *Black and red* traces correspond to the regions as indicated on the *left*. Myocytes in **a** and **b** were treated with isoprenaline (10 nM) for several minutes in order to increase CaTs and the likelihood of spontaneous SR Ca^{2+} release events. Linescans were acquired at 3.02 ms per line, all other recording conditions as in Fig. 3.2c. Images were processed with ImageJ. Lookup table (*colour coding*) was changed in order to facilitate visualisation of small Ca^{2+} release events (Recordings kindly provided by T. Kiess)

Ca^{2+} sparks are caused by the opening of a functional cluster of RyR. Current estimates suggest that the Ca^{2+} flux through ≈ 4–20 RyR underlies a Ca^{2+} spark (Wang et al. 2001; Bers 2008). Ca^{2+} sparks can occur spontaneously. In resting cardiomyocytes, their frequency is low reflecting the low Ca^{2+} sensitivity of RyR. Physiologically, Ca^{2+} sparks are triggered by Ca^{2+} influx through LTCC, and the unitary Ca^{2+} influx through a single LTCC is sufficient to activate a Ca^{2+} spark (Wang et al. 2001). Ca^{2+} influx through LTCC increases local $[Ca^{2+}]$ within the dyadic cleft sufficiently (to ≈ 10–20 µM) to activate the RyR cluster to generate a Ca^{2+} spark, which increases cleft $[Ca^{2+}]$ to >100 µM (Fig. 3.1c, left) (Bers 2008). The spatial and temporal summation of Ca^{2+} sparks during an AP makes up the whole cell CaT (Cheng et al. 1995). Thus, Ca^{2+} sparks are the elementary building blocks of the electrically stimulated CaT in cardiomyocytes.

3.4.4 New Boys in Town: Alternative Mechanisms of Sarcolemmal Ca^{2+} Influx and SR Ca^{2+} Release

Although CICR elicited by L-type Ca^{2+} influx and RyR-mediated SR Ca^{2+} release is the dominant mechanism underlying the systolic $[Ca^{2+}]_i$ increase in ventricular myocytes, there are alternative mechanisms for both sarcolemmal Ca^{2+} influx and SR Ca^{2+} release in cardiomyocytes.

3.4.4.1 NCX-Triggered SR Ca^{2+} Release

In 1990, Ca^{2+} release from ryanodine-sensitive Ca^{2+} stores in ventricular myocytes was described, which was dependent on voltage-dependent Na^+ current and extracellular Ca^{2+} but independent of LTCC (Leblanc and Hume 1990). It was proposed that this SR Ca^{2+} release is mediated by the sarcolemmal NCX running in reverse mode to bring in Ca^{2+}, which then activates RyR to release Ca^{2+} from the SR (Fig. 3.1c, left). Reversal of NCX during the very early phase of the AP is now well accepted (Bers 2002). It is favoured by two mechanisms, which occur during the beginning of the AP: first, the membrane potential is depolarised and, second, Na^+ influx through Na_v channels causes an increase in local subsarcolemmal [Na^+] (Fig. 3.1c, left), thereby reducing the transsarcolemmal [Na^+] gradient. As soon as local [Ca^{2+}] in the cleft starts to rise, however, NCX will operate in forward mode again and remove Ca^{2+} from the cell. Under physiological conditions in a healthy heart (with relatively low $[Na^+]_i$), reversal of NCX occurs for only a few milliseconds during the beginning of the AP and, hence, NCX-triggered SR Ca^{2+} release makes only a small contribution to overall SR Ca^{2+} release (Sipido et al. 1997; Bers 2002). These findings are backed up by structural data suggesting that Na_v channels and NCX – although present in the T-tubules – do colocalise neither with one another nor with LTCC or RyR, but may rather be localised to subdomains within the T-tubules (Scriven et al. 2000). When $[Na^+]_i$ increases, however, NCX-triggered SR Ca^{2+} release can become quantitatively more important. This may be relevant under pathological conditions, e.g. in heart failure, where $[Na^+]_i$ is elevated and also expression of NCX may be increased. Furthermore, treatment with cardiac glycosides will increase $[Na^+]_i$. Even at a high $[Na^+]_i$ of 20 mM, however, SR Ca^{2+} release triggered by LTCC remains the dominant mechanisms of SR Ca^{2+} release in ventricular myocytes (Sipido et al. 1997).

3.4.4.2 IP_3-Mediated SR Ca^{2+} Release

In addition to RyR, the SR in cardiomyocytes also contains IP_3 receptors (IP_3R) albeit at much lower densities (Kockskamper et al. 2008b). Expression of IP_3R is higher in atrial than in ventricular myocytes (Lipp et al. 2000; Domeier et al. 2008). IP_3R can be found on the SR as well as in the perinuclear region including the nuclear envelope (Harzheim et al. 2009; Escobar et al. 2011; Ljubojevic et al. 2014). In atrial myocytes, there is also prominent expression of IP_3R in the subsarcolemmal SR, which partially overlaps with RyR expression (Lipp et al. 2000; Mackenzie et al. 2002). Stimulation of IP_3 signalling via $G\alpha_q$-coupled receptors (e.g. via endothelin-1) increases electrically stimulated CaTs and contractions in both atrial and ventricular myocytes from various species including human (Mackenzie et al. 2002; Zima and Blatter 2004; Proven et al. 2006; Domeier et al. 2008; Kockskamper et al. 2008a; Harzheim et al. 2009; Signore et al. 2013; Ljubojevic et al. 2014). At the subcellular level, stimulation of IP_3R increases the frequency of Ca^{2+} sparks (Zima and Blatter 2004; Domeier et al. 2008). With RyR blocked, however, stimulation of IP_3R (with agonists like IP_3 or adenophostin) elicits smaller Ca^{2+} release events distinct from RyR-mediated Ca^{2+} sparks (Zima and Blatter 2004). Thus, despite their relative paucity, functional IP_3R in atrial and ventricular myocytes release Ca^{2+} from the SR, in concert with neighbouring RyR, to increase CaTs and contractions. One function of IP_3R in cardiomyocytes, therefore, is to elicit a positive inotropic effect upon stimulation of $G\alpha_q$-coupled receptors (e.g. α_1-adrenergic, ET_A and AT_1 receptors). This "booster" function of IP_3 may be physiologically relevant to support cardiac contractile function during conditions of mechanical stress and/or sympathetic activation (when the respective hormone systems become activated). It comes at the expense, however, of a proarrhythmic effect, since the IP_3-induced extra Ca^{2+} release may also trigger Ca^{2+}-dependent arrhythmias (Mackenzie et al. 2002; Zima and Blatter 2004; Proven et al. 2006).

In addition to acting as a booster for EC coupling, IP_3-induced Ca^{2+} release is involved in nuclear Ca^{2+} regulation. The nuclear envelope is contiguous with the SR and acts as a Ca^{2+} store (Wu and Bers 2006; Ljubojevic et al. 2014). It contains IP_3R, which, upon activation, release Ca^{2+} into the perinuclear area and directly into the nucleus (Zima et al. 2007). Stimulation of $G\alpha_q$-coupled receptors increases $[IP_3]$ throughout the cardiomyocyte, and IP_3 is able to diffuse into the nucleus (Remus et al. 2006). IP_3 signalling may cause a preferential or selective increase in nuclear CaTs, which has been implicated in Ca^{2+}-dependent regulation of transcription (Wu et al. 2006; Kockskamper et al. 2008a, b; Higazi et al. 2009; Ljubojevic et al. 2014). In analogy to the EC coupling process, this process has been termed excitation–transcription coupling (Wu et al. 2006). IP_3-mediated cardiomyocyte (nuclear) Ca^{2+} signalling plays an important role for cardiac remodelling, e.g. during development of hypertrophy and heart failure (Bers 2008; Higazi et al. 2009; Nakayama et al. 2010; Ljubojevic et al. 2014).

3.4.4.3 Other Mechanisms of Ca^{2+} Influx and Release

Other mechanisms of Ca^{2+} influx and/or SR Ca^{2+} release have been observed in (some types of) cardiomyocytes (under certain experimental conditions) including but not limited to voltage-dependent SR Ca^{2+} release, Ca^{2+} influx through Na_v channels and Ca^{2+} influx through T-type Ca^{2+} channels (Bers 2001). Some of them have been discussed, at times, very controversially. Their relevance for EC coupling under normal conditions appears to be minor. For further discussion of this issue, see Bers (2001).

3.4.5 Ca^{2+} Removal from the Cytosol: SERCA and NCX Rule

Following the systolic $[Ca^{2+}]_i$ increase, Ca^{2+} has to be removed from the cytosol to allow for diastolic relaxation of cardiomyocytes. Cytosolic Ca^{2+} removal is achieved by four mechanisms: extrusion of Ca^{2+} from the cell via (1) the sarcolemmal NCX and (2) the sarcolemmal Ca^{2+}-ATPase as well as uptake of Ca^{2+} by (3) mitochondria and (4) the SR. The latter is mediated by SERCA (cardiac isoform 2a). Quantitatively, NCX and SERCA are most important accounting for 98–99 % of cytosolic Ca^{2+} removal (Bers 2002). Depending on species, SERCA alone accounts for $\approx$70 % to >90 % of cytosolic Ca^{2+} removal (Bers 2002).

SERCA, like NKA, is a P-type ATPase using the energy derived from ATP hydrolysis to transport Ca^{2+} ions against their electrochemical gradient from the cytosol into the lumen of the SR with an estimated free [Ca^{2+}] of ≈ 1–1.5 mM (Shannon et al. 2003). The predominant isoform in cardiomyocytes is SERCA2a, one of three splice variants of the SERCA2 gene, *ATP2A2* (Kranias and Hajjar 2012). SERCA activity is regulated by PLB (encoded by the *PLN* gene) (Kranias and Hajjar 2012). PLB is a small protein found in the membrane of the SR. In its unphosphorylated state, PLB is bound to SERCA2a and inhibits its activity. Upon phosphorylation, PLB dissociates from SERCA2a to form oligomers and SERCA inhibition is relieved (James et al. 1989; Kranias and Hajjar 2012). Thus, PLB acts like a brake for SERCA activity, which can be loosened by phosphorylation. Phosphorylation of PLB occurs at distinct sites mediated by distinct kinases. In vivo, following β-adrenergic stimulation, PLB is phosphorylated by PKA at S16 and by CaMKII at T17 (Mattiazzi et al. 2005). In vitro, PLB may also be phosphorylated by PKC at S10 and by protein kinase G (PKG) at S16, but the physiological relevance of PKC and PKG phosphorylation is unclear (Kranias and Hajjar 2012).

SERCA2a and PLB are crucial regulators of both diastolic and systolic cardiac function. On the one hand, SERCA activity is responsible, to a large degree, for the decay of the CaT and thus diastolic relaxation. On the other hand, SERCA activity controls SR Ca^{2+} load and hence the amount of Ca^{2+} released in systole. Progressive inhibition of SERCA activity results in progressive reductions of SR Ca^{2+} content and the amplitude and rate of decline of the CaT (Bode et al. 2011). Overexpression of SERCA2a or knockout of PLB both enhances cardiac contraction and relaxation (Luo et al. 1994; He et al. 1997).

NCX is the main sarcolemmal Ca^{2+} extrusion mechanism. Forward mode (Na^+ influx–Ca^{2+} efflux) NCX is activated shortly after the upstroke of the AP upon the systolic $[Ca^{2+}]_i$ increase and persists for the remainder of the CaT (Bers 2002). During CaT decay NCX and SERCA compete for Ca^{2+} removal from the cytosol. NCX activity, however, makes a much smaller contribution to $[Ca^{2+}]_i$ decline than SERCA activity. This can be deduced from the recording shown in Fig. 3.4. It illustrates four electrically stimulated CaTs followed by a caffeine-induced CaT (to estimate SR Ca^{2+} load and fractional release, see below). The grey inset shows an overlay of the last electrically stimulated CaT (black) and the caffeine-induced CaT (dark grey) with amplitudes normalised. During the caffeine-induced CaT, $[Ca^{2+}]_i$ decay is much slower than during the electrically stimulated CaT. In the presence of caffeine, which opens all RyR to release the Ca^{2+} stored in the SR, the SR is short-circuited and only the sarcolemmal NCX mediates $[Ca^{2+}]_i$ decay. Hence, this can be used to derive the rate constant for NCX activity. During the electrically stimulated CaT, on the other hand, both NCX and SERCA contribute to cytosolic Ca^{2+} removal. With the rate constant for NCX activity known, that for SERCA activity can be calculated as well. Finally, with both rate constants for NCX and SERCA activity known, one can calculate their fractional contribution to Ca^{2+} removal during the electrically stimulated CaT (Clarke et al. 2015; Hohendanner et al. 2015). Using this approach, we find a contribution of NCX activity to cytosolic Ca^{2+} removal in the range of ≈ 5–15 % in rat atrial myocytes (6 % in the example shown in Fig. 3.4), in line with results from

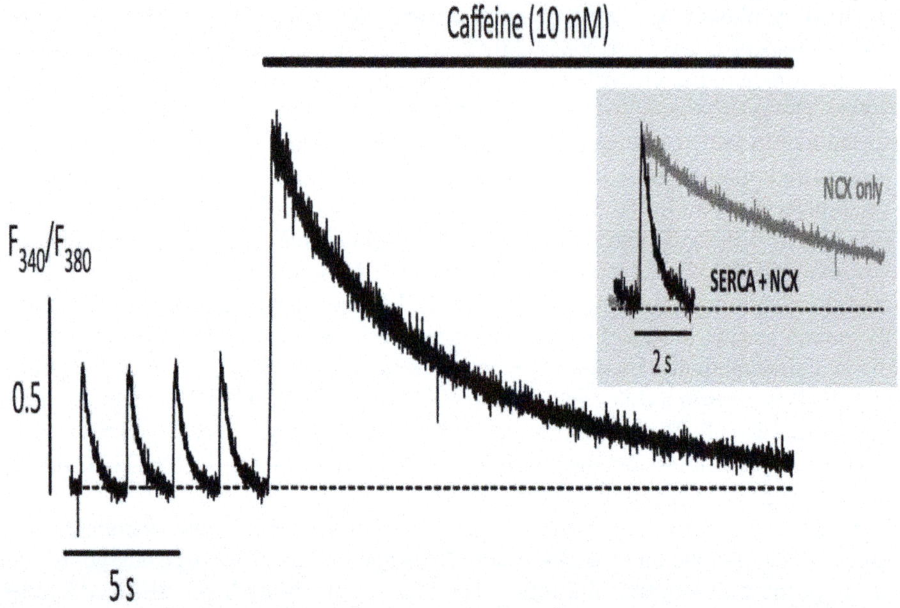

Fig. 3.4 Estimation of SR Ca^{2+} load, fractional SR Ca^{2+} release and SERCA and NCX activities in a rat atrial myocyte. Fura-2 fluorescence (F_{340}/F_{380}) recording of a rat atrial myocyte. Four electrically stimulated CaTs are shown followed by a caffeine-induced CaT. The amplitude of the caffeine-induced CaT is a measure of SR Ca^{2+} load. Dividing the amplitude of the electrically stimulated CaT by the amplitude of the caffeine-induced CaT yields fractional SR Ca^{2+} release (40 % in this case). Inset (*grey*) shows overlay of the last electrically stimulated CaT (*black*) and the caffeine-induced CaT (*dark grey*) with amplitudes normalised. CaT decay is much faster during the electrically stimulated CaT when both SERCA and NCX contribute to cytosolic Ca^{2+} removal. By contrast, during the caffeine-induced CaT when only NCX removes Ca^{2+} from the cytosol, CaT decay is much slower. Assuming that NCX is the only Ca^{2+} removal mechanism during caffeine exposure and that NCX and SERCA are the only Ca^{2+} removal mechanisms during electrically stimulated CaTs, rate constants for SERCA and NCX activities can be derived from the decay kinetics of the electrically stimulated and caffeine-induced CaTs (Clarke et al. 2015). Cell was bathed in Tyrode's solution containing 1.5 mM $[Ca^{2+}]_o$ and loaded with 5 μM Fura-2/AM (20 min); recording conditions as described previously (Pluteanu et al. 2015) (Recording kindly provided by Dr. F. Pluteanu)

rabbit and sheep atrial myocytes (Clarke et al. 2015; Hohendanner et al. 2015). In ventricular myocytes, depending on species, the contribution of NCX activity to cytosolic Ca^{2+} removal varies from $<10\,\%$ to $\approx 30\,\%$ (Bers 2002). Under steady-state conditions, NCX removes the amount of Ca^{2+} that has entered through LTCC, and SERCA removes the amount of Ca^{2+} that has been released from the SR via RyR.

3.4.6 Regulation of the Amplitude of the CaT

The heart must be able to adjust cardiac output to the varying demands of the body. It does so by altering its rate and strength of contraction. The latter is modulated, to a large degree (but not exclusively), by altering the amplitude of the CaT, which is achieved by modulating the activities of the main Ca^{2+}-regulating proteins involved in EC coupling, i.e. LTCC, NCX, SERCA, RyR and – when IP_3 signalling is activated – IP_3R. It is important to note that altering the activity of just one of these proteins may be without or with only limited effect on CaT amplitude in the long term because of Ca^{2+} flux balance (Eisner et al. 2013). Specifically, it is not sufficient to increase RyR P_o to increase CaT amplitude. When RyR P_o is increased experimentally using low concentrations of caffeine (0.1–0.5 mM), an immediate large increase is followed by a quick normalisation of CaT amplitude within a few beats because of "SR autoregulation": the cell loses Ca^{2+} (via sarcolemmal NCX), SR Ca^{2+} load decreases and CaT amplitude comes back to the pre-caffeine level (Eisner et al. 2000; Trafford et al. 2000). In most cases, however, at least under physiological conditions, there is concerted modulation of the activities of several Ca^{2+}-regulating proteins which leads to a long-lasting effect on CaT amplitude. For example, β-adrenergic stimulation increases (1) Ca^{2+} influx through LTCC (to increase trigger Ca^{2+} and to provide extra Ca^{2+} for loading of the SR), (2) RyR P_o (to potentially speed up and further augment SR Ca^{2+} release, but this issue is debated) and (3) the activity of SERCA (to accelerate CaT decay and to increase the load of the SR) (Eisner et al. 2013).

A key determinant for the amplitude of the CaT is SR Ca^{2+} load. SR Ca^{2+} load can be estimated by various means depending on the preparation studied and the methods available. In isolated cardiomyocytes, it is usually estimated by rapid application of high concentrations of caffeine (10–20 mM), which sensitises RyR for activation by ambient cytosolic Ca^{2+} and, thus, opens all RyR to release the Ca^{2+} stored within the SR. SR Ca^{2+} load can then be estimated by the amplitude of the caffeine-induced CaT (in cells loaded with a Ca^{2+}-sensitive fluorescent dye, e.g. Fig. 3.4) or by integration of the caffeine-induced inward current carried by NCX (in patch-clamped cells). Using a low affinity Ca^{2+} dye loaded into the SR, free intra-SR [Ca^{2+}] was estimated in the range of 1–1.5 mM (Shannon et al. 2003).

CaT amplitude is steeply dependent on SR Ca^{2+} load. At a given trigger Ca^{2+}, an increase in SR Ca^{2+} load will cause augmented release due to the increase in the [Ca^{2+}] gradient over the SR membrane and thus the driving force for Ca^{2+} release. However, in addition to this, SR Ca^{2+} load also affects *fractional* SR Ca^{2+} release. Fractional release refers to the fraction of Ca^{2+} that is released from the SR during an electrically stimulated CaT. Figure 3.4 illustrates how fractional release and SR Ca^{2+} load can be determined. A rat atrial myocyte is electrically stimulated before caffeine (10 mM) is applied. The amplitude of the caffeine-induced CaT is a measure for SR Ca^{2+} load. Dividing the amplitude of the electrically stimulated CaT by that of the caffeine-induced CaT yields fractional release. In the example shown, fractional SR Ca^{2+} release amounts to ≈40 %. As shown in a seminal study from the Bers laboratory, fractional release is a steep function of SR Ca^{2+} load (Bassani et al. 1995). Under "normal" conditions (0.5 Hz stimulation, 2 mM $[Ca^{2+}]_o$) in ferret ventricular myocytes, a small increase in SR Ca^{2+} load of ≈4 % results in a large increase of fractional release from ≈35 to ≈60 % (Bassani et al. 1995). The main reason for this steep dependence of fractional release on SR Ca^{2+} load is regulation of RyR by the SR luminal [Ca^{2+}], which increases RyR P_o. Thus, any manoeuvre that increases SR Ca^{2+} load also increases fractional release and, ultimately, CaT amplitude.

Physiologically and pharmacologically relevant manoeuvres to increase SR Ca^{2+} load include β-adrenergic stimulation and exposure to cardiac glycosides. At the extremes, these manoeuvres can even cause SR Ca^{2+} overload and Ca^{2+}-dependent arrhythmias (see below). The mechanisms underlying the increase in SR Ca^{2+} load induced by β-adrenergic stimulation have been outlined above. Cardiac glycosides act mainly via an increase in $[Na^+]_i$ (although additional Na^+-independent modes of action have been proposed) (Altamirano et al. 2006). By inhibiting NKA, cardiac glycosides reduce Na^+ extrusion from the cell. As a result, $[Na^+]_i$ rises and the $[Na^+]$ gradient used by NCX to extrude Ca^{2+} from the cell is diminished. Since less Ca^{2+} is exported from the cell via NCX, more Ca^{2+} can be taken up by the SR via SERCA. This causes an increase in SR Ca^{2+} load and CaT amplitude, which underlies the positive inotropic effect of cardiac glycosides.

Physiologically relevant manoeuvres to increase fractional SR Ca^{2+} release (in addition to increasing SR Ca^{2+} load) encompass stimulation of $G\alpha_s$- and $G\alpha_q$-coupled receptors. $G\alpha_s$-coupled receptors (like β-adrenergic receptors) increase LTCC and thus trigger Ca^{2+}. At a given SR Ca^{2+} load, changes in LTCC are mirrored by concomitant changes in fractional release (Bassani et al. 1995). $G\alpha_q$-coupled receptors (like ET_A or AT_1 receptors), on the other hand, act via increases in $[IP_3]$. IP_3 activates IP_3R to release Ca^{2+} from the SR and this IP_3-induced SR Ca^{2+} release adds to the RyR-mediated Ca^{2+} release to increase fractional release (Kockskamper et al. 2008a).

3.5 Reverse Coupling: Ca^{2+} Affects Membrane Potential and May Elicit Arrhythmias

Membrane potential and $[Ca^{2+}]_i$ are intimately linked. Membrane potential affects $[Ca^{2+}]_i$, e.g. via regulation of LTCC. However, the opposite is also true as Ca^{2+} can affect membrane potential by various means including direct and indirect ones. Direct regulation of membrane potential by Ca^{2+} is mediated by Ca^{2+} ions acting directly on ion channels or electrogenic transporters, whereas indirect regulation occurs through Ca^{2+}-dependent enzymes (or other Ca^{2+}-dependent processes), e.g. phosphorylation of ion channels by CaMKII.

3.5.1 Intracellular Ca^{2+} Directly Affects Ion Channels and Transporters

LTCC are regulated by Ca^{2+} in the vicinity of the channel. Both the Ca^{2+} permeating through the channel as well as the Ca^{2+} released from the SR by nearby RyR cause Ca^{2+}-dependent inactivation of LTCC. This serves as a negative feedback mechanism to limit Ca^{2+} entry and avoid Ca^{2+} overload.

NCX is also regulated by Ca^{2+}. An increase in $[Ca^{2+}]_i$ during the CaT activates forward mode NCX. This mediates membrane depolarisation and contributes to the plateau phase of the AP (see 3.3.2).

Other currents/channels directly regulated by Ca^{2+} include (1) Ca^{2+}-activated non-selective cation current ($I_{NS(Ca)}$), (2) Ca^{2+}-activated Cl^- current ($I_{Cl(Ca)}$) and (3) Ca^{2+}-activated K^+ current ($I_{K(Ca)}$). $I_{NS(Ca)}$ has been described in guinea pig ventricular myocytes (Ehara et al. 1988). Its functional relevance in cardiomyocytes is unclear (Bers 2001). $I_{Cl(Ca)}$ is found in atrial and ventricular myocytes, is activated by the systolic $[Ca^{2+}]_i$ increase and contributes to early repolarisation (Zygmunt and Gibbons 1991, 1992). It is also activated by spontaneous SR Ca^{2+} release under conditions of Ca^{2+} overload and may contribute to delayed afterdepolarisations (Zygmunt et al. 1998). More recently, small conductance (SK) Ca^{2+}-activated K^+ channels have been found in atrial and ventricular myocytes (Xu et al. 2003). Their potential physiological and pathophysiological roles are under investigation. Current evidence suggests that SK channels may contribute to AP repolarisation and exert antiarrhythmic effects, in particular in atrial myocytes (Nattel and Qi 2014; Skibsbye et al. 2014).

3.5.2 First DAD and then DEAD: Spontaneous SR Ca^{2+} Release May Elicit Life-Threatening Arrhythmias

In a normal heart, SR Ca^{2+} release is triggered by LTCC during the AP. It does not (or only rarely so) occur in diastole, when RyR remain closed due to their low Ca^{2+} sensitivity. Under certain conditions, however, when RyR open probability is greatly increased – as in Ca^{2+} overload – spontaneous diastolic Ca^{2+} release occurs more frequently. Ca^{2+} overload may develop as a result of excessive β-adrenergic stimulation (see Fig. 3.3), digitalis action or elevated extracellular $[Ca^{2+}]$ (hypercalcaemia). Under these conditions, spontaneous SR Ca^{2+} release induces a chain of events that may eventually trigger fatal arrhythmias. When Ca^{2+} is released from the SR in diastole, a Ca^{2+} wave can develop (Fig. 3.3). This Ca^{2+} wave activates forward mode NCX to extrude some of the released Ca^{2+} from the cell. As NCX is electrogenic, forward mode NCX elicits an inward current termed I_{ti} (transient inward current). I_{ti} causes a delayed afterdepolarisation or DAD. If the DAD is of sufficient magnitude, it can trigger an extra AP and life-threatening arrhythmias (Lederer and Tsien 1976; Kass et al. 1978; Mechmann and Pott 1986; Fedida et al. 1987; Berlin et al. 1989; Schlotthauer and Bers 2000; Sedej et al. 2010).

3.6 Of Mice and Men: Species Differences in EC Coupling

EC coupling differs substantially between smaller and larger mammals. Regarding cardiac physiology, the most obvious and striking difference between smaller and larger mammals is heart rate, which – as a rule of thumb – increases with decreasing size of the animal species. Resting heart rates differ by a factor of ≈5–10 between small laboratory animals (mice, ≈600 min^{-1}; rats, ≈400 min^{-1}) and humans (≈60–80 min^{-1}). It follows that EC coupling has to proceed much faster in smaller than in larger mammals. APs in mouse and rat ventricular myocytes are considerably shorter than in rabbit and human ventricular myocytes. The shape of mouse and rat

APs is more triangular, a well-defined plateau phase is missing. Hence, there is less time for Ca^{2+} influx through LTCC to trigger Ca^{2+} release from the SR. Peak systolic $[Ca^{2+}]_i$ is reached earlier in mouse and rat ventricular myocytes. CICR is very efficient though. Peak systolic $[Ca^{2+}]_i$ during a CaT is similar in most species reaching values in the range of up to ≈ 1 μM. The gain of EC coupling – defined as the systolic $[Ca^{2+}]_i$ increase over the integrated Ca^{2+} influx through LTCC – is usually higher in smaller mammals. Less Ca^{2+} influx at otherwise comparable systolic $[Ca^{2+}]_i$ increases also means a larger contribution from the SR. The relative contribution of sarcolemmal Ca^{2+} influx versus SR Ca^{2+} release to the systolic CaT amounts to ≈ 10 % versus ≈ 90 % in mouse and rat ventricular myocytes and to ≈ 30 % versus ≈ 70 % in rabbit and human ventricular myocytes (Bassani et al. 1994). Similarly, CaT decay also proceeds faster in smaller mammals largely caused by a higher density of SERCA enabling faster Ca^{2+} reuptake into the SR (Hove-Madsen and Bers 1993). CaT decay is governed mainly by Ca^{2+} reuptake into the SR through SERCA and by Ca^{2+} extrusion from the cell through sarcolemmal NCX. Since under steady-state conditions Ca^{2+} influx must match Ca^{2+} efflux and SR Ca^{2+} release must match SR Ca^{2+} reuptake, it follows that SERCA and NCX contribute ≈ 90 % and ≈ 10 %, respectively, to CaT decay in mouse and rat ventricular myocytes and ≈ 70 % and ≈ 30 %, respectively, in rabbit and human ventricular myocytes. In conclusion, CaTs in mouse and rat ventricular myocytes are fast and largely dependent on SR Ca^{2+} release and reuptake, whereas CaTs in rabbit and human ventricular myocytes are somewhat slower and rely more on sarcolemmal Ca^{2+} fluxes. As CaTs are faster (both upstroke and decay) in smaller mammals, this also allows for faster contraction and relaxation of the heart ultimately enabling the higher heart rates in mice and rats.

3.7 Sick at Heart: EC Coupling Is Altered in Cardiac Disease

Cardiovascular disease still accounts for most deaths in industrialised countries. Many cardiac diseases are associated with alterations in EC coupling. In heart failure and atrial fibrillation, two of the most prevalent and severe cardiac diseases, extensive remodelling of EC coupling has been documented.

In heart failure, in ventricular myocytes, there is altered expression and regulation of sarcolemmal ion channels resulting in increased AP duration as well as greatly disturbed Ca^{2+} handling resulting in elevated diastolic $[Ca^{2+}]_i$ and diminished CaTs and contractions (Beuckelmann et al. 1992). Prolongation of the AP is caused, in part, by reduced K^+ currents (Beuckelmann et al. 1993). Disturbed Ca^{2+} handling includes increased expression and activity of NCX, reduced expression and activity of SERCA and hyperphosphorylation and increased activity of RyR with increased SR Ca^{2+} leak (Wehrens and Marks 2004; Lehnart et al. 2009). Furthermore, there is evidence for remodelling of the T-tubular system and the organisation of LTCC and RyR (Crossman et al. 2011; Wagner et al. 2012).

In atrial fibrillation, in atrial myocytes, there is considerable remodelling of EC coupling including electrical, structural and Ca^{2+} remodelling (Greiser et al. 2011;

Schotten et al. 2011). Electrical remodelling includes greatly reduced LTCC and alterations in several K^+ currents resulting in reduced AP duration (Schotten et al. 2011). Ca^{2+} remodelling consists of reduced Ca^{2+} influx via LTCC, enhanced phosphorylation and activity of RyR with increased SR Ca^{2+} leak and greatly diminished CaTs with impaired centripetal propagation (Greiser et al. 2011, 2014; Voigt et al. 2012).

3.8 Conclusive Remarks

Excitation–contraction (EC) coupling is a complex and fine-tuned mechanism that allows cardiomyocytes to translate electrical activity into mechanical force. As this is the main function of cardiomyocytes, remodelling of cardiomyocyte EC coupling in heart failure and atrial fibrillation contributes to disease progression and represents a potential therapeutic target (Wehrens and Marks 2004; Tamargo and Lopez-Sendon 2011; Dobrev et al. 2012; Nattel and Dobrev 2012).

References

Altamirano J, Li Y, DeSantiago J, Piacentino V 3rd, Houser SR, Bers DM (2006) The inotropic effect of cardioactive glycosides in ventricular myocytes requires Na^+-Ca^{2+} exchanger function. J Physiol 575:845–854

Anderson ME, Braun AP, Schulman H, Premack BA (1994) Multifunctional Ca^{2+}/calmodulin-dependent protein kinase mediates Ca^{2+}-induced enhancement of the L-type Ca^{2+} current in rabbit ventricular myocytes. Circ Res 75:854–861

Armstrong CM, Bezanilla FM, Horowicz P (1972) Twitches in the presence of ethylene glycol bis(-aminoethyl ether)-N, N'-tetracetic acid. Biochim Biophys Acta 267:605–608

Baddeley D, Jayasinghe ID, Lam L, Rossberger S, Cannell MB, Soeller C (2009) Optical single-channel resolution imaging of the ryanodine receptor distribution in rat cardiac myocytes. Proc Natl Acad Sci U S A 106:22275–22280

Bassani JW, Bassani RA, Bers DM (1994) Relaxation in rabbit and rat cardiac cells: species-dependent differences in cellular mechanisms. J Physiol 476:279–293

Bassani JW, Yuan W, Bers DM (1995) Fractional SR Ca release is regulated by trigger Ca and SR Ca content in cardiac myocytes. Am J Physiol 268:C1313–C1319

Belardinelli L, Isenberg G (1983) Isolated atrial myocytes: adenosine and acetylcholine increase potassium conductance. Am J Physiol 244:H734–H737

Benitah JP, Alvarez JL, Gomez AM (2010) L-type Ca^{2+} current in ventricular cardiomyocytes. J Mol Cell Cardiol 48:26–36

Berlin JR (1995) Spatiotemporal changes of Ca^{2+} during electrically evoked contractions in atrial and ventricular cells. Am J Physiol 269:H1165–H1170

Berlin JR, Cannell MB, Lederer WJ (1989) Cellular origins of the transient inward current in cardiac myocytes. Role of fluctuations and waves of elevated intracellular calcium. Circ Res 65:115–126

Bers DM (2001) Excitation-contraction coupling and cardiac contractile force. 2nd edn. Kluwer Academic Publishers. Zipes/Jalife (Editors). Cardiac electrophysiology: from cell to bedside. 6th edn. Elsevier Saunders

Bers DM (2002) Cardiac excitation-contraction coupling. Nature 415:198–205

Bers DM (2008) Calcium cycling and signaling in cardiac myocytes. Annu Rev Physiol 70:23–49

Bers DM (2012) Ryanodine receptor S2808 phosphorylation in heart failure: smoking gun or red herring. Circ Res 110:796–799

Beuckelmann DJ, Nabauer M, Erdmann E (1992) Intracellular calcium handling in isolated ventricular myocytes from patients with terminal heart failure. Circulation 85:1046–1055

Beuckelmann DJ, Nabauer M, Erdmann E (1993) Alterations of K^+ currents in isolated human ventricular myocytes from patients with terminal heart failure. Circ Res 73:379–385

Blatter LA, Kockskamper J, Sheehan KA, Zima AV, Huser J, Lipsius SL (2003) Local calcium gradients during excitation-contraction coupling and alternans in atrial myocytes. J Physiol 546:19–31

Bode EF, Briston SJ, Overend CL, O'Neill SC, Trafford AW, Eisner DA (2011) Changes of SERCA activity have only modest effects on sarcoplasmic reticulum Ca^{2+} content in rat ventricular myocytes. J Physiol 589:4723–4729

Bode F, Sachs F, Franz MR (2001) Tarantula peptide inhibits atrial fibrillation. Nature 409:35–36

Brette F, Despa S, Bers DM, Orchard CH (2005) Spatiotemporal characteristics of SR Ca^{2+} uptake and release in detubulated rat ventricular myocytes. J Mol Cell Cardiol 39:804–812

Brette F, Komukai K, Orchard CH (2002) Validation of formamide as a detubulation agent in isolated rat cardiac cells. Am J Physiol 283:H1720–H1728

Bunemann M, Gerhardstein BL, Gao T, Hosey MM (1999) Functional regulation of L-type calcium channels via protein kinase A-mediated phosphorylation of the beta(2) subunit. J Biol Chem 274:33851–33854

Cheng H, Cannell MB, Lederer WJ (1995) Partial inhibition of Ca^{2+} current by methoxyverapamil (D600) reveals spatial nonuniformities in $[Ca^{2+}]_i$ during excitation-contraction coupling in cardiac myocytes. Circ Res 76:236–241

Cheng H, Lederer MR, Xiao RP, Gomez AM, Zhou YY, Ziman B, Spurgeon H, Lakatta EG, Lederer WJ (1996) Excitation-contraction coupling in heart: new insights from Ca^{2+} sparks. Cell Calcium 20:129–140

Cheng H, Lederer WJ, Cannell MB (1993) Calcium sparks: elementary events underlying excitation-contraction coupling in heart muscle. Science 262:740–744

Clarke JD, Caldwell JL, Horn MA, Bode EF, Richards MA, Hall MC, Graham HK, Briston SJ, Greensmith DJ, Eisner DA, Dibb KM, Trafford AW (2015) Perturbed atrial calcium handling in an ovine model of heart failure: potential roles for reductions in the L-type calcium current. J Mol Cell Cardiol 79:169–179

Cooper PJ, Lei M, Cheng LX, Kohl P (2000) Selected contribution: axial stretch increases spontaneous pacemaker activity in rabbit isolated sinoatrial node cells. J Appl Physiol 89:2099–2104

Crossman DJ, Ruygrok PN, Soeller C, Cannell MB (2011) Changes in the organization of excitation-contraction coupling structures in failing human heart. PLoS One 6:e17901

Dibb KM, Clarke JD, Eisner DA, Richards MA, Trafford AW (2013) A functional role for transverse (t-) tubules in the atria. J Mol Cell Cardiol 58:84–91

Dibb KM, Clarke JD, Horn MA, Richards MA, Graham HK, Eisner DA, Trafford AW (2009) Characterization of an extensive transverse tubular network in sheep atrial myocytes and its depletion in heart failure. Circ Heart Fail 2:482–489

DiFrancesco D (1986) Characterization of single pacemaker channels in cardiac sino-atrial node cells. Nature 324:470–473

DiFrancesco D (2010) The role of the funny current in pacemaker activity. Circ Res 106:434–446

Dobrev D, Carlsson L, Nattel S (2012) Novel molecular targets for atrial fibrillation therapy. Nat Rev Drug Discov 11:275–291

Dolber PC, Bauman RP, Rembert JC, Greenfield JC Jr (1994) Regional changes in myocyte structure in model of canine right atrial hypertrophy. Am J Physiol 267:H1279–H1287

Domeier TL, Zima AV, Maxwell JT, Huke S, Mignery GA, Blatter LA (2008) IP_3 receptor-dependent Ca^{2+} release modulates excitation-contraction coupling in rabbit ventricular myocytes. Am J Physiol 294:H596–H604

Ehara T, Noma A, Ono K (1988) Calcium-activated non-selective cation channel in ventricular cells isolated from adult guinea-pig hearts. J Physiol 403:117–133

Eisner D, Bode E, Venetucci L, Trafford A (2013) Calcium flux balance in the heart. J Mol Cell Cardiol 58:110–117

Eisner DA, Choi HS, Diaz ME, O'Neill SC, Trafford AW (2000) Integrative analysis of calcium cycling in cardiac muscle. Circ Res 87:1087–1094

Escobar M, Cardenas C, Colavita K, Petrenko NB, Franzini-Armstrong C (2011) Structural evidence for perinuclear calcium microdomains in cardiac myocytes. J Mol Cell Cardiol 50:451–459

Fabiato A, Fabiato F (1975) Contractions induced by a calcium-triggered release of calcium from the sarcoplasmic reticulum of single skinned cardiac cells. J Physiol 249:469–495

Farrell EF, Antaramian A, Rueda A, Gomez AM, Valdivia HH (2003) Sorcin inhibits calcium release and modulates excitation-contraction coupling in the heart. J Biol Chem 278:34660–34666

Fedida D, Noble D, Rankin AC, Spindler AJ (1987) The arrhythmogenic transient inward current iTI and related contraction in isolated guinea-pig ventricular myocytes. J Physiol 392:523–542

Fill M, Copello JA (2002) Ryanodine receptor calcium release channels. Physiol Rev 82:893–922

Gaborit N, Le Bouter S, Szuts V, Varro A, Escande D, Nattel S, Demolombe S (2007) Regional and tissue specific transcript signatures of ion channel genes in the non-diseased human heart. J Physiol 582:675–693

Ganesan AN, Maack C, Johns DC, Sidor A, O'Rourke B (2006) Beta-adrenergic stimulation of L-type Ca^{2+} channels in cardiac myocytes requires the distal carboxyl terminus of alpha1C but not serine 1928. Circ Res 98:e11–e18

Glitsch HG (2001) Electrophysiology of the sodium-potassium-ATPase in cardiac cells. Physiol Rev 81:1791–1826

Greiser M, Kerfant BG, Williams GS, Voigt N, Harks E, Dibb KM, Giese A, Meszaros J, Verheule S, Ravens U, Allessie MA, Gammie JS, van der Velden J, Lederer WJ, Dobrev D, Schotten U (2014) Tachycardia-induced silencing of subcellular Ca^{2+} signaling in atrial myocytes. J Clin Invest 124:4759–4772

Greiser M, Lederer WJ, Schotten U (2011) Alterations of atrial Ca^{2+} handling as cause and consequence of atrial fibrillation. Cardiovasc Res 89:722–733

Harzheim D, Movassagh M, Foo RS, Ritter O, Tashfeen A, Conway SJ, Bootman MD, Roderick HL (2009) Increased $InsP_3Rs$ in the junctional sarcoplasmic reticulum augment Ca^{2+} transients and arrhythmias associated with cardiac hypertrophy. Proc Natl Acad Sci U S A 106:11406–11411

He H, Giordano FJ, Hilal-Dandan R, Choi DJ, Rockman HA, McDonough PM, Bluhm WF, Meyer M, Sayen MR, Swanson E, Dillmann WH (1997) Overexpression of the rat sarcoplasmic reticulum Ca^{2+} ATPase gene in the heart of transgenic mice accelerates calcium transients and cardiac relaxation. J Clin Invest 100:380–389

Hibino H, Inanobe A, Furutani K, Murakami S, Findlay I, Kurachi Y (2010) Inwardly rectifying potassium channels: their structure, function, and physiological roles. Physiol Rev 90:291–366

Higazi DR, Fearnley CJ, Drawnel FM, Talasila A, Corps EM, Ritter O, McDonald F, Mikoshiba K, Bootman MD, Roderick HL (2009) Endothelin-1-stimulated $InsP_3$-induced Ca^{2+} release is a nexus for hypertrophic signaling in cardiac myocytes. Mol Cell 33:472–482

Hohendanner F, Walther S, Maxwell JT, Kettlewell S, Awad S, Smith GL, Lonchyna VA, Blatter LA (2015) Inositol-1,4,5-trisphosphate induced Ca^{2+} release and excitation-contraction coupling in atrial myocytes from normal and failing hearts. J Physiol 593:1459–1477

Hove-Madsen L, Bers DM (1993) Sarcoplasmic reticulum Ca^{2+} uptake and thapsigargin sensitivity in permeabilized rabbit and rat ventricular myocytes. Circ Res 73:820–828

Huser J, Blatter LA, Lipsius SL (2000) Intracellular Ca^{2+} release contributes to automaticity in cat atrial pacemaker cells. J Physiol 524:415–422

Huser J, Lipsius SL, Blatter LA (1996) Calcium gradients during excitation-contraction coupling in cat atrial myocytes. J Physiol 494:641–651

Huttlin EL, Jedrychowski MP, Elias JE, Goswami T, Rad R, Beausoleil SA, Villen J, Haas W, Sowa ME, Gygi SP (2010) A tissue-specific atlas of mouse protein phosphorylation and expression. Cell 143:1174–1189

James P, Inui M, Tada M, Chiesi M, Carafoli E (1989) Nature and site of phospholamban regulation of the Ca^{2+} pump of sarcoplasmic reticulum. Nature 342:90–92

Janse MJ, Coronel R, Opthof T, Sosunov EA, Anyukhovsky EP, Rosen MR (2012) Repolarization gradients in the intact heart: transmural or apico-basal? Prog Biophys Mol Biol 109:6–15

Jayasinghe ID, Cannell MB, Soeller C (2009) Organization of ryanodine receptors, transverse tubules, and sodium-calcium exchanger in rat myocytes. Biophys J 97:2664–2673

Kass RS, Lederer WJ, Tsien RW, Weingart R (1978) Role of calcium ions in transient inward currents and aftercontractions induced by strophanthidin in cardiac Purkinje fibres. J Physiol 281:187–208

Kockskamper J, Glitsch HG (1997) Sodium pump of cultured guinea pig atrial myocytes. Ann N Y Acad Sci 834:354–356

Kockskamper J, Seidlmayer L, Walther S, Hellenkamp K, Maier LS, Pieske B (2008a) Endothelin-1 enhances nuclear Ca^{2+} transients in atrial myocytes through Ins(1,4,5)P_3-dependent Ca^{2+} release from perinuclear Ca^{2+} stores. J Cell Sci 121:186–195

Kockskamper J, Sheehan KA, Bare DJ, Lipsius SL, Mignery GA, Blatter LA (2001) Activation and propagation of Ca^{2+} release during excitation-contraction coupling in atrial myocytes. Biophys J 81:2590–2605

Kockskamper J, Zima AV, Blatter LA (2005) Modulation of sarcoplasmic reticulum Ca^{2+} release by glycolysis in cat atrial myocytes. J Physiol 564:697–714

Kockskamper J, Zima AV, Roderick HL, Pieske B, Blatter LA, Bootman MD (2008b) Emerging roles of inositol 1,4,5-trisphosphate signaling in cardiac myocytes. J Mol Cell Cardiol 45:128–147

Kohl P, Nesbitt AD, Cooper PJ, Lei M (2001) Sudden cardiac death by Commotio cordis: role of mechano-electric feedback. Cardiovasc Res 50:280–289

Kohl T, Westphal V, Hell SW, Lehnart SE (2013) Superresolution microscopy in heart – cardiac nanoscopy. J Mol Cell Cardiol 58:13–21

Koval OM, Guan X, Wu Y, Joiner ML, Gao Z, Chen B, Grumbach IM, Luczak ED, Colbran RJ, Song LS, Hund TJ, Mohler PJ, Anderson ME (2010) CaV1.2 beta-subunit coordinates CaMKII-triggered cardiomyocyte death and afterdepolarizations. Proc Natl Acad Sci U S A 107: 4996–5000

Kranias EG, Hajjar RJ (2012) Modulation of cardiac contractility by the phospholamban/ SERCA2a regulatome. Circ Res 110:1646–1660

Kurachi Y, Nakajima T, Sugimoto T (1986) On the mechanism of activation of muscarinic K^+ channels by adenosine in isolated atrial cells: involvement of GTP-binding proteins. Pflugers Arch 407:264–274

Lakatta EG, Maltsev VA, Vinogradova TM (2010) A coupled SYSTEM of intracellular Ca^{2+} clocks and surface membrane voltage clocks controls the timekeeping mechanism of the heart's pacemaker. Circ Res 106:659–673

Leblanc N, Hume JR (1990) Sodium current-induced release of calcium from cardiac sarcoplasmic reticulum. Science 248:372–376

Lederer WJ, Tsien RW (1976) Transient inward current underlying arrhythmogenic effects of cardiotonic steroids in Purkinje fibres. J Physiol 263:73–100

Lee TS, Karl R, Moosmang S, Lenhardt P, Klugbauer N, Hofmann F, Kleppisch T, Welling A (2006) Calmodulin kinase II is involved in voltage-dependent facilitation of the L-type Cav1.2 calcium channel: identification of the phosphorylation sites. J Biol Chem 281:25560–25567

Legato MJ (1973) Ultrastructure of the atrial, ventricular, and Purkinje cell, with special reference to the genesis of arrhythmias. Circulation 47:178–189

Lehnart SE, Maier LS, Hasenfuss G (2009) Abnormalities of calcium metabolism and myocardial contractility depression in the failing heart. Heart Fail Rev 14:213–224

Lehnart SE, Mongillo M, Bellinger A, Lindegger N, Chen BX, Hsueh W, Reiken S, Wronska A, Drew LJ, Ward CW, Lederer WJ, Kass RS, Morley G, Marks AR (2008) Leaky Ca^{2+} release channel/ryanodine receptor 2 causes seizures and sudden cardiac death in mice. J Clin Invest 118:2230–2245

Lehnart SE, Wehrens XH, Reiken S, Warrier S, Belevych AE, Harvey RD, Richter W, Jin SL, Conti M, Marks AR (2005) Phosphodiesterase 4D deficiency in the ryanodine-receptor complex promotes heart failure and arrhythmias. Cell 123:25–35

Lemke T, Welling A, Christel CJ, Blaich A, Bernhard D, Lenhardt P, Hofmann F, Moosmang S (2008) Unchanged beta-adrenergic stimulation of cardiac L-type calcium channels in Ca v 1.2 phosphorylation site S1928A mutant mice. J Biol Chem 283:34738–34744

Lenaerts I, Bito V, Heinzel FR, Driesen RB, Holemans P, D'Hooge J, Heidbuchel H, Sipido KR, Willems R (2009) Ultrastructural and functional remodeling of the coupling between Ca^{2+} influx and sarcoplasmic reticulum Ca^{2+} release in right atrial myocytes from experimental persistent atrial fibrillation. Circ Res 105:876–885

Limberg SH, Netter MF, Rolfes C, Rinne S, Schlichthorl G, Zuzarte M, Vassiliou T, Moosdorf R, Wulf H, Daut J, Sachse FB, Decher N (2011) TASK-1 channels may modulate action potential duration of human atrial cardiomyocytes. Cell Physiol Biochem 28:613–624

Lipp P, Laine M, Tovey SC, Burrell KM, Berridge MJ, Li W, Bootman MD (2000) Functional $InsP_3$ receptors that may modulate excitation-contraction coupling in the heart. Curr Biol 10:939–942

Ljubojevic S, Radulovic S, Leitinger G, Sedej S, Sacherer M, Holzer M, Winkler C, Pritz E, Mittler T, Schmidt A, Sereinigg M, Wakula P, Zissimopoulos S, Bisping E, Post H, Marsche G, Bossuyt J, Bers DM, Kockskamper J, Pieske B (2014) Early remodeling of perinuclear Ca^{2+} stores and nucleoplasmic Ca^{2+} signaling during the development of hypertrophy and heart failure. Circulation 130:244–255

Logothetis DE, Kurachi Y, Galper J, Neer EJ, Clapham DE (1987) The beta gamma subunits of GTP-binding proteins activate the muscarinic K^+ channel in heart. Nature 325: 321–326

Luo W, Grupp IL, Harrer J, Ponniah S, Grupp G, Duffy JJ, Doetschman T, Kranias EG (1994) Targeted ablation of the phospholamban gene is associated with markedly enhanced myocardial contractility and loss of beta-agonist stimulation. Circ Res 75:401–409

Mackenzie L, Bootman MD, Berridge MJ, Lipp P (2001) Predetermined recruitment of calcium release sites underlies excitation-contraction coupling in rat atrial myocytes. J Physiol 530:417–429

Mackenzie L, Bootman MD, Laine M, Berridge MJ, Thuring J, Holmes A, Li WH, Lipp P (2002) The role of inositol 1,4,5-trisphosphate receptors in Ca^{2+} signalling and the generation of arrhythmias in rat atrial myocytes. J Physiol 541:395–409

Maroto R, Raso A, Wood TG, Kurosky A, Martinac B, Hamill OP (2005) TRPC1 forms the stretch-activated cation channel in vertebrate cells. Nat Cell Biol 7:179–185

Marx SO, Reiken S, Hisamatsu Y, Jayaraman T, Burkhoff D, Rosemblit N, Marks AR (2000) PKA phosphorylation dissociates FKBP12.6 from the calcium release channel (ryanodine receptor): defective regulation in failing hearts. Cell 101:365–376

Mattiazzi A, Mundina-Weilenmann C, Guoxiang C, Vittone L, Kranias E (2005) Role of phospholamban phosphorylation on Thr17 in cardiac physiological and pathological conditions. Cardiovasc Res 68:366–375

McDonald TF, Pelzer S, Trautwein W, Pelzer DJ (1994) Regulation and modulation of calcium channels in cardiac, skeletal, and smooth muscle cells. Physiol Rev 74:365–507

Mechmann S, Pott L (1986) Identification of Na-Ca exchange current in single cardiac myocytes. Nature 319:597–599

Nakayama H, Bodi I, Maillet M, DeSantiago J, Domeier TL, Mikoshiba K, Lorenz JN, Blatter LA, Bers DM, Molkentin JD (2010) The IP3 receptor regulates cardiac hypertrophy in response to select stimuli. Circ Res 107:659–666

Nattel S, Dobrev D (2012) The multidimensional role of calcium in atrial fibrillation pathophysiology: mechanistic insights and therapeutic opportunities. Eur Heart J 33:1870–1877

Nattel S, Qi XY (2014) Calcium-dependent potassium channels in the heart: clarity and confusion. Cardiovasc Res 101:185–186

Nerbonne JM, Kass RS (2005) Molecular physiology of cardiac repolarization. Physiol Rev 85:1205–1253

Niggli E (1999) Localized intracellular calcium signaling in muscle: calcium sparks and calcium quarks. Annu Rev Physiol 61:311–335

Niggli E, Lederer WJ (1990) Voltage-independent calcium release in heart muscle. Science 250:565–568

Noma A (1983) ATP-regulated K^+ channels in cardiac muscle. Nature 305:147–148

Peterson BZ, DeMaria CD, Adelman JP, Yue DT (1999) Calmodulin is the Ca^{2+} sensor for Ca^{2+}-dependent inactivation of L-type calcium channels. Neuron 22:549–558

Pluteanu F, Hess J, Plackic J, Nikonova Y, Preisenberger J, Bukowska A, Schotten U, Rinne A, Kienitz MC, Schafer MK, Weihe E, Goette A, Kockskamper J (2015) Early subcellular Ca^{2+} remodelling and increased propensity for Ca^{2+} alternans in left atrial myocytes from hypertensive rats. Cardiovasc Res 106:87–97

Proven A, Roderick HL, Conway SJ, Berridge MJ, Horton JK, Capper SJ, Bootman MD (2006) Inositol 1,4,5-trisphosphate supports the arrhythmogenic action of endothelin-1 on ventricular cardiac myocytes. J Cell Sci 119:3363–3375

Putzke C, Wemhoner K, Sachse FB, Rinne S, Schlichthorl G, Li XT, Jae L, Eckhardt I, Wischmeyer E, Wulf H, Preisig-Muller R, Daut J, Decher N (2007) The acid-sensitive potassium channel TASK-1 in rat cardiac muscle. Cardiovasc Res 75:59–68

Reeves JP, Hale CC (1984) The stoichiometry of the cardiac sodium-calcium exchange system. J Biol Chem 259:7733–7739

Remus TP, Zima AV, Bossuyt J, Bare DJ, Martin JL, Blatter LA, Bers DM, Mignery GA (2006) Biosensors to measure inositol 1,4,5-trisphosphate concentration in living cells with spatiotemporal resolution. J Biol Chem 281:608–616

Richards MA, Clarke JD, Saravanan P, Voigt N, Dobrev D, Eisner DA, Trafford AW, Dibb KM (2011) Transverse tubules are a common feature in large mammalian atrial myocytes including human. Am J Physiol 301:H1996–H2005

Rodriguez P, Bhogal MS, Colyer J (2003) Stoichiometric phosphorylation of cardiac ryanodine receptor on serine 2809 by calmodulin-dependent kinase II and protein kinase A. J Biol Chem 278:38593–38600

Rousseau E, Meissner G (1989) Single cardiac sarcoplasmic reticulum Ca^{2+}-release channel: activation by caffeine. Am J Physiol 256:H328–H333

Sakmann B, Noma A, Trautwein W (1983) Acetylcholine activation of single muscarinic K^+ channels in isolated pacemaker cells of the mammalian heart. Nature 303:250–253

Satoh H, Delbridge LM, Blatter LA, Bers DM (1996) Surface:volume relationship in cardiac myocytes studied with confocal microscopy and membrane capacitance measurements: species-dependence and developmental effects. Biophys J 70:1494–1504

Schlotthauer K, Bers DM (2000) Sarcoplasmic reticulum Ca^{2+} release causes myocyte depolarization. Underlying mechanism and threshold for triggered action potentials. Circ Res 87:774–780

Schmidt C, Wiedmann F, Voigt N, Zhou XB, Heijman J, Lang S, Albert V, Kallenberger S, Ruhparwar A, Szabo G, Kallenbach K, Karck M, Borggrefe M, Biliczki P, Ehrlich JR, Baczko I, Lugenbiel P, Schweizer PA, Donner BC, Katus HA, Dobrev D, Thomas D (2015) Upregulation of K(2P)3.1 K^+ current causes action potential shortening in patients with chronic atrial fibrillation. Circulation 132:82–92

Schotten U, Verheule S, Kirchhof P, Goette A (2011) Pathophysiological mechanisms of atrial fibrillation: a translational appraisal. Physiol Rev 91:265–325

Scriven DR, Asghari P, Moore ED (2013) Microarchitecture of the dyad. Cardiovasc Res 98:169–176

Scriven DR, Asghari P, Schulson MN, Moore ED (2010) Analysis of Cav1.2 and ryanodine receptor clusters in rat ventricular myocytes. Biophys J 99:3923–3929

Scriven DR, Dan P, Moore ED (2000) Distribution of proteins implicated in excitation-contraction coupling in rat ventricular myocytes. Biophys J 79:2682–2691

Sedej S, Heinzel FR, Walther S, Dybkova N, Wakula P, Groborz J, Gronau P, Maier LS, Vos MA, Lai FA, Napolitano C, Priori SG, Kockskamper J, Pieske B (2010) Na^+-dependent SR Ca^{2+} overload induces arrhythmogenic events in mouse cardiomyocytes with a human CPVT mutation. Cardiovasc Res 87:50–59

Shannon TR, Guo T, Bers DM (2003) Ca^{2+} scraps: local depletions of free [Ca^{2+}] in cardiac sarcoplasmic reticulum during contractions leave substantial Ca^{2+} reserve. Circ Res 93:40–45

Shattock MJ, Ottolia M, Bers DM, Blaustein MP, Boguslavskyi A, Bossuyt J, Bridge JH, Chen-Izu Y, Clancy CE, Edwards A, Goldhaber J, Kaplan J, Lingrel JB, Pavlovic D, Philipson K, Sipido KR, Xie ZJ (2015) Na^+/Ca^{2+} exchange and Na^+/K^+-ATPase in the heart. J Physiol 593:1361–1382

Sheehan KA, Zima AV, Blatter LA (2006) Regional differences in spontaneous Ca^{2+} spark activity and regulation in cat atrial myocytes. J Physiol 572:799–809

Signore S, Sorrentino A, Ferreira-Martins J, Kannappan R, Shafaie M, Del Ben F, Isobe K, Arranto C, Wybieralska E, Webster A, Sanada F, Ogorek B, Zheng H, Liu X, Del Monte F, D'Alessandro DA, Wunimenghe O, Michler RE, Hosoda T, Goichberg P, Leri A, Kajstura J, Anversa P, Rota M (2013) Inositol 1, 4, 5-trisphosphate receptors and human left ventricular myocytes. Circulation 128:1286–1297

Sipido KR, Maes M, Van de Werf F (1997) Low efficiency of Ca^{2+} entry through the Na^+-Ca^{2+} exchanger as trigger for Ca^{2+} release from the sarcoplasmic reticulum. A comparison between L-type Ca^{2+} current and reverse-mode Na^+-Ca^{2+} exchange. Circ Res 81:1034–1044

Skibsbye L, Poulet C, Diness JG, Bentzen BH, Yuan L, Kappert U, Matschke K, Wettwer E, Ravens U, Grunnet M, Christ T, Jespersen T (2014) Small-conductance calcium-activated potassium (SK) channels contribute to action potential repolarization in human atria. Cardiovasc Res 103:156–167

Soeller C, Cannell MB (1999) Examination of the transverse tubular system in living cardiac rat myocytes by 2-photon microscopy and digital image-processing techniques. Circ Res 84:266–275

Tamargo J, Lopez-Sendon J (2011) Novel therapeutic targets for the treatment of heart failure. Nat Rev Drug Discov 10:536–555

Terrenoire C, Lauritzen I, Lesage F, Romey G, Lazdunski M (2001) A TREK-1-like potassium channel in atrial cells inhibited by beta-adrenergic stimulation and activated by volatile anesthetics. Circ Res 89:336–342

Trafford AW, Clarke JD, Richards MA, Eisner DA, Dibb KM (2013) Calcium signalling microdomains and the t-tubular system in atrial myocytes: potential roles in cardiac disease and arrhythmias. Cardiovasc Res 98:192–203

Trafford AW, Diaz ME, Sibbring GC, Eisner DA (2000) Modulation of CICR has no maintained effect on systolic Ca^{2+}: simultaneous measurements of sarcoplasmic reticulum and sarcolemmal Ca^{2+} fluxes in rat ventricular myocytes. J Physiol 522:259–270

Varro A, Nanasi PP, Lathrop DA (1993) Potassium currents in isolated human atrial and ventricular cardiocytes. Acta Physiol Scand 149:133–142

Voigt N, Li N, Wang Q, Wang W, Trafford AW, Abu-Taha I, Sun Q, Wieland T, Ravens U, Nattel S, Wehrens XH, Dobrev D (2012) Enhanced sarcoplasmic reticulum Ca^{2+} leak and increased Na^+-Ca^{2+} exchanger function underlie delayed afterdepolarizations in patients with chronic atrial fibrillation. Circ Res 111:402–414

Wagner E, Lauterbach MA, Kohl T, Westphal V, Williams GS, Steinbrecher JH, Streich JH, Korff B, Tuan HT, Hagen B, Luther S, Hasenfuss G, Parlitz U, Jafri MS, Hell SW, Lederer WJ, Lehnart SE (2012) Stimulated emission depletion live-cell super-resolution imaging shows proliferative remodeling of T-tubule membrane structures after myocardial infarction. Circ Res 111:402–414

Walden AP, Dibb KM, Trafford AW (2009) Differences in intracellular calcium homeostasis between atrial and ventricular myocytes. J Mol Cell Cardiol 46:463–473

Wang SQ, Song LS, Lakatta EG, Cheng H (2001) Ca^{2+} signalling between single L-type Ca^{2+} channels and ryanodine receptors in heart cells. Nature 410:592–596

Wang Z, Fermini B, Nattel S (1993) Delayed rectifier outward current and repolarization in human atrial myocytes. Circ Res 73:276–285
Wehrens XH, Lehnart SE, Reiken SR, Deng SX, Vest JA, Cervantes D, Coromilas J, Landry DW, Marks AR (2004a) Protection from cardiac arrhythmia through ryanodine receptor-stabilizing protein calstabin2. Science 304:292–296
Wehrens XH, Lehnart SE, Reiken SR, Marks AR (2004b) Ca^{2+}/calmodulin-dependent protein kinase II phosphorylation regulates the cardiac ryanodine receptor. Circ Res 94:e61–e70
Wehrens XH, Marks AR (2004) Novel therapeutic approaches for heart failure by normalizing calcium cycling. Nat Rev Drug Discov 3:565–573
Workman AJ, Kane KA, Rankin AC (2003) Characterisation of the Na, K pump current in atrial cells from patients with and without chronic atrial fibrillation. Cardiovasc Res 59:593–602
Wu X, Bers DM (2006) Sarcoplasmic reticulum and nuclear envelope are one highly interconnected Ca^{2+} store throughout cardiac myocyte. Circ Res 99:283–291
Wu X, Zhang T, Bossuyt J, Li X, McKinsey TA, Dedman JR, Olson EN, Chen J, Brown JH, Bers DM (2006) Local $InsP_3$-dependent perinuclear Ca^{2+} signaling in cardiac myocyte excitation-transcription coupling. J Clin Invest 116:675–682
Xiao B, Jiang MT, Zhao M, Yang D, Sutherland C, Lai FA, Walsh MP, Warltier DC, Cheng H, Chen SR (2005) Characterization of a novel PKA phosphorylation site, serine-2030, reveals no PKA hyperphosphorylation of the cardiac ryanodine receptor in canine heart failure. Circ Res 96:847–855
Xu Y, Tuteja D, Zhang Z, Xu D, Zhang Y, Rodriguez J, Nie L, Tuxson HR, Young JN, Glatter KA, Vazquez AE, Yamoah EN, Chiamvimonvat N (2003) Molecular identification and functional roles of a Ca^{2+}-activated K^+ channel in human and mouse hearts. J Biol Chem 278: 49085–49094
Zima AV, Bare DJ, Mignery GA, Blatter LA (2007) IP_3-dependent nuclear Ca^{2+} signalling in the mammalian heart. J Physiol 584:601–611
Zima AV, Blatter LA (2004) Inositol-1,4,5-trisphosphate-dependent Ca^{2+} signalling in cat atrial excitation-contraction coupling and arrhythmias. J Physiol 555:607–615
Zima AV, Kockskamper J, Blatter LA (2006) Cytosolic energy reserves determine the effect of glycolytic sugar phosphates on sarcoplasmic reticulum Ca^{2+} release in cat ventricular myocytes. J Physiol 577:281–293
Zygmunt AC, Gibbons WR (1991) Calcium-activated chloride current in rabbit ventricular myocytes. Circ Res 68:424–437
Zygmunt AC, Gibbons WR (1992) Properties of the calcium-activated chloride current in heart. J Gen Physiol 99:391–414
Zygmunt AC, Goodrow RJ, Weigel CM (1998) I_{NaCa} and $I_{Cl(Ca)}$ contribute to isoproterenol-induced delayed after depolarizations in midmyocardial cells. Am J Physiol 275:H1979–H1992

Cardiac Metabolism and Energetic Control 4

Susanne Rohrbach and Bernd Niemann

Abstract

Cardiomyocytes are incessantly contracting; thus, they also require a continuous energy supply to fuel an optimal function. Fatty acids are the predominant substrate used in the adult heart for ATP production. However, the heart is metabolically versatile in using other substrates such as glucose, lactate, amino acids and ketone bodies when they become abundantly available. In addition to this versatile substrate utilization, complex regulatory mechanisms such as transcriptional regulation and post-translational modifications of key metabolic protein as well as allosteric regulation by metabolites or substrates within the metabolic network contribute to the metabolic flexibility of the heart. This flexibility confers the advantage of adequately supplying ATP in response to a variety of physiological conditions including developmental or nutritional changes. However, the failing heart is characterized by an energetic deficit and a loss of metabolic flexibility. Here, we will summarize typical metabolic features of the healthy heart, the foetal heart, the hypertrophied or failing heart, the aged heart and the obese or diabetic heart. In addition, strategies used as metabolic therapies in heart failure as well as the impact of well-established heart failure therapeutics on cardiac metabolism will be discussed.

S. Rohrbach (✉)
Institute of Physiology, Justus-Liebig-Universität Gießen,
Aulweg 129, Giessen 35392, Germany
e-mail: Susanne.Rohrbach@physiologie.med.uni-giessen.de

B. Niemann
Klinik für Herz-, Kinderherz und Gefäßchirurgie, Justus-Liebig-Universität Gießen,
Rudolf-Buchheim-Sr. 7, Giessen 35392, Germany
e-mail: bernd.niemann@chiru.med.uni-giessen.de

K.-D. Schlüter (ed.), *Cardiomyocytes – Active Players in Cardiac Disease*,
DOI 10.1007/978-3-319-31251-4_4

4.1 Metabolism of the Adult Healthy Heart

The maintenance of metabolic homeostasis, which is among the most highly regulated processes in our organisms, is challenging due to the variation in diet composition, caloric content of the diet and food availability. Accordingly, nutrient sensing and coordination of nutrient utilization are mandatory to meet the need of the different organs of an organism. ATP is the major energy intermediate for all functions of organelles and organisms. A human produces per day nearly the same amount of ATP as the own body weight (Christian and Spremulli 2012). This impressive relation objectifies the central importance of the main source for ATP, the mitochondria, which produce about 90 % of the ATP needed. Overall, the cytosolic concentration of ATP remains stable at 3–4 mM, representing an amount of ~50 g ATP/body, a hydrolysis of 50 g ATP/min and thus the need of repetitive molecular ATP hydrolysis and ATP synthesis up to 1000 times/day.

The contractile performance of cardiomyocytes requires a highly efficient and adaptable system of energy generation, energy transfer and energy sensing. The heart utilizes an enormous amount of ATP each day (approximately 6 kg), multiple times its own organ weight, and varying activity levels require a rapid adaptation of energy supply (Ashrafian et al. 2007). However, the ATP reserve is relatively low and lasts only for approximately ten contractions. Therefore, a robust constant ATP synthesis is required. The healthy, adult heart derives 60–90 % of ATP from the fatty acid oxidation (FAO) and up to 30 % of myocardial ATP is generated by glucose and lactate, with smaller contributions from ketones and amino acids. Oxidation carbohydrates can become the major energy source under certain conditions such as a high-carbohydrate meal glucose (Mori et al. 2012; Lopaschuk et al. 1988). Cardiac extraction and oxidation of lactate are strongly increased during exercise when skeletal muscle lactate production increases (Kaijser and Berglund 1992). Complex regulatory mechanisms such as transcriptional regulation and post-translational modifications of key metabolic protein as well as allosteric regulation by metabolites or substrates within the metabolic network contribute to the metabolic flexibility of the heart. To understand metabolic alterations in the diseased heart, we will first provide an overview of metabolic processes in the healthy heart (Fig. 4.1).

4.1.1 Fatty Acid Metabolism

Fatty acids cross the plasma membrane by diffusion or by protein-mediated transport (Schaffer 2002). The latter mechanism was suggested to account for 70–80 % of fatty acid uptake (Luiken et al. 2004). In particular, long-chain fatty acids (LCFA) cross the cell membrane via fatty acid transporters such as the fatty acid translocase (FAT/CD36), fatty acid transport proteins (FATP) and the plasma membrane-bound fatty acid-binding protein ($FABP_{pm}$) (Su and Abumrad 2009; Koonen et al. 2005; Schaffer 2002; Chabowski et al. 2008). Insulin and contractions induce the translocation of intracellularly stored FAT/CD36 to the plasma membrane to increase cellular LCFA uptake (Koonen et al. 2005). Fatty acyl-CoA synthases (FACS) add a

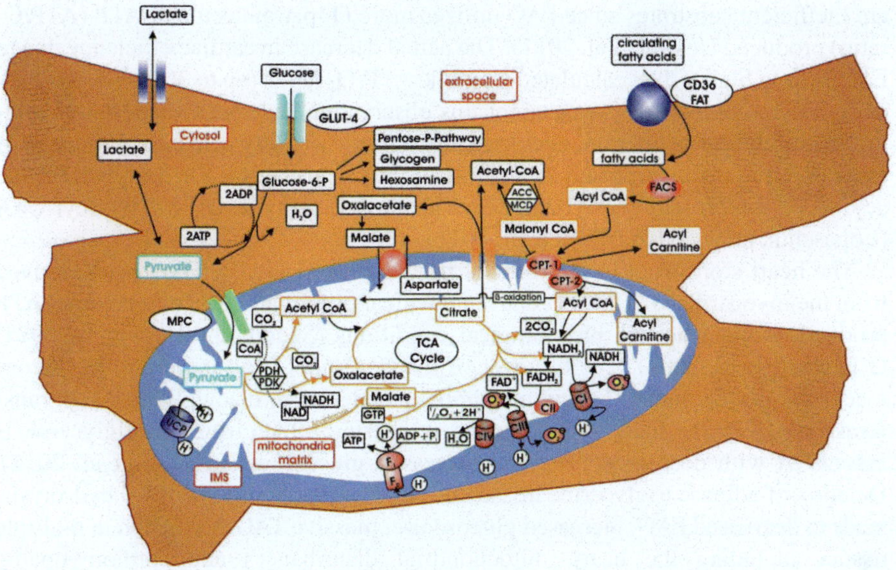

Fig. 4.1 Schematic representation of classic pathways of cardiac substrate metabolism. *ACC* acetyl-CoA carboxylase, *ADP* adenosine diphosphate, *CD36* cluster of differentiation 36, *CoA* coenzyme A, *CPT* carnitine palmitoyltransferase, *FACS* fatty acyl-CoA synthase, *FADH2* flavin adenine dinucleotide, *FAT* fatty acid translocase, *IMS* intermembrane space, *LCAD* long-chain acyl-CoA dehydrogenase, *MCD* malonyl-CoA decarboxylase, *MPC* mitochondrial pyruvate carrier, *TCA* tricarboxylic acid, *GTP* guanosine triphosphate, *ATP* adenosine triphosphate, *PDH* pyruvate dehydrogenase, *PDK* pyruvate dehydrogenase kinase, *NADH* nicotinamide adenine dinucleotide (reduced form), *NAD⁺* nicotinamide adenine dinucleotide (oxidized form), *FAD* flavin adenine dinucleotide, *MPC* mitochondrial pyruvate carrier, *CI* complex I of the respiratory chain, *CII* complex II of the respiratory chain, *CIII* complex III of the respiratory chain, *CIV* complex IV of the respiratory chain

CoA group to the fatty acids through esterification. Acyl-CoA synthases vary in their affinities for fatty acids with different chain length and show also differences in tissue localization (Ellis et al. 2010). The carnitine palmitoyltransferase 1 (CPT-1) converts the fatty acyl-CoA and free carnitine into long-chain fatty acylcarnitine. There are two isoforms of CPT-1: CPT-1α predominates in the liver, and CPT-1β is the main isoform in the heart (McGarry and Brown 1997). CPT-1β is 30-fold more sensitive to malonyl-CoA inhibition than is CPT-1α (McGarry and Brown 1997; Weis et al. 1994; Bonnefont et al. 2004). The carnitine acylcarnitine translocase (CAT) transports the fatty acid across the inner mitochondrial membrane, where CPT-2 converts the acylcarnitine into acyl-CoA, which will then enter the fatty acid oxidation (FAO) spiral to produce acetyl-CoA. One acetyl-CoA, NADH and $FADH_2$ are produced during each cycle of FAO in the mitochondrial matrix. Acetyl-CoA enters the Krebs cycle, and NADH and $FADH_2$ from FAO as well as from TCA cycle (see below) are utilized by the mitochondrial respiratory chain to produce the final products ATP and H_2O. The oxidation of palmitate ($C_{16}H_{32}O_2$) requires 23 O_2 and produces 105 ATP. Compared to glucose, fatty acids are inefficient substrates since FAO utilizes more O_2 per molecule of ATP (ATP/O_2 ratio) produced (Jaswal et al. 2011). The actual decrease in cardiac efficiency due to FAO is even higher than calculated from the ATP/O_2 ratios (up to 30 %) (Lopaschuk et al. 2010). A number of reasons for this observation have been suggested including the inhibition of adenine nucleotide transferase by fatty acids, the influence of mitochondrial uncoupling and the cycling of fatty acids between triacylglycerol and acyl-CoA, which requires ATP for esterification of fatty acids to fatty acyl-CoA (Lopaschuk et al. 2010; Jaswal et al. 2011).

The heart stores fuel in the form of triacylglycerol (TAG). Fatty acids derived from the myocardial TAG pool contribute with approximately 10 % to the total ATP production under normal physiological conditions (Saddik and Lopaschuk 1991; O'Donnell et al. 2008). The turnover of the TAG pool is an important regulatory mechanism of cardiac fatty acid metabolism. Deletion of diacylglycerol acyltransferase (DGAT), the final enzyme in the synthesis of TAG from diacylglycerol, is associated with decreased FAO and increased glucose uptake (Liu et al. 2011). Deletion of adipose triglyceride lipase, the enzyme responsible for TAG hydrolysis, leads to decreased FAO, increased glucose use, massive TAG deposition in multiple tissues including the heart, mitochondrial disturbances and cardiomyopathy (Haemmerle et al. 2006, 2011).

4.1.2 Carbohydrate Metabolism

Glucose uptake in cardiomyocytes is regulated by the translocation of the glucose transporters GLUT1 and GLUT4 to the cell membrane (Becker et al. 2001). GLUT4 is the most abundant glucose transporter in cardiomyocytes. The translocation of GLUT4 follows the release of insulin or in response to metabolic or contraction-induced activation of the AMPK-dependent protein kinase (AMPK). The heart is constantly contracting, and therefore, it is likely that contraction-mediated GLUT4 translocation represents an important glucose uptake mechanism (Abel 2004). In comparison, many muscles are often at rest and insulin-mediated GLUT4 translocation represents a quantitatively more important mechanism regulating skeletal muscle glucose uptake than in the heart. Cardiomyocytes also show a strong expression of GLUT1, which is responsible for a significant component of basal cardiac glucose uptake (Abel 2004).

After uptake, intracellular glucose is converted to glucose-6-phosphate by hexokinase II (Postic et al. 1994) and can either enter the glycolytic pathway or be stored as glycogen. The phosphofructokinase (PFK), a glycolytic enzyme that catalyzes the irreversible transfer of a phosphate, exists in two isoforms: PFK1 and PFK2. PFK1 catalyzes the conversion of fructose-6-phosphate and ATP to fructose 1,6-bisphosphate and ADP, which is a key regulatory and rate-limiting step of glycolysis. PFK2 phosphorylates fructose-6-phosphate to form fructose-2,6-bisphosphate. PFK1 is allosterically inhibited by high levels of ATP, but AMP reverses the inhibitory action of ATP. PFK1 is also inhibited by low pH, phosphoenolpyruvic acid and citrate. PFK1 is allosterically activated by a high concentration of AMP, but the most potent activator is fructose-2,6-bisphosphate. Fructose-2,6-bisphosphate is produced by PFK2 and increases the glycolytic flux. The precise regulation of PFK1 prevents uncoupling of glycolysis/glucose oxidation and prevents glycolysis and gluconeogenesis from occurring simultaneously. During glycolysis, glucose is converted to pyruvate, which is either oxidized and converted entirely to CO_2 or converted to lactate. When pyruvate does not enter glucose oxidation, it is converted to lactate by lactate dehydrogenase, which occurs under anaerobic conditions or when mitochondrial pyruvate oxidation is inhibited. The conversion of glucose into lactate produces 2 ATP. When glycolysis is coupled to glucose oxidation, pyruvate needs to be shuttled across the impermeable inner mitochondrial membrane by a mitochondrial pyruvate carrier (Herzig et al. 2012). Glucose oxidation is regulated via the activity of the pyruvate dehydrogenase (PDH) complex, which catalyzes the oxidative decarboxylation of pyruvate, linking glycolysis to the tricarboxylic acid cycle. PDH is a giant multienzyme complex that also contains a PDH phosphatase (PDP) and a PDH kinase (PDK) (Holness and Sugden 2003). It is regulated by phosphorylation and by the levels of its products and substrates. PDK inhibits PDH, while dephosphorylation by PDP reverses this inhibition (Holness and Sugden 2003). Products of FAO (NADH and acetyl-CoA) activate PDK, which results in a PDH phosphorylation and inhibition. Thus, when fatty acids are available and oxidized, the capacity to utilize glucose is low. There are different PDK isoforms. PDK2 and PDK4 are expressed in most tissues, PDK1 is limited to the heart and pancreatic islets and PDK3 is limited to the kidney, brain and testis (Sugden and Holness 2002). Elevated pyruvate levels result in an increased PDH activity (Denton et al. 1996). Mediators that increase cytoplasmic calcium can also activate pyruvate dehydrogenase (Denton et al. 1996). Sympathetic stimulation or insulin increases glycolytic flux via phosphorylation and activation of PFK2 (Rider et al. 2004; Mouton et al. 2010) and activation of PDH (Mazumder et al. 2004; McCormack and Denton 1981). The final stage in the oxidation of glucose entails a set of nine reactions in which the acetyl group of acetyl-CoA is oxidized to CO_2. These reactions operate in a cycle that is referred to by several names: citric acid cycle, tricarboxylic acid (TCA) cycle or Krebs cycle. Most enzymes involved in the TCA cycle are localized to the mitochondrial matrix. However, succinate dehydrogenase and α-ketoglutarate dehydrogenase are localized to the inner mitochondrial membrane. The full oxidation of one molecule of glucose requires 6 O_2 and produces 31 ATP. Glucose is therefore a more efficient substrate than fatty acids (see above). Since Krebs cycle intermediates can be used for many biosynthetic pathways, they are constantly removed from the cycle and need to be replenished. This process of replenishment of Krebs cycle through pathways independent of acetyl-CoA is called anaplerosis. An example of such an anaplerotic reaction is the carboxylation of pyruvate to malate, which requires nicotinamide adenine dinucleotide phosphate (NADPH) and is therefore linked to other pathways that consume (e.g. antioxidative defence, lipogenesis) or produce (pentose phosphate pathway) NADPH.

Glucose metabolism through glycolysis yields pyruvate, which is subsequently oxidized in the mitochondrial compartment as described above. However, glycolytic intermediates such as glucose-6-phosphate (G6P) and lactate participate in additional pathways. G6P enters the pentose phosphate pathway (PPP), resulting in the generation of NADPH during the oxidative phase and the generation of pentose during the non-oxidative phase of the PPP. NADPH is required for maintaining the level of reduced glutathione and has therefore a major role in the antioxidant defence. Furthermore, NADPH is used in reductive biosynthesis reactions within cells such as fatty acid synthesis or biosynthesis of hormones. The pentose ribose 5-phosphate is utilized in the synthesis of nucleotides.

G6P can also be converted to sorbitol by the enzyme aldose reductase, in the polyol pathway. An increased flux through this pathway has been associated with cardiac dysfunction in diabetic patients (Johnson et al. 2004) and an impairment of functional and metabolic recovery after ischaemia (Hwang et al. 2004). Treatment with an aldose reductase inhibitor improved LV systolic function (Johnson et al. 2004). Fructose-6-phosphate can be converted into uridine diphosphate-N-acetylglucosamine, a substrate for the O-linked GlcNAc transferase, and is thus involved in the O-GlcNAcylation of proteins (Wells et al. 2001). Increased protein O-linked GlcNAcylation was suggested to represent a cardioprotective mechanism against ischaemia-reperfusion injury (Liu et al. 2007). The heart stores fuel in the form of glycogen, although the turnover rate of cardiac glycogen is low in the adult heart under resting conditions. Myocardial glycogen metabolism in the hypertrophied heart is similar to that observed in the normal heart (Allard et al. 1997). However, an acute increase in workload or ischaemia has been suggested to result in an increased utilization of glycogen (Goodwin et al. 1998; Schaefer and Ramasamy 1997).

Glucose and free fatty acid (FFA) metabolic pathways are subject to complex reciprocal regulation, where acetyl-CoA derived from glycolysis inhibits FAO and vice versa (Randle cycle or glucose fatty acid cycle) (Randle et al. 1963). Acetyl-CoA and NADH produced by FAO inhibits PDH, while acetyl-CoA produced by PDH inhibits the FAO enzyme 3-ketoacyl-CoA thiolase and NADH from glucose oxidation inhibits the FAO enzyme 3-hydroxyacyl-CoA dehydrogenase and acyl-CoA dehydrogenase (Jaswal et al. 2011). In addition, citrate inhibits PFK1 (Randle 1998). Malonyl-CoA, which can arise from acetyl-CoA, inhibits the carnitine palmitoyltransferase (CPT) that controls the entry and oxidation of LCFA. The Randle cycle controls fuel selection and adapts the substrate supply and demand in normal tissues through a nutrient-mediated fine-tuning of metabolism in addition to its hormonal control. Therefore, utilization of FFA and glucose is well balanced in the healthy heart and enables the heart to switch between energy sources according to their availability and in response to environmental stimuli. This provides the heart with a high flexibility in substrate utilization.

4.1.3 Amino Acid Metabolism

Although most catabolic and metabolic pathways of amino acids are found in the liver, the branched-chain amino acid (BCAA) catabolism is localized mainly in non-hepatic tissues (Harper et al. 1984). The heart is also capable to metabolize the branched-chain amino acids leucine, isoleucine and valine. BCAA are essential amino acids acquired from external food and play a major role in protein synthesis and turnover. Myocardial utilization of BCAAs is insensitive to the circulating insulin level and is regulated instead by their availability in arterial plasma (McNulty et al. 2000). An excess amount of free BCAA or their products, as observed in genetic disorders resulting from defects in BCAA catabolic pathway, are associated with cardiomyopathies and defects in mitochondrial respiratory chain (Romano et al. 2010; Arn and Funanage 2006).

BCAA are metabolized first to branched-chain keto acids (BCKAs) by the branched-chain aminotransferase (BCAT) and then oxidized to propionyl-CoA by the BCKA dehydrogenase complex (BCKD), the rate-limiting step in BCAA metabolism (Harper et al. 1984). Propionyl-CoA can be metabolized to acetyl-CoA or to succinyl-CoA and then enter the TCA cycle. Activity and expression of the BCKD complex are tightly regulated in response to nutrient and growth signals through the phosphorylation status of the regulatory subunit of the E1α subunit, similar to the regulation of the PDH complex (Arn and Funanage 2006; Harris et al. 2004). Phosphorylation of the BCKD complex by BCKD kinase and dephosphorylation by a mitochondrial matrix-targeted protein phosphatase 2C family member (PP2Cm) regulate BCKD activity and, thus, the oxidation of BCAA (Lu et al. 2009). When BCAA level is high, E1α is dephosphorylated by the BCKD phosphatase (PP2Cm), which leads to BCKD activation and reduction of BCAA levels. Recent investigations suggest that PP2Cm, which is highly expressed in the heart, regulates mitochondrial membrane permeability transition pore opening and is essential for cell survival, ROS defence, embryonic development and cardiac function (Lu et al. 2007). This suggests that PP2Cm may have other targets in addition to BCKD. The contribution of amino acids to overall cardiac oxidative metabolism is considered minor due to the low availability of these substrates. However, BCAA are not only an important nutrient source, they are also important signalling molecules. They activate the mTOR signalling pathway, which promotes cardiac hypertrophy and suppresses autophagy (Nicklin et al. 2009). This suggests that increased cardiac BCAA could lead to a chronic mTOR activation, which promotes hypertrophy and inhibits potentially cardioprotective autophagy. However, another study also suggested that dietary supplementation of BCAA promotes cardiac mitochondrial biogenesis, reduces ROS damage and increases life span in mice. The underlying mechanisms appear to involve mTOR, eNOS and Sirt1 (D'Antona et al. 2010).

4.1.4 Ketone Body Metabolism

Ketone bodies such as acetoacetic acid and β-hydroxybutyric acid are produced from fatty acids during periods of fasting, and metabolism of ketone bodies yields acetyl-CoA. Ketone metabolism in the healthy heart under non-starving conditions is rather modest. However, the contribution of ketone bodies to overall cardiac oxidative metabolism under conditions of starvation was shown to be considerable (Jeffrey et al. 1995). Ketotic nutrient environments increase ketone bodies in the blood and result in an enhanced use by the heart (Wentz et al. 2010). In the isolated perfused heart, ketone bodies reduce fatty acid uptake and oxidation, an effect that is not due to elevated myocardial malonyl-CoA content (Jeffrey et al. 1995; Stanley et al. 2003). Blood ketone bodies are elevated in CHF in proportion to the severity of cardiac dysfunction and neurohormonal activation (Lommi et al. 1996). A moderate ketosis was suggested as a compensatory mechanism for the defects in mitochondrial energy transduction associated with acute insulin deficiency or mitochondrial senescence (Sato et al. 1995).

4.1.5 Respiratory Chain

Under physiological conditions, mitochondria consume large amounts of oxygen to produce ATP at complex V of the respiratory chain. The mitochondrial matrix homes the PDH complex, enzymes of the Krebs cycle or FAO and most enzymes of the urea cycle and heme biosynthesis. These metabolic pathways directly feed electrons via NADH and FADH into the respiratory chain. Oxidation releases electrons which are stepwise transferred to less energetic acceptors thus enabling controlled energy release for ATP synthesis by generating a highly energetic phosphoric acid anhydride bond (Gautheron 1984; Hatefi 1985). Terminal redox acceptor is oxygen thus defining the name of the process as oxidative phosphorylation (oxphos). Hydrogen is transferred to the respiratory chain by $NADH/H^+$ and $FADH_2$. Central metabolic processors are the four respiratory chain complexes associated together with the F_1/F_0 ATP synthase which is often also called complex V of the respiratory chain (Gautheron 1984; Hatefi 1985). Traditional understanding defines separate complexes (complex I, NADH-ubiquinone-oxidoreductase; complex II, succinate ubiquinone oxidoreductase; complex III, cytochrome bc-complex; complex IV, cytochrome C oxidase), but current knowledge assumes the existence of respiratory super-complexes, so-called respisomes (Chance and Williams 1955; Schagger and Pfeiffer 2000; Lenaz et al. 2010; Genova and Lenaz 2014). Transfer of electrons between complexes is provided by two mobile substrates, cytochrome C and ubiquinone. Energetic output is a gradient of ten protons per $NADH/H^+$ (complexes I–IV) and six protons per succinate (complexes II–IV). The P/O ratio describes how many ATP can be produced per oxygen. For $NADH_2$-dependent respiration P/O is 2,7 ATP/O_2, while succinate-dependent respiration exhibits a P/O of 1,6 ATP/O_2 when assuming energy-consuming transmembrane transport mechanisms. As functional status of the respiratory chain depends on availability of substrates and integrity of complexes, lack of ADP or Pi reduces respiratory flow. This phenomenon is called "respiratory control" or "regulation of oxidative phosphorylation". Dynamic equilibria known as steady states have been defined in five classes (Chance and Williams 1956). Flux in respiration is controlled by availability of different substrates as shown in Table 4.1. During state 3 respiration, the respiratory chain works at maximum activity, depending on the transmembrane potential. State 4, also termed as "controlled state", exhibits limited function of respiratory chain due to ADP deprivation. Uncoupling the respiratory chain by protonophores disturbs respiratory

Table 4.1 Metabolic states of mitochondria (Chance and Williams 1956)

State	Excess	Respiration rate	Limiting factor
1	O_2	slow	ADP and substrate
2	O_2, ADP	slow	substrate
3	O_2, ADP, substrate	fast	ΔpH, respiratory chain
4	O_2, substrate	slow	ADP
5	ADP, substrate	0	O_2

control. Because passive proton reflux into the matrix abolishes the mitochondrial membrane potential, the respiratory chain is working unimpeded as long as substrates are available. While unregulated uncoupling leads to a complete breakdown of cellular energy supply and thereafter cellular death, controlled uncoupling is needed to survive. Uncoupling proteins such as uncoupling protein 2 (UCP2) directly reduce transmembrane potential in states of reduced electron flux to avoid redox-associated injury to proteome, genome or lipidome of mitochondria and the entire cell (Negre-Salvayre et al. 1997; Arsenijevic et al. 2000).

4.2 Metabolism of the Foetal and Newborn Heart

In utero, the foetus receives blood rich in carbohydrates (especially glucose and lactate) but poor in fatty acids (Battaglia and Meschia 1978; Girard et al. 1992). Although FFA are transferred across the placenta also in humans, fatty acids are transformed into TAG and stored in liver and adipose tissue rather than oxidized by the heart or other foetal organs (Girard et al. 1985). Accordingly, the rates of FAO are low and the foetus oxidizes mainly glucose, lactate and amino acids. In the foetal heart, glucose uptake is regulated by the low-affinity transporter GLUT1 (Postic et al. 1994). The activities of enzymes in the glycolytic pathway are high during foetal development (Girard et al. 1992; Hoerter and Opie 1978; Ferre et al. 1986; Lopaschuk et al. 1992), which may also be related to the stronger expression of HIF-1α in the foetal heart (Nau et al. 2002). PFK1, the rate-limiting enzyme of glycolysis, was described to be less sensitive to the inhibitory effects of cytosolic ATP and citrate than the adult heart (Bristow et al. 1987). Instead, PFK1 may be more sensitive to the activating effects of fructose-2,6-bisphosphate (Bristow et al. 1987). The circulating levels of lactate can exceed 10 mM in the foetus compared to 1 mM in the adult (Girard et al. 1992). Lactate influx and efflux are controlled by monocarboxylate cotransporters, which work according to the transmembrane proton gradient (Poole and Halestrap 1993). The foetal heart extracts lactate, and lactate oxidation accounts for the majority of cardiac oxygen consumption (Girard et al. 1992; Fisher et al. 1980). Since glycolytic rates are higher than glucose oxidation, the intracellular pH is low and can thus enable an effective influx of lactate into the foetal cardiomyocyte (Lopaschuk et al. 1991; Onay-Besikci 2006). During foetal life, insulin levels are high and glucagon is low, while the immediate postnatal period is characterized by an increase in plasma glucagon and a decrease in plasma insulin in humans (Blazquez et al. 1974). This change in plasma concentrations may be related to the activation of the sympathetic nervous system (Girard et al. 1974). Foetal cardiomyocytes are also characterized by a lower mitochondrial content and a lower activity of enzymes of the Krebs cycle (Goodwin et al. 1976). While in the adult heart CPT-1α represents only a very minor constituent, the foetal heart shows a much greater proportion of this isoform (Brown et al. 1995), which is less sensitive to malonyl-CoA inhibition (McGarry and Brown 1997; Weis et al. 1994; Bonnefont et al. 2004).

The adaptation to the extrauterine life is characterized by an increasing workload for the heart and the demand for increased energy production. In the immediate newborn period, glycolytic rates are high but glucose oxidation is low (Lopaschuk et al. 1991; Lopaschuk et al. 1992). However, within a few days after birth, glycolysis decreases and provides less than 10% of ATP production (Lopaschuk and Spafford 1990; Lopaschuk et al. 1991). Shortly after birth, glucose transporter molecules demonstrate a shift from GLUT1 towards a higher expression of GLUT4 (Postic et al. 1994). As the newborn heart matures, the capacity to oxidize glucose increases, FAO increases and fatty acids become the main source of energy within the first weeks after birth (Lopaschuk and Spafford 1990; Lopaschuk et al. 1991). However, plasma concentrations of FFA in newborns reach adult levels soon after birth due to mobilization of endogenous fat stores (Callikan et al. 1979). Thus, the rise in plasma FFA occurs before the increase in FAO, suggesting that other factors than FFA supply are involved. Plasma levels of the adipokine adiponectin increase after birth and adiponectin was shown to stimulate FAO by activating AMPK (Onay-Besikci et al. 2004; Yamauchi et al. 2002). Expression and activity of AMPK increase after birth, which results in a decrease in acetyl-CoA carboxylase (ACC) activity and thus lower malonyl-CoA but increased FAO (Makinde et al. 1997; Lopaschuk et al. 1994; Dyck et al. 1998). The postnatal increase in adiponectin and AMPK may thus contribute to the stimulation of FAO (Onay-Besikci et al. 2004; Combs et al. 2003; Makinde et al. 1997; Lopaschuk et al. 1994). The main pathway to remove malonyl-CoA in the heart is via degradation by malonyl-CoA decarboxylase (Dyck et al. 1998), which shows an increase in expression and activity during cardiac maturation and may thus also contribute to the increased FAO (Onay-Besikci et al. 2003).

4.3 Metabolism of the Hypertrophied or Failing Heart

Irrespective of cause, heart failure is associated with an energy deficit, resulting from the higher energy demand of the hypertrophied cardiomyocytes along with impairments in substrate utilization. It has therefore been called "an engine out of fuel", since the phosphocreatine to ATP (PCr/ATP) ratio is reduced compared to healthy hearts (Neubauer 2007; Hardy et al. 1991). Cardiac hypertrophy is associated with maintained or moderately reduced ATP levels but strongly reduced phosphocreatine (Liao et al. 1996; Tian et al. 1997; Ingwall 2009), resulting in a reduced PCr/ATP ratio. In end-stage heart failure, also a significant decrease in ATP is observed (Ingwall and Weiss 2004; Ingwall 2009; Neubauer 2007). Furthermore,

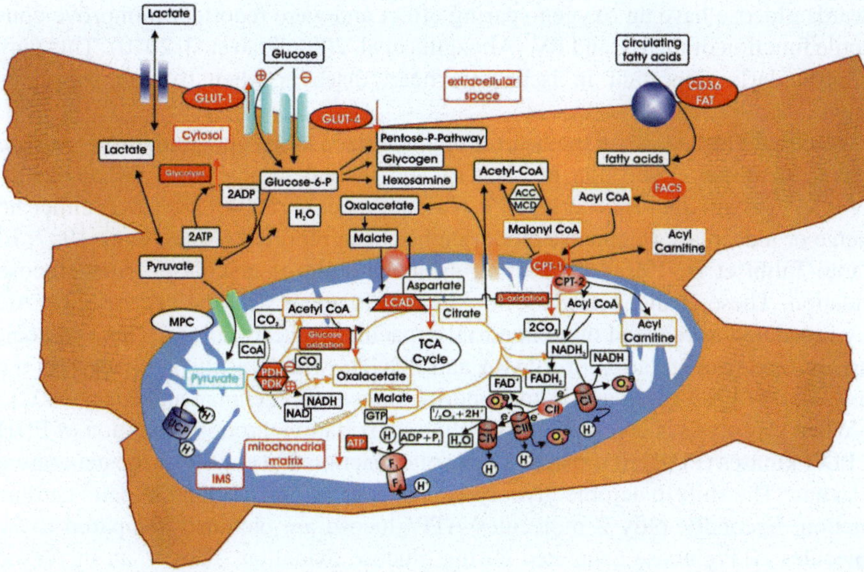

Fig. 4.2 Alterations in myocardial substrate metabolisms observed in heart failure. The affected molecules are highlighted in *red* and the direction of the change is marked with an *arrow*. Abbreviations as described in Fig. 4.1

the myocardial PCr/ATP ratio is a predictor of both total and cardiovascular mortality in patients with dilated cardiomyopathy (Neubauer et al. 1997). Accordingly, enhancing ATP synthesis or FAO was shown to preserve cardiac energetic as well as cardiac function in preclinical models (Gupta et al. 2012; Kolwicz et al. 2012).

In patient with heart failure, the above-described high metabolic flexibility of the healthy heart is impaired and contributes significantly to disease progression. However, the metabolic phenotype (Fig. 4.2) shows significant differences depending on the pathogenesis of heart failure and the disease stage (compensated hypertrophy vs. end-stage heart failure). Moderate heart failure is characterized by a shift from FAO towards glucose oxidation. At this stage, FFA uptake, enzymes involved in fatty acid oxidation and markers of mitochondrial biogenesis are reduced (Abdurrachim et al. 2015). Metabolic disturbances such as decrease in FAO were shown to be evident already 2 weeks after transverse aortic constriction (TAC) and thus may precede the onset of congestive heart failure, while cardiac dilatation and impaired LV ejection fraction were evident only after 20 weeks in this rat model of pressure overload (Doenst et al. 2010). However, others have shown that no decrease in FAO occurs during compensated HF and that the downregulation of FAO enzymes and the switch to carbohydrate oxidation are only phenomena of late-stage heart failure (Chandler et al. 2004). This shift in substrate utilization seems to be an adaptive response, since glucose oxidation uses 12% less oxygen to produce the same amount of ATP (Korvald et al. 2000). Accordingly, so-called metabolic agents such as PDK inhibitors that cause a shift in the myocardial substrate use from FFA towards glucose have an oxygen-sparing effect and were reported to improve contractile function of the LV and RV (Abozguia et al. 2006; Piao et al. 2010). This may be particularly important in ischaemic heart diseases when oxygen supply is limited.

Increased glucose use in the heart is characterized by an upregulation of glucose uptake, increased expression of GLUT1 and increased glycolysis (Allard et al. 1994; Montessuit and Thorburn 1999; Nascimben et al. 2004). However, either no change or a decrease in glucose oxidation has been reported (Allard et al. 1997; El Alaoui-Talibi et al. 1997), suggesting an uncoupling of glycolysis from glucose oxidation. These changes in glucose oxidation together with the decrease in FAO may be related to reduced mitochondrial capacities in heart failure. This is associated with a downregulation of PPARα and PGC1α, which results in changes in the expression of FAO genes and mitochondrial biogenesis (Lehman and Kelly 2002).

When glycolysis is uncoupled from glucose oxidation through inhibition of PDH by PDH kinases (PDK), it remains a cytosolic reaction that ends with the generation of lactate. The shift to aerobic glycolysis results in acidosis, which impairs cardiac function. Secondly, only 2 molecules ATP/glucose are obtained compared to 32 molecules ATP/glucose generated during glucose oxidation. Thirdly, an increased glucose uptake is required to accomplish the increased energy demand of the hypertrophied myocardium (Lundgrin et al. 2013). All these changes in metabolism alter cardiac susceptibility to hypertrophy and fibrosis, rates of apoptosis and proliferation and cardiac energy balance, calcium balance and thus also contractility. Many studies providing evidence of energy deficit and metabolic disturbances in heart failure have been performed in models of LV pressure or volume overload. However, there are also studies showing comparable changes in the RV in response to increased RV afterload. Hypertrophied RV cardiomyocytes in pulmonary arterial hypertension (PAH) demonstrate an inhibition of mitochondrial respiration, reliance on glycolysis and glutaminolysis and reduced energy production (Paulin and Michelakis 2014). While some forms of right ventricular hypertrophy (RVH) are well tolerated (adaptive RVH), other forms rapidly result in RV failure (maladaptive RVH). Increased glycolysis along with PDH inhibition is more evident in maladaptive RVH (Piao et al. 2010). Interestingly, these RV changes extend to the LV in maladaptive RVH.

Recent data suggest that heart failure may itself promote metabolic changes such as insulin resistance, in part through neurohumoral activation (Ashrafian et al. 2007; Nikolaidis et al. 2004b). Insulin resistance limits glucose uptake and can thus results in an impairment of ATP synthesis (Nikolaidis et al. 2004b; Swan et al. 1997). Cardiac-specific overexpression of GLUT1, which results in an increased insulin-independent uptake of glucose and recapitulates the increases in basal glucose uptake observed in hypertrophied hearts, delays the development of heart failure in mice (Liao et al. 2002). The long-term increase in basal glucose uptake does not induce glucotoxicity but confers protections against ageing-associated increase of susceptibility to ischaemic injury (Luptak et al. 2007). The development of cardiac insulin resistance and decreased mitochondrial oxidative metabolism have been demonstrated to be early metabolic changes in the development of cardiac hypertrophy in a mouse model of abdominal aortic constriction (Zhang et al. 2013). This suggests that these changes in energy metabolism precede the development of heart failure and may actually contribute to the progression from hypertrophy to heart failure. Indeed, cardiac mitochondria are affected early and undergo progressive remodelling during cardiac hypertrophy and the progression to end-stage heart failure. Among the diverse mitochondrial changes in cardiac failure in humans as well as in preclinical models are decreased state 3 respiration, changes in mitochondrial morphology, decreased protein and mRNA levels of ETC proteins and markers of mitochondrial biogenesis, increased mitochondrial ROS release and reduced enzyme activity of selected respiratory chain complexes (Rosca and Hoppel 2013; Turrens 2003; Bugger et al. 2010; Doenst et al. 2010; Riehle et al. 2011; Scheubel et al. 2002; Kato et al. 2010; Amorim et al. 2010).

A factor that may influence the interpretation of glucose oxidation is the fact that pyruvate is channelled into anaplerotic pathways. An increase in anaplerosis, the replenishment of Krebs cycle through pathways independent of acetyl-CoA, may account for the mismatch between increased glycolysis and decreased glucose oxidation in cardiac hypertrophy. It has been reported that anaplerosis increases in cardiac hypertrophy in rats and mice (Sorokina et al. 2007; Pound et al. 2009; Kolwicz et al. 2012). Anaplerosis maintains the flux through the Krebs cycle and accounts for a great portion of glucose oxidation. The shift away from acetyl-CoA and NADPH production towards anaplerosis bypasses energy, which may also contribute to a less energy-efficient heart. Activation of PDH on the other hand reduces anaplerosis, results in normalized TAG levels and improves ex vivo contractility (Pound et al. 2009). However, cardiac hypertrophy results in an increased demand for amino acids, derived from precursors of the Krebs cycle. Therefore, activation of anaplerotic pathways could also be envisioned as a countermeasure to maintain Krebs cycle and thus a protective mechanism. Indeed, anaplerosis plays an important role in the maintenance of contractile function and impairment in anaplerosis causes contractile dysfunction (Russell and Taegtmeyer 1991). Currently, the role of anaplerosis under conditions of cardiac hypertrophy remains incompletely understood (des Rosiers et al. 2011).

4.4 Metabolism of the Diabetic or Obese Heart

The metabolic phenotype of the diabetic and the obese heart shows many similarities. In both diseases, fatty acid uptake and oxidation are increased, intramyocardial TAGs show increased levels, glucose uptake and glucose oxidation are reduced, malonyl-CoA levels are decreased and malonyl-CoA decarboxylase expression is induced (Lopaschuk et al. 2010). Preclinical models of obesity or diabetes and studies in obese patients have also shown that circulating fatty acids and TAGs are increased (Koutsari and Jensen 2006; Young et al. 2002; Luiken et al. 2001; Mazumder et al. 2004). Although glucose uptake is reduced, an increased flux of the accessory pathways of glucose metabolism such as the polyol pathway or the hexosamine biosynthetic pathway has been identified in the diabetic myocardium

(Kolwicz et al. 2013). The inability of the heart to appropriately utilize glucose results in a reliance on FAO and a metabolic inflexibility compared to the healthy heart. This metabolic inflexibility has been associated with marked changes in the insulin-signalling cascade including reduced insulin-mediated phosphorylation of IRS-1, decreased PKB phosphorylation and reduced GLUT4 translocation, which can partially be mimicked with short-term high-fat diet (Park et al. 2005; Wright et al. 2009). Within 1 day of high-fat diet, cardiac mitochondria demonstrate a reduced ability to utilize pyruvate, which is induced by an increase in the expression of PDK4 and inhibition of PDH (Crewe et al. 2013). Interestingly, the increased expression of PDK4 precedes the diminished insulin signalling (Crewe et al. 2013). This suggests that fatty acid-induced inhibition of PDH may be an initiating event in decreased metabolic flexibility of the heart. PDK4 upregulation does not only occur in response to high-fat diet but is also with food restriction (Wu et al. 2001; Holness et al. 2002), both conditions with high levels of circulating fatty acids. When food is limited, the PDK4-mediated inhibition of cardiac PDH may be a protective mechanism to preserve limited glucose for organs highly dependent on glucose such as the brain. However, the same mechanism may contribute significantly to a deterioration of cardiac function in the context of obesity or diabetes.

Mitochondria respiring fatty acids produce more ROS than mitochondria respiring pyruvate. Attenuation of mitochondrial ROS release results in completely preserved insulin sensitivity despite a high-fat diet (Anderson et al. 2009; Seifert et al. 2010). Accordingly, mice with overexpression of a mitochondrially targeted catalase show reduced ROS release and do not develop insulin resistance despite high-fat diet (Anderson et al. 2009). In addition to allosteric interactions that modulate enzyme activities, metabolic intermediates such as acetyl-CoA but also ROS are involved in controlling the balance between glucose and fatty acid oxidation. Post-translational modifications such as acetylation modulate enzyme activities and thus metabolism (Still et al. 2013; Guan and Xiong 2011). In vitro acetylation of cardiac mitochondria increases ROS production and inhibits pyruvate oxidation (Vadvalkar et al. 2013), suggesting that acetylation of mitochondrial proteins is also involved in the induction of metabolic inflexibility. Proteins involved in pyruvate oxidation may be particularly sensitive to acetylation, possibly due to their proximity to acetyl-CoA or a resistance to deacetylation, e.g. by SIRT3.

In the obese or diabetic heart, the supply of substrates exceeds the need for ATP synthesis. This results in a significant accumulation of lipids despite increased FAO (van de Weijer et al. 2011). Ectopic lipid accumulation in the heart, a process termed lipotoxicity, is associated with cardiac hypertrophy, cardiac dysfunction and apoptosis (van de Weijer et al. 2011; Chiu et al. 2001). The strong association between increased cardiac fatty acid uptake and cardiomyopathy has been demonstrated in various animal models (Yagyu et al. 2003; Chiu et al. 2005). However, the evidence that patients without a comparable genetic defect will accumulate cardiac lipids due to elevated circulating fatty acids and TAGs may not be as strong although numerous studies have shown that excess lipid accumulation occurs in obese or insulin-resistant patients (Marfella et al. 2009; Sharma et al. 2004). Indeed, patients with type 2 diabetes mellitus and heart failure adapt to increases in circulating lipids and do not display features of acute myocardial lipotoxicity (Nielsen et al. 2013). Furthermore, lipid accumulation appears to be reversible in humans as recently shown. Mechanical unloading was able to correct metabolic derangements and myocardial lipotoxicity in advanced heart failure and reverse insulin resistance (Chokshi et al. 2012).

Obesity or type 2 diabetes mellitus induces an increase in fatty acid uptake and oxidation together with a reduced glucose oxidation and cardiac efficiency in animal models as well as in patients (Buchanan et al. 2005; Peterson et al. 2004; Mazumder et al. 2004; Rijzewijk et al. 2009). The decreased cardiac efficiency may be related to the low O_2 efficiency of FAO as well as to increased mitochondrial uncoupling induced by fatty acids (Burkhoff et al. 1991; Cole et al. 2011). This uncoupling effect, which was suggested to be due to the protonophoric actions of fatty acids or the induction of uncoupling protein (Schonfeld and Wojtczak 2008; Millet et al. 1997), results in a decreased ATP production despite fuel oxidation. Increased delivery of reducing equivalents from FAO to the respiratory chain but decreased oxidative phosphorylation capacity of the respiratory chain can result in increased ROS production. Initially, the mitochondrial uncoupling may therefore represent an adaptation to increased FA-induced ROS production. However, this does not completely compensate for the increased ROS production as deduced from increased ROS damage in obesity or diabetes (Boudina et al. 2007). On a long-term basis, the uncoupling significantly contributes to the energetic deficit with a decreased PCr/ATP ratio in diabetic hearts (Scheuermann-Freestone et al. 2003).

Whereas oxidation of pyruvate and pyruvate dehydrogenase activity are significantly reduced in obese mice, mitochondrial respiration with palmitate as a substrate is preserved (Boudina et al. 2005). This shows that obesity rather affects carbohydrate oxidation in a manner similar to diabetes

Obesity and diabetes also affect the secretion of a large number of adipokines, which regulate multiple metabolic processes as well as insulin sensitivity in different organs. Some of these adipokines such as adiponectin or leptin have profound cardiac effects. While adiponectin acutely stimulates glucose uptake and oxidation in cardiomyocytes, long-term effects of adiponectin include increased uptake and oxidation of fatty acids and decreased glucose oxidation (Palanivel et al. 2007). Leptin on the other hand does not alter the uptake and oxidation of glucose or insulin signalling in cardiomyocytes but also increases fatty acid uptake. While short-term stimulation with leptin increases FA oxidation, it decreased over time leading to intracellular lipid accumulation (Palanivel et al. 2006). Finally, the same group also reported that conditioned medium derived from adipocytes of wild-type and streptozotocin-induced diabetic rats mediates differential effects on cardiomyocyte metabolism. The conditioned medium from wild-type rats increased cardiomyocyte glucose and fatty acid uptake and oxidation, whereas the medium from diabetic rats stimulated non-oxidative glucose metabolism and suppressed fatty acid oxidation (Palanivel et al. 2008). This was associated with significant differences in adiponectin and leptin levels in conditioned medium, suggesting that adipokines secreted from diabetic rats have a deleterious influence on cardiomyocyte metabolism. However, a detailed description of the diverse cardiac effects of these two adipokines or the impact of other adipokines on cardiac metabolic homeostasis and substrate utilization is beyond the scope of this article.

4.5 The Aged Heart

The ageing human heart shows a decline in myocardial fatty acid uptake and oxidation and an increase in the relative contribution of glucose to myocardial substrate metabolism but glucose uptake is not changed under resting conditions (Kates et al. 2003). Under inotropic stimulation and increased oxygen consumption, myocardial fatty acid uptake and oxidation increases by a similar extent in younger and older volunteers, but glucose uptake is only increased in young volunteers (Soto et al. 2003). Aged myocardium has a decreased GLUT4 protein expression, which is not reversed by chronic treadmill running (Hall et al. 1994). Old animals show a significant lowering of cardiac CPT-1 activity and in the rates of palmitoylcarnitine oxidation, probably resulting in reduced delivery of fatty acyl units to FAO (McMillin et al. 1993). Cardiomyocytes from old rats demonstrate higher ceramide content and reduced viability after palmitate (C16:0) in vitro, while no major deleterious effects are observed in young cardiomyocytes. Aged but not young cardiomyocytes are unable to respond to higher palmitate with increased FAO (Aurich et al. 2013). Furthermore, a significant reduction in heart mitochondrial cardiolipin and carnitine content has been reported (McMillin et al. 1993). The suppression of glucose oxidation by fatty acids seen in the adult rat heart is not seen in old hearts (McMillin et al. 1993). PPARα, a key regulator of many FAO genes, is decreased in aged animals and may also contribute to the decreased myocardial FAO (Iemitsu et al. 2002).

Respiratory chain function decreases with age while oxidative damage and mutations in mtDNA increase (Krishnan et al. 2007). Furthermore, cardiomyocytes from old rats demonstrate significantly reduced AMPK activation and impaired mitochondrial biogenesis (Aurich et al. 2013). The mitochondrial theory of ageing, a variant of the free radical theory of aging, is based on the idea of a vicious cycle, in which mtDNA mutations cause respiratory chain dysfunction, enhancing the production of DNA-damaging ROS, finally resulting in cell death and an ageing phenotype. Expression of genes coding for proteins involved in oxidative phosphorylation and for transcription factors involved in the regulation of mitochondrial metabolism and biogenesis is downregulated in aged ventricles, which is associated with a decline in the activities of oxidative phosphorylation complexes I and IV (Preston et al. 2008; Niemann et al. 2010; Rohrbach et al. 2006). Some of these mitochondrial changes are reversed in response to caloric restriction or exercise (Iemitsu et al. 2002; Niemann et al. 2010; Rohrbach et al. 2006). Mutant mice with a deficit in the proofreading function of the mtDNA polymerase γ (POLG) show an increase in somatic mtDNA mutations (Trifunovic et al. 2004). In the heart, these mutant mice demonstrate a disproportional respiratory chain with depressed activities of complex I and complex IV, reduced ATP-synthesizing capacity and accumulation of

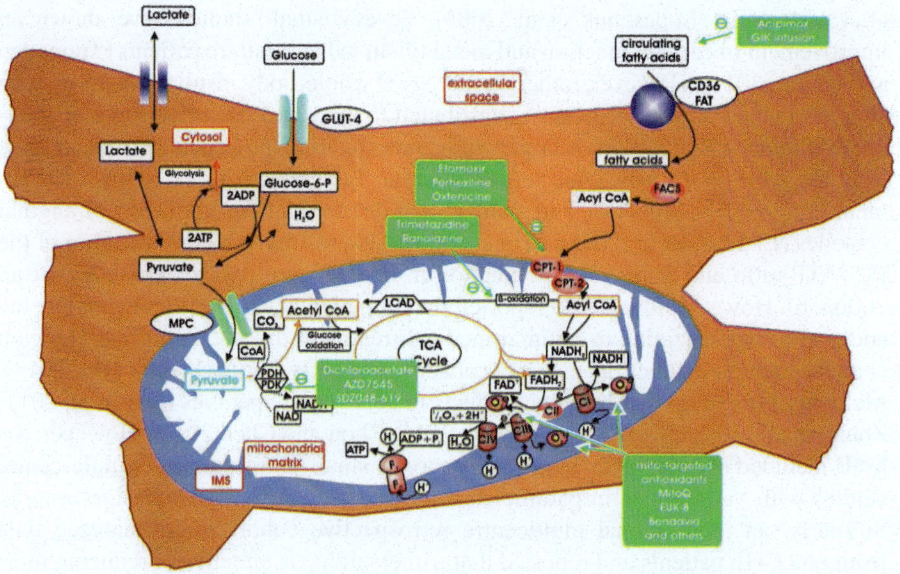

Fig. 4.3 Therapeutics used to target cardiac substrate metabolism for the treatment of heart failure. The mechanism of actions is with a *green arrow*. Abbreviations as described in Fig. 4.1

abnormal mitochondria, and they die prematurely with dilated cardiomyopathy. Although these data suggested a causative link between mtDNA mutations and ageing phenotypes in mammals, the relevance of these findings for ageing mechanisms in wild-type mammals with intact POLG has been debated (Hasty and Vijg 2004; Miller 2004; Khrapko et al. 2006).

4.6 Metabolic Therapies

Many metabolic approaches rely on the hypothesis that a shift towards the more energy-efficient substrate glucose may have beneficial effects for the diseased heart through an increase in ATP production per oxygen utilized. This switch can be achieved through (a) inhibition of FAO, (b) interference with circulating substrate levels, (c) CPT-1 inhibition and (d) promotion of carbohydrate oxidation as shown in Fig. 4.3.

4.6.1 Inhibition of FAO

Trimetazidine is an anti-anginal agent, inhibits FAO by blocking long-chain 3-ketoacyl-CoA thiolase and enhances glucose oxidation. Evidence suggests that trimetazidine confers protection in heart failure. Beneficial effects of trimetazidine in ex vivo working ischaemic hearts involve a decrease in FAO and stimulation of glucose oxidation, resulting in an improvement in cardiac function and efficiency after ischaemia (Lopaschuk et al. 2003). Several small studies have shown an improvement of cardiac function and metabolism in heart failure patients (Tuunanen and Knuuti 2011). However, rather an increased whole-body insulin sensitivity than changes in cardiac FAO may have contributed to the protective effects of trimetazidine (Tuunanen et al. 2008). A larger study, investigating the effects of trimetazidine in 200 patients with ischaemic left ventricular dysfunction, reported an improvement in myocardial function and a significant decrease of the frequency of anginal episodes (El-Kady et al. 2005). Trimetazidine also results in a normalization of the PCr/ATP ratio and improves LV function in heart failure patients (Fragasso et al. 2006a, b). However, it was also reported that in patients with nonischaemic HF, the addition of trimetazidine to optimal medical treatment did not result in significant changes in cardiac function, exercise capacity or quality of life (Winter et al. 2014). Meta-analyses suggested beneficial effects in heart failure patients (Gao et al. 2011; Zhang et al. 2012; Grajek and Michalak 2015; Zhou and Chen 2014). However, the trials included in these meta-analyses are mostly small, short-duration, single-centre studies with variabilities in parameters such as study design or inclusion criteria. Similarly, an international multicentre retrospective cohort study analyzed data from 669 CHF patients and reported that trimetazidine is effective in reducing mortality and event-free survival in patients with CHF (Fragasso et al. 2013). In general, trimetazidine is well tolerated, although extrapyramidal disorders have been reported (Masmoudi et al. 2012). This puts even more emphasis on the need of a prospective multicentre study in order to investigate the cardiac effects and potential side effects of trimetazidine and to define the subgroup of patients (e.g. stage of heart failure) with highest benefits from this additional therapy.

Ranolazine is an FDA-approved anti-anginal agent and is a partial FAO inhibitor, which increases glucose oxidation and PDH activity and improves coupling between glycolysis and glucose oxidation (McCormack et al. 1996; Clarke et al. 1996). In preclinical models, ranolazine was shown to reduce infarct size, to improve LV function and to prevent the progression of remodelling and contractile failure (Hale and Kloner 2006; Chandler et al. 2002; Rastogi et al. 2008; Hale and Kloner 2014). Short-term clinical trials showed that ranolazine significantly improves functional capacity and decreases anginal episodes in patients with chronic stable angina. However, long-term clinical trials showed that ranolazine did not improve cardiovascular mortality or affect incidence of myocardial infarction (Codolosa et al. 2014; Savarese et al. 2013). A prospective, randomized, double-blind, placebo-controlled small proof-of-concept study revealed that ranolazine improved measures of haemodynamics but did not improve relaxation parameters (Maier et al. 2013). In addition, ranolazine has recently been shown to have additional beneficial antiarrhythmic actions on supraventricular and ventricular tachyarrhythmias, attributed to its inhibitory effect on both peak and late sodium current (Gupta et al. 2015).

However, strategies of enhancing glucose oxidation by reducing FAO may not be a promising approach in all patients with heart failure. In advanced-stage heart failure, mitochondrial oxidative capacity is depressed and inhibition of FAO may mediate deleterious effects at this disease stage (Stanley et al. 2005). Indeed, recent preclinical investigations suggest that sustaining FAO in the hypertrophied heart through high-fat diets preserves LV energetics and function (Chess et al. 2009; Okere et al. 2006).

4.6.2 Interference with Circulating Substrate Levels

Acipimox is a nicotinic acid derivate that acutely inhibits lipolysis in adipose tissue and thus decreases FFA in plasma. Although FFA uptake is decreased and glucose uptake is enhanced in heart failure patients, acipimox is unable to improve myocardial function (Halbirk et al. 2010b). It was also reported that acipimox results in a decline in cardiac efficiency and work in heart failure patient but not in healthy controls, suggesting that failing hearts are more dependent on FFA availability (Tuunanen et al. 2006). Recently, it was shown that FFA deprivation (acipimox), but not FFA excess, impairs LV energy metabolism and LV function in patients with heart failure within hours (Salerno et al. 2015).

Glucose-insulin-potassium (GIK) infusion has been shown to increase glycolysis and decrease circulating FFA, which is associated with reduced infarct size and improved postischaemic function (Jonassen et al. 2000; Zhang et al. 2006), but also a lack of infarct size reduction has been reported (Bellows and Kloner 1998). A reduction in mortality was shown in some patients following myocardial infarction (Malmberg et al. 1996; van der Horst et al. 2003), but also no effect on mortality has been suggested to occur after GIK therapy (Ceremuzynski et al. 1999; Rasoul et al. 2007). Thus, there is no clear consensus whether GIK therapy is beneficial or not.

4.6.3 CPT-1 Inhibition

Studies on the effects of CPT-1 inhibitors such as *perhexiline, etomoxir* or *oxfenicine* have yielded conflicting results in heart failure patients. Similar to the results from preclinical models (Lopaschuk et al. 1988; Turcani and Rupp 1997), the first human trial evaluating etomoxir in heart failure patients shows an improved LV function in those patients (Schmidt-Schweda and Holubarsch 2000). Due to liver toxicity, etomoxir is not a candidate for further development of treatment of heart failure patients (Holubarsch et al. 2007).

Perhexiline was originally used as an anti-anginal agent in the 1970s, but hepatic toxicity and neuropathy led to a decline of its use. A double-blind, placebo-controlled study demonstrated that perhexiline improves VO_2 max and cardiac function (Lee et al. 2005). Furthermore, it corrects cardiac energy deficiency and diastolic dysfunction and increases exercise capacity in patients with symptomatic hypertrophic cardiomyopathy (Abozguia et al. 2010). However, long-term administration of perhexiline does not result in a significant functional improvement in patients with ischaemic LV dysfunction (Bansal et al. 2010). Finally, oxfenicine was shown to prevent ventricular remodelling and delay the time to end-stage heart failure in pacing-induced heart failure in dogs (Lionetti et al. 2005). However, oxfenicine was also reported to induce myocardial hypertrophy and even to induce cardiotoxic and mitochondriotoxic effects (Greaves et al. 1984; Bachmann and Weber 1988).

4.6.4 Promotion of Carbohydrate Oxidation

Dichloroacetate (DCA) which increases PDH activity by inhibiting PDK has been shown to improve the functional recovery after cardiac ischaemia already a while ago but also in more recent animal studies (McVeigh and Lopaschuk 1990; Ussher et al. 2012). A study in patients with coronary artery disease revealed an increased stroke volume and increased myocardial efficiency after DCA infusion (Wargovich et al. 1988). The effects of DCA have also been tested in heart failure studies in animals and humans. DCA was shown to increase energy reserves and glucose uptake in rats with compensated LV hypertrophy or LV failure. This is associated with improved cardiac function and the survival of the animals. Interestingly, DCA also activates the pentose phosphate pathway and reduces oxidative stress (Kato et al. 2010). In patients with class III to IV congestive heart failure, DCA stimulates myocardial lactate consumption, improves left ventricular mechanical efficiency and increases cardiac work (Bersin et al. 1994). However, another study performed in patients with chronic heart failure did not show an improvement in left ventricular function (Lewis et al. 1998). DCA treatment also results in a partial inhibition of an anaplerotic pathway (flux through malic enzyme), which led to normalization of the myocardial triglyceride pool and improved contractile function (Pound et al. 2009). Furthermore, there are also a number of studies investigating the effects of DCA in right ventricular failure. DCA prevents and reverses PAH, improves mortality and restores right ventricular repolarization in different models of right ventricular hypertrophy (McMurtry et al. 2004; Piao et al. 2010). Recent experimental studies also suggest novel mechanisms involved in DCA actions. DCA was shown to induce epigenetic remodelling in the heart (Matsuhashi et al. 2015), to mitigate cardiac contractile dysfunction following ischaemia-induced ventricular fibrillation (Azam et al. 2015) and to prevent restenosis in preclinical animal models of vessel injury (Deuse et al. 2014). Although short-term parenteral administration of DCA appears to be safe and capable of significantly improving myocardial function, human studies are limited because of the chronic neurotoxicity of the drug in long-term studies (Brandsma et al. 2010). The cardiovascular impact in humans and the potential side effects of the novel PDK inhibitors AZD7545 or SDZ048-619, which have been tested in preclinical models (Mayers et al. 2003; Aicher et al. 2000; Bebernitz et al. 2000), remain to be elucidated.

Glucagon-like peptide (GLP-1) is a gut hormone and incretin derived from the intestinal L cells, increases myocardial glucose uptake, improves LV performance in dogs with dilated cardiomyopathy (Nikolaidis et al. 2005) and mediates beneficial effects in ischaemia and reperfusion injury (Ravassa et al. 2012; Zhao et al. 2006). Chronic GLP-1 treatment prolongs survival, preserves LV function, increases myocardial glucose uptake and reduces myocyte apoptosis in spontaneously hypertensive, heart failure-prone rats (Poornima et al. 2008), which is associated with increased insulin levels. However, some of the cardioprotective effects of GLP-1 have also been suggested to be independent from changes in plasma insulin and thus beyond the incretin effect (Doenst et al. 2013; Nystrom et al. 2004; Noyan-Ashraf et al. 2013). For example, GLP-1 agonists also improve mitochondrial function in vitro and reduce mitochondrial oxidative stress (Chang et al. 2014). The activation of AMPK, which enhances insulin-dependent signalling and increases glucose uptake, may be implicated in some of the cardioprotective effects of GLP-1 or GLP-1 analogs. However, only a few clinical trials with a limited number of patients suffering from heart failure have been performed so far. Short-term (3 days) or long-term (5 weeks) GLP-1 infusion improves LV function in patients with chronic heart failure or acute myocardial infarction (Thrainsdottir et al. 2004; Sokos et al. 2006; Nikolaidis et al. 2004a). Others have reported that short-term GLP-1 increases circulating insulin levels and reduces plasma glucose concentration but has no major cardiovascular effects in NYHA II–III heart failure patients without diabetes (Halbirk et al. 2010a). Nathanson et al. reported in a double-blinded, clinical trial in 20 diabetic patients with NYHA III–IV that short-term infusion of the GLP-1 agonist improves haemodynamic function in diabetic patients with congestive heart failure (Nathanson et al. 2012). In addition to GLP-1 or GLP-1 agonists, inhibitors of the dipeptidyl peptidase-4 (DPP-4), which inactivates GLP-1, represent another class of incretin-based therapies. Although small-scale clinical trials suggested cardioprotective effects of DPP-4 inhibitors, several large-scale clinical trials did not prove the beneficial effects of DPP-4 inhibitors (Scirica et al. 2013; White et al. 2013). Ongoing clinical studies such as the FIGHT, the TECOS or the CAROLINA trial need to clarify whether GLP-1 and DPP-4 inhibitors are effective and/or safe for patients with heart failure and who mainly receives their benefits among the patients with cardiovascular diseases and/or DM.

In summary, there are promising data available for some metabolic therapeutics. However, a number of questions remain before they can be integrated into clinical practice. One question is which agent should be used? At what stage of heart failure should patients be treated? How does the underlying aetiology of heart failure affect the efficacy of metabolic therapeutics? These questions need to be answered in comprehensive clinical trials before consideration of a broad clinical use.

In addition to these metabolic approaches, a number of classical therapeutic regimens in heart failure patients are known to affect cardiac metabolism. *Beta-blockers* have been described to inhibit CPT-1, increase glucose oxidation and increase cardiac efficiency (Eichhorn et al. 1994; Panchal et al. 1998; Podbregar and Voga 2002). Some beta-blockers decrease circulating fatty acids through a reduction of catecholamine-induced lipolysis in adipose tissue (Jaswal et al. 2011). Carvedilol was suggested to reduce myocardial FFA use without a significant change in glucose use (Wallhaus et al. 2001) and cause a reduction in FFA uptake, an effect not observed under metoprolol (Al-Hesayen et al. 2005).

Angiotensin II (Ang II) is an important mediator of increased neurohumoral activation in heart failure patients and induces mitochondrial dysfunction (de Cavanagh et al. 2009), and the existence of a mitochondrial renin-angiotensin system, which is coupled to mitochondrial nitric oxide (NO) production, has been reported (Abadir et al. 2011). NO is an important modulator of mitochondrial biogenesis (Nisoli et al. 2003) and blockade of the angiotensin II type 1 receptor (AT1R) enhances NO production (Abadir et al. 2003).

Ang II also results in reduced glucose oxidation through an upregulation of pyruvate dehydrogenase kinase (Mori et al. 2012). This decrease in glucose oxidation is not accompanied by a reduced glucose uptake or glycolysis, which results in the production of lactate in Ang II-treated hearts. However, Ang II also results in the downregulation of PPARα, a key regulator of many FAO genes, and reduced protein expression of FAO enzymes (Pellieux et al. 2006). Activation of the renin-angiotensin system in heart failure has been associated with insulin resistance, an effect also observed in cardiomyocytes under Ang II stimulation (Doehner et al. 2005; Wong et al. 2008; Velloso et al. 1996; Mori et al. 2013a). However, the development of insulin resistance in heart failure is likely multifactorial, which involves chronic activation of the renin-angiotensin system, increased sympathetic nervous system activity and a number of other factors. In addition to the classic renin-angiotensin system, alternative pathways involving Ang 1–7, ACE2 or the Mas receptors have been described (Santos et al. 2013). While activation of the ACE/AngII/AT1R axis leads to systemic insulin resistance (Mori et al. 2013a), the ACE2/Ang1-7/Mas axis may mediate opposite effects on insulin signalling in the heart (Giani et al. 2007; Mori et al. 2013b). Accordingly, blockade of the renin-angiotensin system with ACE inhibitors or AT1R blockers can be envisioned to mediate profound metabolic effects in the heart and new components of the system may be targets for the optimization of cardiac energy metabolism in the diseased heart. Future studies, investigating the relationship between neurohumoral activation and cardiac energy metabolism, are expected to develop metabolic pharmacological interventions.

Thiazolidinedione which activates PPARγ decreases circulating levels of fatty acids and increases myocardial glucose and lactate oxidation (Zhu et al. 2000; Yue et al. 2005). However, the clear metabolic benefits of thiazolidinedione are overshadowed by the risk for fluid retention and congestive heart failure, and safety concerns have been raised by large clinical trials testing thiazolidinedione in heart failure patients (Kung and Henry 2012; Komajda et al. 2010; Robinson 2007).

Metformin is a widely used insulin-sensitizing drug, has been shown to increase glucose uptake with or without insulin in cardiomyocytes (Bertrand et al. 2006), improves LV function and reduces myocardial lipid accumulation (Cittadini et al. 2012; Wang et al. 2011). A retrospective analysis in patients with heart failure and diabetes suggests that thiazolidinediones and metformin are not associated with increased mortality and may improve outcomes in older patients (Masoudi et al. 2005). However, a more recent study showed that metformin has no effect on several surrogate markers of cardiovascular disease (Preiss et al. 2014).

Although heart failure is associated with mitochondrial dysfunction and increased mitochondrial ROS production, there are only a few therapies directly targeted to improve mitochondrial function. Several *mitochondria-targeted antioxidants* have demonstrated cardioprotection in animal models. Mitoquinone (MitoQ) was shown to mediate cardioprotective effects in experimental models of ischaemia and reperfusion, cardiac hypertrophy and doxorubicin-induced cardiomyopathy (Adlam et al. 2005;

Chandran et al. 2009; Dikalova et al. 2010), MitoTEMPO demonstrated cardioprotective effects in hypertension (Graham et al. 2009) and EUK-8 improved cardiac function in a model of pressure overload-induced heart failure (van Empel et al. 2006). Other interesting molecules in this field are synthetic peptides called Szeto-Schiller peptides (SS). SS-02 and SS-31 were shown to be protective against cardiac ischaemia-reperfusion injury, in different heart failure models, but also in a number of noncardiac diseases (Szeto 2008; Cho et al. 2007; Szeto 2014; Dai et al. 2011, 2014). SS-31 entered into clinical development in 2010 using a clinical formulation named Bendavia. Based on the extensive preclinical data and the results of phase I studies assessing the safety, tolerability and pharmacokinetics of Bendavia, phase II clinical studies have been initiated. One of these studies with Bendavia is focussing on reperfusion injury (Chakrabarti et al. 2013) and another one will investigate patients with congestive heart failure (Szeto 2014). However, general antioxidants such as vitamin E have not shown cardioprotection in clinical trials, but demonstrated an increased risk of developing heart failure after myocardial infarction (Lonn et al. 2005; Marchioli et al. 2006).

Conclusions

There are major differences in the uptake and utilization of substrates between the healthy adult heart, the foetal heart and the diseased heart. Furthermore, different aetiologies of heart disease such as ischaemia, diabetes, ageing, obesity or mechanical overload result in distinct but disease-typical changes in cardiac metabolism. Although many of these alterations have been first described many years ago, the importance of metabolic events and metabolic flexibility in disease progression has long gone unappreciated. In addition to its role in ATP synthesis, substrate metabolism also has a major impact on various aspects of cardiac physiology including growth, proliferation or survival (Doenst et al. 2013). Recent investigations have clearly shown that a number of substrates or metabolites are also potent signalling molecules, suggesting an important role of these molecules in the progression of heart diseases independent from ATP supply. Despite the high prevalence of heart failure, drug therapy has evolved little in recent years. Novel therapeutic strategies are therefore needed and should include drugs targeting substrate metabolism. Because of the complexity of cardiac substrate metabolism and disease-specific differences, additional experimental work and conclusive clinical trials are needed.

References

Abadir PM, Carey RM, Siragy HM (2003) Angiotensin AT2 receptors directly stimulate renal nitric oxide in bradykinin B2-receptor-null mice. Hypertension 42:600–604

Abadir PM, Foster DB, Crow M, Cooke CA, Rucker JJ, Jain A, Smith BJ, Burks TN, Cohn RD, Fedarko NS, Carey RM, O'Rourke B, Walston JD (2011) Identification and characterization of a functional mitochondrial angiotensin system. Proc Natl Acad Sci U S A 108:14849–14854

Abdurrachim D, Luiken JJ, Nicolay K, Glatz JF, Prompers JJ, Nabben M (2015) Good and bad consequences of altered fatty acid metabolism in heart failure: evidence from mouse models. Cardiovasc Res 106:194–205

Abel ED (2004) Glucose transport in the heart. Front Biosci J Virt Libr 9:201–215
Abozguia K, Clarke K, Lee L, Frenneaux M (2006) Modification of myocardial substrate use as a therapy for heart failure. Nat Clin Pract Cardiovasc Med 3:490–498
Abozguia K, Elliott P, McKenna W, Phan TT, Nallur-Shivu G, Ahmed I, Maher AR, Kaur K, Taylor J, Henning A, Ashrafian H, Watkins H, Frenneaux M (2010) Metabolic modulator perhexiline corrects energy deficiency and improves exercise capacity in symptomatic hypertrophic cardiomyopathy. Circulation 122:1562–1569
Adlam VJ, Harrison JC, Porteous CM, James AM, Smith RA, Murphy MP, Sammut IA (2005) Targeting an antioxidant to mitochondria decreases cardiac ischemia-reperfusion injury. FASEB J 19:1088–1095
Aicher TD, Anderson RC, Gao J, Shetty SS, Coppola GM, Stanton JL, Knorr DC, Sperbeck DM, Brand LJ, Vinluan CC, Kaplan EL, Dragland CJ, Tomaselli HC, Islam A, Lozito RJ, Liu X, Maniara WM, Fillers WS, DelGrande D, Walter RE, Mann WR (2000) Secondary amides of (R)-3,3,3-trifluoro-2-hydroxy-2-methylpropionic acid as inhibitors of pyruvate dehydrogenase kinase. J Med Chem 43:236–249
Al-Hesayen A, Azevedo ER, Floras JS, Hollingshead S, Lopaschuk GD, Parker JD (2005) Selective versus nonselective beta-adrenergic receptor blockade in chronic heart failure: differential effects on myocardial energy substrate utilization. Eur J Heart Fail 7:618–623
Allard MF, Schonekess BO, Henning SL, English DR, Lopaschuk GD (1994) Contribution of oxidative metabolism and glycolysis to ATP production in hypertrophied hearts. Am J Physiol 267:H742–H750
Allard MF, Henning SL, Wambolt RB, Granleese SR, English DR, Lopaschuk GD (1997) Glycogen metabolism in the aerobic hypertrophied rat heart. Circulation 96:676–682
Amorim PA, Nguyen TD, Shingu Y, Schwarzer M, Mohr FW, Schrepper A, Doenst T (2010) Myocardial infarction in rats causes partial impairment in insulin response associated with reduced fatty acid oxidation and mitochondrial gene expression. J Thorac Cardiovasc Surg 140:1160–1167
Anderson EJ, Lustig ME, Boyle KE, Woodlief TL, Kane DA, Lin CT, Price JW 3rd, Kang L, Rabinovitch PS, Szeto HH, Houmard JA, Cortright RN, Wasserman DH, Neufer PD (2009) Mitochondrial H2O2 emission and cellular redox state link excess fat intake to insulin resistance in both rodents and humans. J Clin Invest 119:573–581
Arn P, Funanage VL (2006) 3-methylglutaconic aciduria disorders: the clinical spectrum increases. J Ped Hemat Oncol 28:62–63
Arsenijevic D, Onuma H, Pecqueur C, Raimbault S, Manning BS, Miroux B, Couplan E, Alves-Guerra MC, Goubern M, Surwit R, Bouillaud F, Richard D, Collins S, Ricquier D (2000) Disruption of the uncoupling protein-2 gene in mice reveals a role in immunity and reactive oxygen species production. Nat Genet 26:435–439
Ashrafian H, Frenneaux MP, Opie LH (2007) Metabolic mechanisms in heart failure. Circulation 116:434–448
Aurich AC, Niemann B, Pan R, Gruenler S, Issa H, Silber RE, Rohrbach S (2013) Age-dependent effects of high fat-diet on murine left ventricles: role of palmitate. Bas Res Cardiol 108:369
Azam MA, Wagg CS, Masse S, Farid T, Lai PF, Kusha M, Asta J, Jaimes R 3rd, Kuzmiak-Glancy S, Kay MW, Lopaschuk GD, Nanthakumar K (2015) Feeding the fibrillating feart: dichloroacetate improves cardiac contractile dysfunction following VF. Am J Physiol. doi:10.1152/ajpheart.00404.2015
Bachmann E, Weber E (1988) Biochemical mechanisms of oxfenicine cardiotoxicity. Pharmacology 36:238–248
Bansal M, Chan J, Leano R, Pillans P, Horowitz J, Marwick TH (2010) Effects of perhexiline on myocardial deformation in patients with ischaemic left ventricular dysfunction. Int J Cardiol 139:107–112
Battaglia FC, Meschia G (1978) Principal substrates of fetal metabolism. Physiol Rev 58:499–527
Bebernitz GR, Aicher TD, Stanton JL, Gao J, Shetty SS, Knorr DC, Strohschein RJ, Tan J, Brand LJ, Liu C, Wang WH, Vinluan CC, Kaplan EL, Dragland CJ, DelGrande D, Islam A, Lozito RJ,

Liu X, Maniara WM, Mann WR (2000) Anilides of (R)-trifluoro-2-hydroxy-2-methylpropionic acid as inhibitors of pyruvate dehydrogenase kinase. J Med Chem 43:2248–2257

Becker C, Sevilla L, Tomas E, Palacin M, Zorzano A, Fischer Y (2001) The endosomal compartment is an insulin-sensitive recruitment site for GLUT4 and GLUT1 glucose transporters in cardiac myocytes. Endocrinology 142:5267–5276

Bellows SD, Kloner RA (1998) Glucose-insulin-potassium does not reduce myocardial infarct size in an ischemic/reperfusion rabbit model. J Thromb Thrombolysis 5:25–27

Bersin RM, Wolfe C, Kwasman M, Lau D, Klinski C, Tanaka K, Khorrami P, Henderson GN, de Marco T, Chatterjee K (1994) Improved hemodynamic function and mechanical efficiency in congestive heart failure with sodium dichloroacetate. J Am Coll Cardiol 23:1617–1624

Bertrand L, Ginion A, Beauloye C, Hebert AD, Guigas B, Hue L, Vanoverschelde JL (2006) AMPK activation restores the stimulation of glucose uptake in an in vitro model of insulin-resistant cardiomyocytes via the activation of protein kinase B. Am J Physiol 291:H239–H250

Blazquez E, Sugase T, Blazquez M, Foa PP (1974) Neonatal changes in the concentration of rat liver cyclic AMP and of serum glucose, free fatty acids, insulin, pancreatic, and total glucagon in man and in the rat. J Lab Clin Chem 83:957–967

Bonnefont JP, Djouadi F, Prip-Buus C, Gobin S, Munnich A, Bastin J (2004) Carnitine palmitoyltransferases 1 and 2: biochemical, molecular and medical aspects. Mol Aspects Med 25:495–520

Boudina S, Sena S, O'Neill BT, Tathireddy P, Young ME, Abel ED (2005) Reduced mitochondrial oxidative capacity and increased mitochondrial uncoupling impair myocardial energetics in obesity. Circulation 112:2686–2695

Boudina S, Sena S, Theobald H, Sheng X, Wright JJ, Hu XX, Aziz S, Johnson JI, Bugger H, Zaha VG, Abel ED (2007) Mitochondrial energetics in the heart in obesity-related diabetes: direct evidence for increased uncoupled respiration and activation of uncoupling proteins. Diabetes 56:2457–2466

Brandsma D, Dorlo TP, Haanen JH, Beijnen JH, Boogerd W (2010) Severe encephalopathy and polyneuropathy induced by dichloroacetate. J Neurol 257:2099–2100

Bristow J, Bier DM, Lange LG (1987) Regulation of adult and fetal myocardial phosphofructokinase. Relief of cooperativity and competition between fructose 2,6-bisphosphate, ATP, and citrate. J Biol Chem 262:2171–2175

Brown NF, Weis BC, Husti JE, Foster DW, McGarry JD (1995) Mitochondrial carnitine palmitoyltransferase I isoform switching in the developing rat heart. J Biol Chem 270:8952–8957

Buchanan J, Mazumder PK, Hu P, Chakrabarti G, Roberts MW, Yun UJ, Cooksey RC, Litwin SE, Abel ED (2005) Reduced cardiac efficiency and altered substrate metabolism precedes the onset of hyperglycemia and contractile dysfunction in two mouse models of insulin resistance and obesity. Endocrinology 146:5341–5349

Bugger H, Schwarzer M, Chen D, Schrepper A, Amorim PA, Schoepe M, Nguyen TD, Mohr FW, Khalimonchuk O, Weimer BC, Doenst T (2010) Proteomic remodelling of mitochondrial oxidative pathways in pressure overload-induced heart failure. Cardiovasc Res 85:376–384

Burkhoff D, Weiss RG, Schulman SP, Kalil-Filho R, Wannenburg T, Gerstenblith G (1991) Influence of metabolic substrate on rat heart function and metabolism at different coronary flows. Am J Physiol 261:H741–H750

Callikan S, Ferre P, Pegorier JP, Girard JR, Marliss EB, Assan R (1979) Fuel metabolism in fasted newborn rabbits. J Develop Physiol 1:267–281

Ceremuzynski L, Budaj A, Czepiel A, Burzykowski T, Achremczyk P, Smielak-Korombel W, Maciejewicz J, Dziubinska J, Nartowicz E, Kawka-Urbanek T, Piotrowski W, Hanzlik J, Cieslinski A, Kawecka-Jaszcz K, Gessek J, Wrabec K (1999) Low-dose glucose-insulin-potassium is ineffective in acute myocardial infarction: results of a randomized multicenter Pol-GIK trial. Cardiovasc Drugs Ther 13:191–200

Chabowski A, Gorski J, Glatz JF, Luiken JJ P, Bonen A (2008) Protein-mediated fatty acid uptake in the heart. Curr Cardiol Rev 4:12–21

Chakrabarti AK, Feeney K, Abueg C, Brown DA, Czyz E, Tendera M, Janosi A, Giugliano RP, Kloner RA, Weaver WD, Bode C, Godlewski J, Merkely B, Gibson CM (2013) Rationale and design of the EMBRACE STEMI study: a phase 2a, randomized, double-blind, placebo-controlled trial to evaluate the safety, tolerability and efficacy of intravenous Bendavia on reperfusion injury in patients treated with standard therapy including primary percutaneous coronary intervention and stenting for ST-segment elevation myocardial infarction. Am Heart J 165:509–514

Chance B, Williams GR (1955) A method for the localization of sites for oxidative phosphorylation. Nature 176:250–254

Chance B, Williams GR (1956) The respiratory chain and oxidative phosphorylation. Adv Enzymol Relat Subj Biochem 17:65–134

Chandler MP, Stanley WC, Morita H, Suzuki G, Roth BA, Blackburn B, Wolff A, Sabbah HN (2002) Short-term treatment with ranolazine improves mechanical efficiency in dogs with chronic heart failure. Circ Res 91:278–280

Chandler MP, Kerner J, Huang H, Vazquez E, Reszko A, Martini WZ, Hoppel CL, Imai M, Rastogi S, Sabbah HN, Stanley WC (2004) Moderate severity heart failure does not involve a downregulation of myocardial fatty acid oxidation. Am J Physiol 287:H1538–H1543

Chandran K, Aggarwal D, Migrino RQ, Joseph J, McAllister D, Konorev EA, Antholine WE, Zielonka J, Srinivasan S, Avadhani NG, Kalyanaraman B (2009) Doxorubicin inactivates myocardial cytochrome c oxidase in rats: cardioprotection by Mito-Q. Biophys J 96:1388–1398

Chang G, Zhang D, Liu J, Zhang P, Ye L, Lu K, Duan Q, Zheng A, Qin S (2014) Exenatide protects against hypoxia/reoxygenation-induced apoptosis by improving mitochondrial function in H9c2 cells. Exptl Biol Med 239:414–422

Chess DJ, Khairallah RJ, O'Shea KM, Xu W, Stanley WC (2009) A high-fat diet increases adiposity but maintains mitochondrial oxidative enzymes without affecting development of heart failure with pressure overload. Am J Physiol 297:H1585–H1593

Chiu HC, Kovacs A, Ford DA, Hsu FF, Garcia R, Herrero P, Saffitz JE, Schaffer JE (2001) A novel mouse model of lipotoxic cardiomyopathy. J Clin Invest 107:813–822

Chiu HC, Kovacs A, Blanton RM, Han X, Courtois M, Weinheimer CJ, Yamada KA, Brunet S, Xu H, Nerbonne JM, Welch MJ, Fettig NM, Sharp TL, Sambandam N, Olson KM, Ory DS, Schaffer JE (2005) Transgenic expression of fatty acid transport protein 1 in the heart causes lipotoxic cardiomyopathy. Circ Res 96:225–233

Cho J, Won K, Wu D, Soong Y, Liu S, Szeto HH, Hong MK (2007) Potent mitochondria-targeted peptides reduce myocardial infarction in rats. Coron Artery Dis 18:215–220

Chokshi A, Drosatos K, Cheema FH, Ji R, Khawaja T, Yu S, Kato T, Khan R, Takayama H, Knoll R, Milting H, Chung CS, Jorde U, Naka Y, Mancini DM, Goldberg IJ, Schulze PC (2012) Ventricular assist device implantation corrects myocardial lipotoxicity, reverses insulin resistance, and normalizes cardiac metabolism in patients with advanced heart failure. Circulation 125:2844–2853

Christian BE, Spremulli LL (2012) Mechanism of protein biosynthesis in mammalian mitochondria. Biochim Biophys Acta 1819:1035–1054

Cittadini A, Napoli R, Monti MG, Rea D, Longobardi S, Netti PA, Walser M, Sama M, Aimaretti G, Isgaard J, Sacca L (2012) Metformin prevents the development of chronic heart failure in the SHHF rat model. Diabetes 61:944–953

Clarke B, Wyatt KM, McCormack JG (1996) Ranolazine increases active pyruvate dehydrogenase in perfused normoxic rat hearts: evidence for an indirect mechanism. J Mol Cell Cardiol 28:341–350

Codolosa JN, Acharjee S, Figueredo VM (2014) Update on ranolazine in the management of angina. Vasc Health Risk Manag 10:353–362

Cole MA, Murray AJ, Cochlin LE, Heather LC, McAleese S, Knight NS, Sutton E, Jamil AA, Parassol N, Clarke K (2011) A high fat diet increases mitochondrial fatty acid oxidation and uncoupling to decrease efficiency in rat heart. Basic Res Cardiol 106:447–457

Combs TP, Berg AH, Rajala MW, Klebanov S, Iyengar P, Jimenez-Chillaron JC, Patti ME, Klein SL, Weinstein RS, Scherer PE (2003) Sexual differentiation, pregnancy, calorie restriction, and aging affect the adipocyte-specific secretory protein adiponectin. Diabetes 52:268–276

Crewe C, Kinter M, Szweda LI (2013) Rapid inhibition of pyruvate dehydrogenase: an initiating event in high dietary fat-induced loss of metabolic flexibility in the heart. PLoS One 8:e77280

D'Antona G, Ragni M, Cardile A, Tedesco L, Dossena M, Bruttini F, Caliaro F, Corsetti G, Bottinelli R, Carruba MO, Valerio A, Nisoli E (2010) Branched-chain amino acid supplementation promotes survival and supports cardiac and skeletal muscle mitochondrial biogenesis in middle-aged mice. Cell Metab 12:362–372

Dai DF, Chen T, Szeto H, Nieves-Cintron M, Kutyavin V, Santana LF, Rabinovitch PS (2011) Mitochondrial targeted antioxidant Peptide ameliorates hypertensive cardiomyopathy. J Am Coll Cardiol 58:73–82

Dai W, Shi J, Gupta RC, Sabbah HN, Hale SL, Kloner RA (2014) Bendavia, a mitochondria-targeting peptide, improves postinfarction cardiac function, prevents adverse left ventricular remodeling, and restores mitochondria-related gene expression in rats. J Cardiovasc Pharmacol 64:543–553

de Cavanagh EM, Ferder M, Inserra F, Ferder L (2009) Angiotensin II, mitochondria, cytoskeletal, and extracellular matrix connections: an integrating viewpoint. Am J Physiol 296:H550–H558

Denton RM, McCormack JG, Rutter GA, Burnett P, Edgell NJ, Moule SK, Diggle TA (1996) The hormonal regulation of pyruvate dehydrogenase complex. Adv Enzyme Regul 36:183–198

Des Rosiers C, Labarthe F, Lloyd SG, Chatham JC (2011) Cardiac anaplerosis in health and disease: food for thought. Cardiovasc Res 90:210–219

Deuse T, Hua X, Wang D, Maegdefessel L, Heeren J, Scheja L, Bolanos JP, Rakovic A, Spin JM, Stubbendorff M, Ikeno F, Langer F, Zeller T, Schulte-Uentrop L, Stoehr A, Itagaki R, Haddad F, Eschenhagen T, Blankenberg S, Kiefmann R, Reichenspurner H, Velden J, Klein C, Yeung A, Robbins RC, Tsao PS, Schrepfer S (2014) Dichloroacetate prevents restenosis in preclinical animal models of vessel injury. Nature 509:641–644

Dikalova AE, Bikineyeva AT, Budzyn K, Nazarewicz RR, McCann L, Lewis W, Harrison DG, Dikalov SI (2010) Therapeutic targeting of mitochondrial superoxide in hypertension. Circ Res 107:106–116

Doehner W, Rauchhaus M, Ponikowski P, Godsland IF, von Haehling S, Okonko DO, Leyva F, Proudler AJ, Coats AJ, Anker SD (2005) Impaired insulin sensitivity as an independent risk factor for mortality in patients with stable chronic heart failure. J Am Coll Cardiol 46: 1019–1026

Doenst T, Pytel G, Schrepper A, Amorim P, Farber G, Shingu Y, Mohr FW, Schwarzer M (2010) Decreased rates of substrate oxidation ex vivo predict the onset of heart failure and contractile dysfunction in rats with pressure overload. Cardiovasc Res 86:461–470

Doenst T, Nguyen TD, Abel ED (2013) Cardiac metabolism in heart failure: implications beyond ATP production. Circ Res 113:709–724

Dyck JR, Barr AJ, Barr RL, Kolattukudy PE, Lopaschuk GD (1998) Characterization of cardiac malonyl-CoA decarboxylase and its putative role in regulating fatty acid oxidation. Am J Physiol 275:H2122–H2129

Eichhorn EJ, Heesch CM, Barnett JH, Alvarez LG, Fass SM, Grayburn PA, Hatfield BA, Marcoux LG, Malloy CR (1994) Effect of metoprolol on myocardial function and energetics in patients with nonischemic dilated cardiomyopathy: a randomized, double-blind, placebo-controlled study. J Am Coll Cardiol 24:1310–1320

El Alaoui-Talibi Z, Guendouz A, Moravec M, Moravec J (1997) Control of oxidative metabolism in volume-overloaded rat hearts: effect of propionyl-L-carnitine. Am J Physiol 272: H1615–H1624

El-Kady T, El-Sabban K, Gabaly M, Sabry A, Abdel-Hady S (2005) Effects of trimetazidine on myocardial perfusion and the contractile response of chronically dysfunctional myocardium in ischemic cardiomyopathy: a 24-month study. Am J Cardiovasc Drugs 5:271–278

Ellis JM, Frahm JL, Li LO, Coleman RA (2010) Acyl-coenzyme A synthetases in metabolic control. Curr Opin Lipidol 21:212–217

Ferre P, Decaux JF, Issad T, Girard J (1986) Changes in energy metabolism during the suckling and weaning period in the newborn. Reprod Nutr Dev 26:619–631

Fisher DJ, Heymann MA, Rudolph AM (1980) Myocardial oxygen and carbohydrate consumption in fetal lambs in utero and in adult sheep. Am J Physiol 238:H399–H405

Fragasso G, Palloshi A, Puccetti P, Silipigni C, Rossodivita A, Pala M, Calori G, Alfieri O, Margonato A (2006a) A randomized clinical trial of trimetazidine, a partial free fatty acid oxidation inhibitor, in patients with heart failure. J Am Coll Cardiol 48:992–998

Fragasso G, Perseghin G, De Cobelli F, Esposito A, Palloshi A, Lattuada G, Scifo P, Calori G, Del Maschio A, Margonato A (2006b) Effects of metabolic modulation by trimetazidine on left ventricular function and phosphocreatine/adenosine triphosphate ratio in patients with heart failure. Eur Heart J 27:942–948

Fragasso G, Rosano G, Baek SH, Sisakian H, Di Napoli P, Alberti L, Calori G, Kang SM, Sahakyan L, Sanosyan A, Vitale C, Marazzi G, Margonato A, Belardinelli R (2013) Effect of partial fatty acid oxidation inhibition with trimetazidine on mortality and morbidity in heart failure: results from an international multicentre retrospective cohort study. Int J Cardiol 163:320–325

Gao D, Ning N, Niu X, Hao G, Meng Z (2011) Trimetazidine: a meta-analysis of randomised controlled trials in heart failure. Heart 97:278–286

Gautheron DC (1984) Mitochondrial oxidative phosphorylation and respiratory chain: review. J Inherit Metab Dis 7(Suppl 1):57–61

Genova ML, Lenaz G (2014) Functional role of mitochondrial respiratory supercomplexes. Biochim Biophys Acta 1837:427–443

Giani JF, Gironacci MM, Munoz MC, Pena C, Turyn D, Dominici FP (2007) Angiotensin-(1 7) stimulates the phosphorylation of JAK2, IRS-1 and Akt in rat heart in vivo: role of the AT1 and Mas receptors. Am J Physiol 293:H1154–H1163

Girard JR, Kervran A, Soufflet E, Assan R (1974) Factors affecting the secretion of insulin and glucagon by the rat fetus. Diabetes 23:310–317

Girard J, Duee PH, Ferre P, Pegorier JP, Escriva F, Decaux JF (1985) Fatty acid oxidation and ketogenesis during development. Reprod Nutr Dev 25:303–319

Girard J, Ferre P, Pegorier JP, Duee PH (1992) Adaptations of glucose and fatty acid metabolism during perinatal period and suckling-weaning transition. Physiol Rev 72:507–562

Goodwin CW, Mela L, Deutsch C, Forster RE, Miller LD, Kelivoria-Papadopoulos M (1976) Development and adaptation of heart mitochondrial respiratory chain function in fetus and in newborn. Adv Exp Med Biol 75:713–719

Goodwin GW, Taylor CS, Taegtmeyer H (1998) Regulation of energy metabolism of the heart during acute increase in heart work. J Biol Chem 273:29530–29539

Graham D, Huynh NN, Hamilton CA, Beattie E, Smith RA, Cocheme HM, Murphy MP, Dominiczak AF (2009) Mitochondria-targeted antioxidant MitoQ10 improves endothelial function and attenuates cardiac hypertrophy. Hypertension 54:322–328

Grajek S, Michalak M (2015) The effect of trimetazidine added to pharmacological treatment on all-cause mortality in patients with systolic heart failure. Cardiology 131:22–29

Greaves P, Martin J, Michel MC, Mompon P (1984) Cardiac hypertrophy in the dog and rat induced by oxfenicine, an agent which modifies muscle metabolism. Arch Toxicol Suppl 7:488–493

Guan KL, Xiong Y (2011) Regulation of intermediary metabolism by protein acetylation. Trends Biochem Sci 36:108–116

Gupta A, Akki A, Wang Y, Leppo MK, Chacko VP, Foster DB, Caceres V, Shi S, Kirk JA, Su J, Lai S, Paolocci N, Steenbergen C, Gerstenblith G, Weiss RG (2012) Creatine kinase-mediated improvement of function in failing mouse hearts provides causal evidence the failing heart is energy starved. J Clin Invest 122:291–302

Gupta T, Khera S, Kolte D, Aronow WS, Iwai S (2015) Antiarrhythmic properties of ranolazine: a review of the current evidence. Int J Cardiol 187:66–74

Haemmerle G, Lass A, Zimmermann R, Gorkiewicz G, Meyer C, Rozman J, Heldmaier G, Maier R, Theussl C, Eder S, Kratky D, Wagner EF, Klingenspor M, Hoefler G, Zechner R (2006) Defective lipolysis and altered energy metabolism in mice lacking adipose triglyceride lipase. Science 312:734–737

Haemmerle G, Moustafa T, Woelkart G, Buttner S, Schmidt A, van de Weijer T, Hesselink M, Jaeger D, Kienesberger PC, Zierler K, Schreiber R, Eichmann T, Kolb D, Kotzbeck P,

Schweiger M, Kumari M, Eder S, Schoiswohl G, Wongsiriroj N, Pollak NM, Radner FP, Preiss-Landl K, Kolbe T, Rulicke T, Pieske B, Trauner M, Lass A, Zimmermann R, Hoefler G, Cinti S, Kershaw EE, Schrauwen P, Madeo F, Mayer B, Zechner R (2011) ATGL-mediated fat catabolism regulates cardiac mitochondrial function via PPAR-alpha and PGC-1. Nat Med 17:1076–1085

Halbirk M, Norrelund H, Moller N, Holst JJ, Schmitz O, Nielsen R, Nielsen-Kudsk JE, Nielsen SS, Nielsen TT, Eiskjaer H, Botker HE, Wiggers H (2010a) Cardiovascular and metabolic effects of 48-h glucagon-like peptide-1 infusion in compensated chronic patients with heart failure. Am J Physiol 298:H1096–H1102

Halbirk M, Norrelund H, Moller N, Schmitz O, Gotzsche L, Nielsen R, Nielsen-Kudsk JE, Nielsen SS, Nielsen TT, Eiskjaer H, Botker HE, Wiggers H (2010b) Suppression of circulating free fatty acids with acipimox in chronic heart failure patients changes whole body metabolism but does not affect cardiac function. Am J Physiol 299:H1220–H1225

Hale SL, Kloner RA (2006) Ranolazine, an inhibitor of the late sodium channel current, reduces postischemic myocardial dysfunction in the rabbit. J Cardiovasc Pharmacol Therap 11:249–255

Hale SL, Kloner RA (2014) Ranolazine treatment for myocardial infarction? Effects on the development of necrosis, left ventricular function and arrhythmias in experimental models. Cardiovasc Drugs Ther 28:469–475

Hall JL, Mazzeo RS, Podolin DA, Cartee GD, Stanley WC (1994) Exercise training does not compensate for age-related decrease in myocardial GLUT-4 content. J Appl Physiol 76:328–332

Hardy CJ, Weiss RG, Bottomley PA, Gerstenblith G (1991) Altered myocardial high-energy phosphate metabolites in patients with dilated cardiomyopathy. Am Heart J 122:795–801

Harper AE, Miller RH, Block KP (1984) Branched-chain amino acid metabolism. Annu Rev Nutr 4:409–454

Harris RA, Joshi M, Jeoung NH (2004) Mechanisms responsible for regulation of branched-chain amino acid catabolism. Biochem Biophys Res Commun 313:391–396

Hasty P, Vijg J (2004) Rebuttal to Miller: 'Accelerated aging': a primrose path to insight?'. Aging Cell 3:67–69

Hatefi Y (1985) The mitochondrial electron transport and oxidative phosphorylation system. Ann Rev Biochem 54:1015–1069

Herzig S, Raemy E, Montessuit S, Veuthey JL, Zamboni N, Westermann B, Kunji ER, Martinou JC (2012) Identification and functional expression of the mitochondrial pyruvate carrier. Science 337:93–96

Hoerter JA, Opie LH (1978) Perinatal changes in glycolytic function in response to hypoxia in the incubated or perfused rat heart. Biol Neonate 33:144–161

Holness MJ, Sugden MC (2003) Regulation of pyruvate dehydrogenase complex activity by reversible phosphorylation. Biochem Soc Trans 31:1143–1151

Holness MJ, Smith ND, Bulmer K, Hopkins T, Gibbons GF, Sugden MC (2002) Evaluation of the role of peroxisome-proliferator-activated receptor alpha in the regulation of cardiac pyruvate dehydrogenase kinase 4 protein expression in response to starvation, high-fat feeding and hyperthyroidism. Biochem J 364:687–694

Holubarsch CJ, Rohrbach M, Karrasch M, Boehm E, Polonski L, Ponikowski P, Rhein S (2007) A double-blind randomized multicentre clinical trial to evaluate the efficacy and safety of two doses of etomoxir in comparison with placebo in patients with moderate congestive heart failure: the ERGO (etomoxir for the recovery of glucose oxidation) study. Clin Sci 113:205–212

Hwang YC, Kaneko M, Bakr S, Liao H, Lu Y, Lewis ER, Yan S, Ii S, Itakura M, Rui L, Skopicki H, Homma S, Schmidt AM, Oates PJ, Szabolcs M, Ramasamy R (2004) Central role for aldose reductase pathway in myocardial ischemic injury. FASEB J 18:1192–1199

Iemitsu M, Miyauchi T, Maeda S, Tanabe T, Takanashi M, Irukayama-Tomobe Y, Sakai S, Ohmori H, Matsuda M, Yamaguchi I (2002) Aging-induced decrease in the PPAR-alpha level in hearts is improved by exercise training. Am J Physiol 283:H1750–H1760

Ingwall JS (2009) Energy metabolism in heart failure and remodelling. Cardiovasc Res 81: 412–419

Ingwall JS, Weiss RG (2004) Is the failing heart energy starved? On using chemical energy to support cardiac function. Circ Res 95:135–145

Jaswal JS, Keung W, Wang W, Ussher JR, Lopaschuk GD (2011) Targeting fatty acid and carbohydrate oxidation–a novel therapeutic intervention in the ischemic and failing heart. Biochim Biophys Acta 1813:1333–1350

Jeffrey FM, Diczku V, Sherry AD, Malloy CR (1995) Substrate selection in the isolated working rat heart: effects of reperfusion, afterload, and concentration. Basic Res Cardiol 90:388–396

Johnson BF, Nesto RW, Pfeifer MA, Slater WR, Vinik AI, Chyun DA, Law G, Wackers FJ, Young LH (2004) Cardiac abnormalities in diabetic patients with neuropathy: effects of aldose reductase inhibitor administration. Diabetes Care 27:448–454

Jonassen AK, Aasum E, Riemersma RA, Mjos OD, Larsen TS (2000) Glucose-insulin-potassium reduces infarct size when administered during reperfusion. Cardiovasc Drugs Ther 14: 615–623

Kaijser L, Berglund B (1992) Myocardial lactate extraction and release at rest and during heavy exercise in healthy men. Acta Physiol Scand 144:39–45

Kates AM, Herrero P, Dence C, Soto P, Srinivasan M, Delano DG, Ehsani A, Gropler RJ (2003) Impact of aging on substrate metabolism by the human heart. J Am Coll Cardiol 41:293–299

Kato T, Niizuma S, Inuzuka Y, Kawashima T, Okuda J, Tamaki Y, Iwanaga Y, Narazaki M, Matsuda T, Soga T, Kita T, Kimura T, Shioi T (2010) Analysis of metabolic remodeling in compensated left ventricular hypertrophy and heart failure. Circ Heart Fail 3:420–430

Khrapko K, Kraytsberg Y, de Grey AD, Vijg J, Schon EA (2006) Does premature aging of the mtDNA mutator mouse prove that mtDNA mutations are involved in natural aging? Aging Cell 5:279–282

Kolwicz SC Jr, Olson DP, Marney LC, Garcia-Menendez L, Synovec RE, Tian R (2012) Cardiac-specific deletion of acetyl CoA carboxylase 2 prevents metabolic remodeling during pressure-overload hypertrophy. Circ Res 111:728–738

Kolwicz SC Jr, Purohit S, Tian R (2013) Cardiac metabolism and its interactions with contraction, growth, and survival of cardiomyocytes. Circ Res 113:603–616

Komajda M, McMurray JJ, Beck-Nielsen H, Gomis R, Hanefeld M, Pocock SJ, Curtis PS, Jones NP, Home PD (2010) Heart failure events with rosiglitazone in type 2 diabetes: data from the RECORD clinical trial. Eur Heart J 31:824–831

Koonen DP, Glatz JF, Bonen A, Luiken JJ (2005) Long-chain fatty acid uptake and FAT/CD36 translocation in heart and skeletal muscle. Biochim Biophys Acta 1736:163–180

Korvald C, Elvenes OP, Myrmel T (2000) Myocardial substrate metabolism influences left ventricular energetics in vivo. Am J Physiol 278:H1345–H1351

Koutsari C, Jensen MD (2006) Thematic review series: patient-oriented research. Free fatty acid metabolism in human obesity. J Lipid Res 47:1643–1650

Krishnan KJ, Greaves LC, Reeve AK, Turnbull DM (2007) Mitochondrial DNA mutations and aging. Ann N Y Acad Sci 1100:227–240

Kung J, Henry RR (2012) Thiazolidinedione safety. Expert Opin Drug Saf 11:565–579

Lee L, Campbell R, Scheuermann-Freestone M, Taylor R, Gunaruwan P, Williams L, Ashrafian H, Horowitz J, Fraser AG, Clarke K, Frenneaux M (2005) Metabolic modulation with perhexiline in chronic heart failure: a randomized, controlled trial of short-term use of a novel treatment. Circulation 112:3280–3288

Lehman JJ, Kelly DP (2002) Gene regulatory mechanisms governing energy metabolism during cardiac hypertrophic growth. Heart Fail Rev 7:175–185

Lenaz G, Baracca A, Barbero G, Bergamini C, Dalmonte ME, Del Sole M, Faccioli M, Falasca A, Fato R, Genova ML, Sgarbi G, Solaini G (2010) Mitochondrial respiratory chain super-complex I-III in physiology and pathology. Biochim Biophys Acta 1797:633–640

Lewis JF, DaCosta M, Wargowich T, Stacpoole P (1998) Effects of dichloroacetate in patients with congestive heart failure. Clin Cardiol 21:888–892

Liao R, Nascimben L, Friedrich J, Gwathmey JK, Ingwall JS (1996) Decreased energy reserve in an animal model of dilated cardiomyopathy. Relationship to contractile performance. Circ Res 78:893–902

Liao R, Jain M, Cui L, D'Agostino J, Aiello F, Luptak I, Ngoy S, Mortensen RM, Tian R (2002) Cardiac-specific overexpression of GLUT1 prevents the development of heart failure attributable to pressure overload in mice. Circulation 106:2125–2131

Lionetti V, Linke A, Chandler MP, Young ME, Penn MS, Gupte S, d'Agostino C, Hintze TH, Stanley WC, Recchia FA (2005) Carnitine palmitoyl transferase-I inhibition prevents ventricular remodeling and delays decompensation in pacing-induced heart failure. Cardiovasc Res 66:454–461

Liu J, Marchase RB, Chatham JC (2007) Glutamine-induced protection of isolated rat heart from ischemia/reperfusion injury is mediated via the hexosamine biosynthesis pathway and increased protein O-GlcNAc levels. J Mol Cell Cardiol 42:177–185

Liu L, Yu S, Khan RS, Ables GP, Bharadwaj KG, Hu Y, Huggins LA, Eriksson JW, Buckett LK, Turnbull AV, Ginsberg HN, Blaner WS, Huang LS, Goldberg IJ (2011) DGAT1 deficiency decreases PPAR expression and does not lead to lipotoxicity in cardiac and skeletal muscle. J Lipid Res 52:732–744

Lommi J, Kupari M, Koskinen P, Naveri H, Leinonen H, Pulkki K, Harkonen M (1996) Blood ketone bodies in congestive heart failure. J Am Coll Cardiol 28:665–672

Lonn E, Bosch J, Yusuf S, Sheridan P, Pogue J, Arnold JM, Ross C, Arnold A, Sleight P, Probstfield J, Dagenais GR, Hope, Investigators H-TT (2005) Effects of long-term vitamin E supplementation on cardiovascular events and cancer: a randomized controlled trial. JAMA 293: 1338–1347

Lopaschuk GD, Spafford MA (1990) Energy substrate utilization by isolated working hearts from newborn rabbits. Am J Physiol 258:H1274–H1280

Lopaschuk GD, Wall SR, Olley PM, Davies NJ (1988) Etomoxir, a carnitine palmitoyltransferase I inhibitor, protects hearts from fatty acid-induced ischemic injury independent of changes in long chain acylcarnitine. Circ Res 63:1036–1043

Lopaschuk GD, Spafford MA, Marsh DR (1991) Glycolysis is predominant source of myocardial ATP production immediately after birth. Am J Physiol 261:H1698–H1705

Lopaschuk GD, Collins-Nakai RL, Itoi T (1992) Developmental changes in energy substrate use by the heart. Cardiovasc Res 26:1172–1180

Lopaschuk GD, Witters LA, Itoi T, Barr R, Barr A (1994) Acetyl-CoA carboxylase involvement in the rapid maturation of fatty acid oxidation in the newborn rabbit heart. J Biol Chem 269:25871–25878

Lopaschuk GD, Barr R, Thomas PD, Dyck JR (2003) Beneficial effects of trimetazidine in ex vivo working ischemic hearts are due to a stimulation of glucose oxidation secondary to inhibition of long-chain 3-ketoacyl coenzyme a thiolase. Circ Res 93:e33–e37

Lopaschuk GD, Ussher JR, Folmes CD, Jaswal JS, Stanley WC (2010) Myocardial fatty acid metabolism in health and disease. Physiol Rev 90:207–258

Lu G, Ren S, Korge P, Choi J, Dong Y, Weiss J, Koehler C, Chen JN, Wang Y (2007) A novel mitochondrial matrix serine/threonine protein phosphatase regulates the mitochondria permeability transition pore and is essential for cellular survival and development. Genes Dev 21:784–796

Lu G, Sun H, She P, Youn JY, Warburton S, Ping P, Vondriska TM, Cai H, Lynch CJ, Wang Y (2009) Protein phosphatase 2Cm is a critical regulator of branched-chain amino acid catabolism in mice and cultured cells. J Clin Invest 119:1678–1687

Luiken JJ, Arumugam Y, Dyck DJ, Bell RC, Pelsers MM, Turcotte LP, Tandon NN, Glatz JF, Bonen A (2001) Increased rates of fatty acid uptake and plasmalemmal fatty acid transporters in obese Zucker rats. J Biol Chem 276:40567–40573

Luiken JJ, Coort SL, Koonen DP, van der Horst DJ, Bonen A, Zorzano A, Glatz JF (2004) Regulation of cardiac long-chain fatty acid and glucose uptake by translocation of substrate transporters. Pflugers Archiv Eur J Physiol 448:1–15

Lundgrin EL, Park MM, Sharp J, Tang WH, Thomas JD, Asosingh K, Comhair SA, DiFilippo FP, Neumann DR, Davis L, Graham BB, Tuder RM, Dostanic I, Erzurum SC (2013) Fasting 2-deoxy-2-[18F]fluoro-D-glucose positron emission tomography to detect metabolic changes in pulmonary arterial hypertension hearts over 1 year. Ann Am Thorac Soc 10:1–9

Luptak I, Yan J, Cui L, Jain M, Liao R, Tian R (2007) Long-term effects of increased glucose entry on mouse hearts during normal aging and ischemic stress. Circulation 116:901–909

Maier LS, Layug B, Karwatowska-Prokopczuk E, Belardinelli L, Lee S, Sander J, Lang C, Wachter R, Edelmann F, Hasenfuss G, Jacobshagen C (2013) RAnoLazIne for the treatment of diastolic heart failure in patients with preserved ejection fraction: the RALI-DHF proof-of-concept study. JACC Heart Fail 1:115–122

Makinde AO, Gamble J, Lopaschuk GD (1997) Upregulation of 5′-AMP-activated protein kinase is responsible for the increase in myocardial fatty acid oxidation rates following birth in the newborn rabbit. Circ Res 80:482–489

Malmberg K, Ryden L, Hamsten A, Herlitz J, Waldenstrom A, Wedel H (1996) Effects of insulin treatment on cause-specific one-year mortality and morbidity in diabetic patients with acute myocardial infarction. DIGAMI Study Group. Diabetes insulin-glucose in acute myocardial infarction. Eur Heart J 17:1337–1344

Marchioli R, Levantesi G, Macchia A, Marfisi RM, Nicolosi GL, Tavazzi L, Tognoni G, Valagussa F, Investigators GI-P (2006) Vitamin E increases the risk of developing heart failure after myocardial infarction: results from the GISSI-Prevenzione trial. J Cardiovasc Med 7:347–350

Marfella R, Di Filippo C, Portoghese M, Barbieri M, Ferraraccio F, Siniscalchi M, Cacciapuoti F, Rossi F, D'Amico M, Paolisso G (2009) Myocardial lipid accumulation in patients with pressure-overloaded heart and metabolic syndrome. J Lipid Res 50:2314–2323

Masmoudi K, Masson H, Gras V, Andrejak M (2012) Extrapyramidal adverse drug reactions associated with trimetazidine: a series of 21 cases. Fundam Clin Pharmacol 26:198–203

Masoudi FA, Inzucchi SE, Wang Y, Havranek EP, Foody JM, Krumholz HM (2005) Thiazolidinediones, metformin, and outcomes in older patients with diabetes and heart failure: an observational study. Circulation 111:583–590

Matsuhashi T, Hishiki T, Zhou H, Ono T, Kaneda R, Iso T, Yamaguchi A, Endo J, Katsumata Y, Atsushi A, Yamamoto T, Shirakawa K, Yan X, Shinmura K, Suematsu M, Fukuda K, Sano M (2015) Activation of pyruvate dehydrogenase by dichloroacetate has the potential to induce epigenetic remodeling in the heart. J Mol Cell Cardiol 82:116–124

Mayers RM, Butlin RJ, Kilgour E, Leighton B, Martin D, Myatt J, Orme JP, Holloway BR (2003) AZD7545, a novel inhibitor of pyruvate dehydrogenase kinase 2 (PDHK2), activates pyruvate dehydrogenase in vivo and improves blood glucose control in obese (fa/fa) Zucker rats. Biochem Soc Trans 31:1165–1167

Mazumder PK, O'Neill BT, Roberts MW, Buchanan J, Yun UJ, Cooksey RC, Boudina S, Abel ED (2004) Impaired cardiac efficiency and increased fatty acid oxidation in insulin-resistant ob/ob mouse hearts. Diabetes 53:2366–2374

McCormack JG, Denton RM (1981) The activation of pyruvate dehydrogenase in the perfused rat heart by adrenaline and other inotropic agents. Biochem J 194:639–643

McCormack JG, Barr RL, Wolff AA, Lopaschuk GD (1996) Ranolazine stimulates glucose oxidation in normoxic, ischemic, and reperfused ischemic rat hearts. Circulation 93:135–142

McGarry JD, Brown NF (1997) The mitochondrial carnitine palmitoyltransferase system. From concept to molecular analysis. Eur J Biochem 244:1–14

McMillin JB, Taffet GE, Taegtmeyer H, Hudson EK, Tate CA (1993) Mitochondrial metabolism and substrate competition in the aging Fischer rat heart. Cardiovasc Res 27:2222–2228

McMurtry MS, Bonnet S, Wu X, Dyck JR, Haromy A, Hashimoto K, Michelakis ED (2004) Dichloroacetate prevents and reverses pulmonary hypertension by inducing pulmonary artery smooth muscle cell apoptosis. Circ Res 95:830–840

McNulty PH, Jacob R, Deckelbaum LI, Young LH (2000) Effect of hyperinsulinemia on myocardial amino acid uptake in patients with coronary artery disease. Metab Clin Exp 49: 1365–1369

McVeigh JJ, Lopaschuk GD (1990) Dichloroacetate stimulation of glucose oxidation improves recovery of ischemic rat hearts. Am J Physiol 259:H1079–H1085

Miller RA (2004) 'Accelerated aging': a primrose path to insight? Aging Cell 3:47–51

Millet L, Vidal H, Andreelli F, Larrouy D, Riou JP, Ricquier D, Laville M, Langin D (1997) Increased uncoupling protein-2 and −3 mRNA expression during fasting in obese and lean humans. J Clin Invest 100:2665–2670

Montessuit C, Thorburn A (1999) Transcriptional activation of the glucose transporter GLUT1 in ventricular cardiac myocytes by hypertrophic agonists. J Biol Chem 274:9006–9012

Mori J, Basu R, McLean BA, Das SK, Zhang L, Patel VB, Wagg CS, Kassiri Z, Lopaschuk GD, Oudit GY (2012) Agonist-induced hypertrophy and diastolic dysfunction are associated with selective reduction in glucose oxidation: a metabolic contribution to heart failure with normal ejection fraction. Circ Heart Fail 5:493–503

Mori J, Alrob OA, Wagg CS, Harris RA, Lopaschuk GD, Oudit GY (2013a) ANG II causes insulin resistance and induces cardiac metabolic switch and inefficiency: a critical role of PDK4. Am J Physiol 304:H1103–H1113

Mori J, Zhang L, Oudit GY, Lopaschuk GD (2013b) Impact of the renin-angiotensin system on cardiac energy metabolism in heart failure. J Mol Cell Cardiol 63:98–106

Mouton V, Toussaint L, Vertommen D, Gueuning MA, Maisin L, Havaux X, Sanchez-Canedo C, Bertrand L, Dequiedt F, Hemmings BA, Hue L, Rider MH (2010) Heart 6-phosphofructo-2-kinase activation by insulin requires PKB (protein kinase B), but not SGK3 (serum- and glucocorticoid-induced protein kinase 3). Biochem J 431:267–275

Nascimben L, Ingwall JS, Lorell BH, Pinz I, Schultz V, Tornheim K, Tian R (2004) Mechanisms for increased glycolysis in the hypertrophied rat heart. Hypertens 44:662–667

Nathanson D, Ullman B, Lofstrom U, Hedman A, Frick M, Sjoholm A, Nystrom T (2012) Effects of intravenous exenatide in type 2 diabetic patients with congestive heart failure: a double-blind, randomised controlled clinical trial of efficacy and safety. Diabetologia 55:926–935

Nau PN, Van Natta T, Ralphe JC, Teneyck CJ, Bedell KA, Caldarone CA, Segar JL, Scholz TD (2002) Metabolic adaptation of the fetal and postnatal ovine heart: regulatory role of hypoxia-inducible factors and nuclear respiratory factor-1. Pediatr Res 52:269–278

Negre-Salvayre A, Hirtz C, Carrera G, Cazenave R, Troly M, Salvayre R, Penicaud L, Casteilla L (1997) A role for uncoupling protein-2 as a regulator of mitochondrial hydrogen peroxide generation. FASEB J 11:809–815

Neubauer S (2007) The failing heart–an engine out of fuel. New Engl J Med 356:1140–1151

Neubauer S, Horn M, Cramer M, Harre K, Newell JB, Peters W, Pabst T, Ertl G, Hahn D, Ingwall JS, Kochsiek K (1997) Myocardial phosphocreatine-to-ATP ratio is a predictor of mortality in patients with dilated cardiomyopathy. Circulation 96:2190–2196

Nicklin P, Bergman P, Zhang B, Triantafellow E, Wang H, Nyfeler B, Yang H, Hild M, Kung C, Wilson C, Myer VE, MacKeigan JP, Porter JA, Wang YK, Cantley LC, Finan PM, Murphy LO (2009) Bidirectional transport of amino acids regulates mTOR and autophagy. Cell 136:521–534

Nielsen R, Norrelund H, Kampmann U, Kim WY, Ringgaard S, Schar M, Moller N, Botker HE, Wiggers H (2013) Failing heart of patients with type 2 diabetes mellitus can adapt to extreme short-term increases in circulating lipids and does not display features of acute myocardial lipotoxicity. Circ Heart Fail 6:845–852

Niemann B, Chen Y, Issa H, Silber RE, Rohrbach S (2010) Caloric restriction delays cardiac ageing in rats: role of mitochondria. Cardiovasc Res 88:267–276

Nikolaidis LA, Mankad S, Sokos GG, Miske G, Shah A, Elahi D, Shannon RP (2004a) Effects of glucagon-like peptide-1 in patients with acute myocardial infarction and left ventricular dysfunction after successful reperfusion. Circulation 109:962–965

Nikolaidis LA, Sturzu A, Stolarski C, Elahi D, Shen YT, Shannon RP (2004b) The development of myocardial insulin resistance in conscious dogs with advanced dilated cardiomyopathy. Cardiovasc Res 61:297–306

Nikolaidis LA, Elahi D, Shen YT, Shannon RP (2005) Active metabolite of GLP-1 mediates myocardial glucose uptake and improves left ventricular performance in conscious dogs with dilated cardiomyopathy. Am J Physiol 289:H2401–H2408

Nisoli E, Clementi E, Paolucci C, Cozzi V, Tonello C, Sciorati C, Bracale R, Valerio A, Francolini M, Moncada S, Carruba MO (2003) Mitochondrial biogenesis in mammals: the role of endogenous nitric oxide. Science 299:896–899

Noyan-Ashraf MH, Shikatani EA, Schuiki I, Mukovozov I, Wu J, Li RK, Volchuk A, Robinson LA, Billia F, Drucker DJ, Husain M (2013) A glucagon-like peptide-1 analog reverses the molecular pathology and cardiac dysfunction of a mouse model of obesity. Circulation 127:74–85

Nystrom T, Gutniak MK, Zhang Q, Zhang F, Holst JJ, Ahren B, Sjoholm A (2004) Effects of glucagon-like peptide-1 on endothelial function in type 2 diabetes patients with stable coronary artery disease. Am J Physiol 287:E1209–E1215

O'Donnell JM, Fields AD, Sorokina N, Lewandowski ED (2008) The absence of endogenous lipid oxidation in early stage heart failure exposes limits in lipid storage and turnover. J Mol Cell Cardiol 44:315–322

Okere IC, Young ME, McElfresh TA, Chess DJ, Sharov VG, Sabbah HN, Hoit BD, Ernsberger P, Chandler MP, Stanley WC (2006) Low carbohydrate/high-fat diet attenuates cardiac hypertrophy, remodeling, and altered gene expression in hypertension. Hypertension 48:1116–1123

Onay-Besikci A (2006) Regulation of cardiac energy metabolism in newborn. Mol Cell Biochem 287:1–11

Onay-Besikci A, Campbell FM, Hopkins TA, Dyck JR, Lopaschuk GD (2003) Relative importance of malonyl CoA and carnitine in maturation of fatty acid oxidation in newborn rabbit heart. Am J Physiol 284:H283–H289

Onay-Besikci A, Altarejos JY, Lopaschuk GD (2004) gAd-globular head domain of adiponectin increases fatty acid oxidation in newborn rabbit hearts. J Biol Chem 279:44320–44326

Palanivel R, Eguchi M, Shuralyova I, Coe I, Sweeney G (2006) Distinct effects of short- and long-term leptin treatment on glucose and fatty acid uptake and metabolism in HL-1 cardiomyocytes. Metab Clini Exp 55:1067–1075

Palanivel R, Fang X, Park M, Eguchi M, Pallan S, De Girolamo S, Liu Y, Wang Y, Xu A, Sweeney G (2007) Globular and full-length forms of adiponectin mediate specific changes in glucose and fatty acid uptake and metabolism in cardiomyocytes. Cardiovasc Res 75:148–157

Palanivel R, Vu V, Park M, Fang X, Sweeney G (2008) Differential impact of adipokines derived from primary adipocytes of wild-type versus streptozotocin-induced diabetic rats on glucose and fatty acid metabolism in cardiomyocytes. J Endocrinol 199:389–397

Panchal AR, Stanley WC, Kerner J, Sabbah HN (1998) Beta-receptor blockade decreases carnitine palmitoyl transferase I activity in dogs with heart failure. J Cardiac Fail 4:121–126

Park SY, Cho YR, Kim HJ, Higashimori T, Danton C, Lee MK, Dey A, Rothermel B, Kim YB, Kalinowski A, Russell KS, Kim JK (2005) Unraveling the temporal pattern of diet-induced insulin resistance in individual organs and cardiac dysfunction in C57BL/6 mice. Diabetes 54:3530–3540

Paulin R, Michelakis ED (2014) The metabolic theory of pulmonary arterial hypertension. Circ Res 115:148–164

Pellieux C, Aasum E, Larsen TS, Montessuit C, Papageorgiou I, Pedrazzini T, Lerch R (2006) Overexpression of angiotensinogen in the myocardium induces downregulation of the fatty acid oxidation pathway. J Mol Cell Cardiol 41:459–466

Peterson LR, Herrero P, Schechtman KB, Racette SB, Waggoner AD, Kisrieva-Ware Z, Dence C, Klein S, Marsala J, Meyer T, Gropler RJ (2004) Effect of obesity and insulin resistance on myocardial substrate metabolism and efficiency in young women. Circulation 109:2191–2196

Piao L, Fang YH, Cadete VJ, Wietholt C, Urboniene D, Toth PT, Marsboom G, Zhang HJ, Haber I, Rehman J, Lopaschuk GD, Archer SL (2010) The inhibition of pyruvate dehydrogenase kinase improves impaired cardiac function and electrical remodeling in two models of right ventricular hypertrophy: resuscitating the hibernating right ventricle. J Mol Med 88:47–60

Podbregar M, Voga G (2002) Effect of selective and nonselective beta-blockers on resting energy production rate and total body substrate utilization in chronic heart failure. J Cardiac Fail 8:369–378

Poole RC, Halestrap AP (1993) Transport of lactate and other monocarboxylates across mammalian plasma membranes. Am J Physiol 264:C761–C782

Poornima I, Brown SB, Bhashyam S, Parikh P, Bolukoglu H, Shannon RP (2008) Chronic glucagon-like peptide-1 infusion sustains left ventricular systolic function and prolongs survival in the spontaneously hypertensive, heart failure-prone rat. Circ Heart Fail 1:153–160

Postic C, Leturque A, Printz RL, Maulard P, Loizeau M, Granner DK, Girard J (1994) Development and regulation of glucose transporter and hexokinase expression in rat. Am J Physiol 266:E548–E559

Pound KM, Sorokina N, Ballal K, Berkich DA, Fasano M, Lanoue KF, Taegtmeyer H, O'Donnell JM, Lewandowski ED (2009) Substrate-enzyme competition attenuates upregulated anaplerotic flux through malic enzyme in hypertrophied rat heart and restores triacylglyceride content: attenuating upregulated anaplerosis in hypertrophy. Circ Res 104:805–812

Preiss D, Lloyd SM, Ford I, McMurray JJ, Holman RR, Welsh P, Fisher M, Packard CJ, Sattar N (2014) Metformin for non-diabetic patients with coronary heart disease (the CAMERA study): a randomised controlled trial. Lancet Diabetes Endocrinol 2:116–124

Preston CC, Oberlin AS, Holmuhamedov EL, Gupta A, Sagar S, Syed RH, Siddiqui SA, Raghavakaimal S, Terzic A, Jahangir A (2008) Aging-induced alterations in gene transcripts and functional activity of mitochondrial oxidative phosphorylation complexes in the heart. Mech Ageing Dev 129:304–312

Randle PJ (1998) Regulatory interactions between lipids and carbohydrates: the glucose fatty acid cycle after 35 years. Diabetes Metab Rev 14:263–283

Randle PJ, Garland PB, Hales CN, Newsholme EA (1963) The glucose fatty-acid cycle. Its role in insulin sensitivity and the metabolic disturbances of diabetes mellitus. Lancet 1:785–789

Rasoul S, Ottervanger JP, Timmer JR, Svilaas T, Henriques JP, Dambrink JH, van der Horst IC, Zijlstra F (2007) One year outcomes after glucose-insulin-potassium in ST elevation myocardial infarction. The glucose-insulin-potassium study II. Int J Cardiol 122:52–55

Rastogi S, Sharov VG, Mishra S, Gupta RC, Blackburn B, Belardinelli L, Stanley WC, Sabbah HN (2008) Ranolazine combined with enalapril or metoprolol prevents progressive LV dysfunction and remodeling in dogs with moderate heart failure. Am J Physiol 295:H2149–H2155

Ravassa S, Zudaire A, Diez J (2012) GLP-1 and cardioprotection: from bench to bedside. Cardiovasc Res 94:316–323

Rider MH, Bertrand L, Vertommen D, Michels PA, Rousseau GG, Hue L (2004) 6-phosphofructo-2-kinase/fructose-2,6-bisphosphatase: head-to-head with a bifunctional enzyme that controls glycolysis. Biochem J 381:561–579

Riehle C, Wende AR, Zaha VG, Pires KM, Wayment B, Olsen C, Bugger H, Buchanan J, Wang X, Moreira AB, Doenst T, Medina-Gomez G, Litwin SE, Lelliott CJ, Vidal-Puig A, Abel ED (2011) PGC-1beta deficiency accelerates the transition to heart failure in pressure overload hypertrophy. Circ Res 109:783–793

Rijzewijk LJ, van der Meer RW, Lamb HJ, de Jong HW, Lubberink M, Romijn JA, Bax JJ, de Roos A, Twisk JW, Heine RJ, Lammertsma AA, Smit JW, Diamant M (2009) Altered myocardial substrate metabolism and decreased diastolic function in nonischemic human diabetic cardiomyopathy: studies with cardiac positron emission tomography and magnetic resonance imaging. J Am Coll Cardiol 54:1524–1532

Robinson JG (2007) Update on PPAR agonists: the clinical significance of FIELD and PROACTIVE. Curr Atheroscler Rep 9:64–71

Rohrbach S, Niemann B, Abushouk AM, Holtz J (2006) Caloric restriction and mitochondrial function in the ageing myocardium. Exp Gerontol 41:525–531

Romano S, Valayannopoulos V, Touati G, Jais JP, Rabier D, de Keyzer Y, Bonnet D, de Lonlay P (2010) Cardiomyopathies in propionic aciduria are reversible after liver transplantation. J Pediat 156:128–134

Rosca MG, Hoppel CL (2013) Mitochondrial dysfunction in heart failure. Heart Fail Rev 18: 607–622

Russell RR 3rd, Taegtmeyer H (1991) Changes in citric acid cycle flux and anaplerosis antedate the functional decline in isolated rat hearts utilizing acetoacetate. J Clin Invest 87:384–390

Saddik M, Lopaschuk GD (1991) Myocardial triglyceride turnover and contribution to energy substrate utilization in isolated working rat hearts. J Biol Chem 266:8162–8170

Salerno A, Fragasso G, Esposito A, Canu T, Lattuada G, Manzoni G, Del Maschio A, Margonato A, De Cobelli F, Perseghin G (2015) Effects of short-term manipulation of serum FFA concentrations on left ventricular energy metabolism and function in patients with heart failure: no association with circulating bio-markers of inflammation. Acta Diabetol 52:753–761

Santos RA, Ferreira AJ, Verano-Braga T, Bader M (2013) Angiotensin-converting enzyme 2, angiotensin-(1–7) and Mas: new players of the renin-angiotensin system. J Endocrinol 216:R1–R17

Sato K, Kashiwaya Y, Keon CA, Tsuchiya N, King MT, Radda GK, Chance B, Clarke K, Veech RL (1995) Insulin, ketone bodies, and mitochondrial energy transduction. FASEB J 9:651–658
Savarese G, Rosano G, D'Amore C, Musella F, Della Ratta GL, Pellegrino AM, Formisano T, Vitagliano A, Cirillo A, Cice G, Fimiani L, del Guercio L, Trimarco B, Perrone-Filardi P (2013) Effects of ranolazine in symptomatic patients with stable coronary artery disease. A systematic review and meta-analysis. Int J Cardiol 169:262–270
Schaefer S, Ramasamy R (1997) Glycogen utilization and ischemic injury in the isolated rat heart. Cardiovasc Res 35:90–98
Schaffer JE (2002) Fatty acid transport: the roads taken. Am J Physiol 282:E239–E246
Schagger H, Pfeiffer K (2000) Supercomplexes in the respiratory chains of yeast and mammalian mitochondria. EMBO J 19:1777–1783
Scheubel RJ, Tostlebe M, Simm A, Rohrbach S, Prondzinsky R, Gellerich FN, Silber RE, Holtz J (2002) Dysfunction of mitochondrial respiratory chain complex I in human failing myocardium is not due to disturbed mitochondrial gene expression. J Am Coll Cardiol 40:2174–2181
Scheuermann-Freestone M, Madsen PL, Manners D, Blamire AM, Buckingham RE, Styles P, Radda GK, Neubauer S, Clarke K (2003) Abnormal cardiac and skeletal muscle energy metabolism in patients with type 2 diabetes. Circulation 107:3040–3046
Schmidt-Schweda S, Holubarsch C (2000) First clinical trial with etomoxir in patients with chronic congestive heart failure. Clin Sci 99:27–35
Schonfeld P, Wojtczak L (2008) Fatty acids as modulators of the cellular production of reactive oxygen species. Free Rad Bio Med 45:231–241
Scirica BM, Bhatt DL, Braunwald E, Steg PG, Davidson J, Hirshberg B, Ohman P, Frederich R, Wiviott SD, Hoffman EB, Cavender MA, Udell JA, Desai NR, Mosenzon O, McGuire DK, Ray KK, Leiter LA, Raz I, Committee S-TS, Investigators (2013) Saxagliptin and cardiovascular outcomes in patients with type 2 diabetes mellitus. New Engl J Med 369:1317–1326
Seifert EL, Estey C, Xuan JY, Harper ME (2010) Electron transport chain-dependent and -independent mechanisms of mitochondrial H_2O_2 emission during long-chain fatty acid oxidation. J Biol Chem 285:5748–5758
Sharma S, Adrogue JV, Golfman L, Uray I, Lemm J, Youker K, Noon GP, Frazier OH, Taegtmeyer H (2004) Intramyocardial lipid accumulation in the failing human heart resembles the lipotoxic rat heart. FASEB J 18:1692–1700
Sokos GG, Nikolaidis LA, Mankad S, Elahi D, Shannon RP (2006) Glucagon-like peptide-1 infusion improves left ventricular ejection fraction and functional status in patients with chronic heart failure. J Cardiac Fail 12:694–699
Sorokina N, O'Donnell JM, McKinney RD, Pound KM, Woldegiorgis G, LaNoue KF, Ballal K, Taegtmeyer H, Buttrick PM, Lewandowski ED (2007) Recruitment of compensatory pathways to sustain oxidative flux with reduced carnitine palmitoyltransferase I activity characterizes inefficiency in energy metabolism in hypertrophied hearts. Circulation 115:2033–2041
Soto PF, Herrero P, Kates AM, Dence CS, Ehsani AA, Davila-Roman V, Schechtman KB, Gropler RJ (2003) Impact of aging on myocardial metabolic response to dobutamine. Am J Physiol 285:H2158–H2164
Stanley WC, Meadows SR, Kivilo KM, Roth BA, Lopaschuk GD (2003) beta-Hydroxybutyrate inhibits myocardial fatty acid oxidation in vivo independent of changes in malonyl-CoA content. Am J Physiol 285:H1626–H1631
Stanley WC, Recchia FA, Lopaschuk GD (2005) Myocardial substrate metabolism in the normal and failing heart. Physiol Rev 85:1093–1129
Still AJ, Floyd BJ, Hebert AS, Bingman CA, Carson JJ, Gunderson DR, Dolan BK, Grimsrud PA, Dittenhafer-Reed KE, Stapleton DS, Keller MP, Westphall MS, Denu JM, Attie AD, Coon JJ, Pagliarini DJ (2013) Quantification of mitochondrial acetylation dynamics highlights prominent sites of metabolic regulation. J Biol Chem 288:26209–26219
Su X, Abumrad NA (2009) Cellular fatty acid uptake: a pathway under construction. Trends Endocrinol Metab 20:72–77
Sugden MC, Holness MJ (2002) Therapeutic potential of the mammalian pyruvate dehydrogenase kinases in the prevention of hyperglycaemia. Curr Drug Targets Immune Endocr Metabol Disord 2:151–165

Swan JW, Anker SD, Walton C, Godsland IF, Clark AL, Leyva F, Stevenson JC, Coats AJ (1997) Insulin resistance in chronic heart failure: relation to severity and etiology of heart failure. J Am Coll Cardiol 30:527–532

Szeto HH (2008) Mitochondria-targeted cytoprotective peptides for ischemia-reperfusion injury. Antiox Redox Signal 10:601–619

Szeto HH (2014) First-in-class cardiolipin-protective compound as a therapeutic agent to restore mitochondrial bioenergetics. Br J Pharmacol 171:2029–2050

Thrainsdottir I, Malmberg K, Olsson A, Gutniak M, Ryden L (2004) Initial experience with GLP-1 treatment on metabolic control and myocardial function in patients with type 2 diabetes mellitus and heart failure. Diab Vasc Dis Res 1:40–43

Tian R, Nascimben L, Ingwall JS, Lorell BH (1997) Failure to maintain a low ADP concentration impairs diastolic function in hypertrophied rat hearts. Circulation 96:1313–1319

Trifunovic A, Wredenberg A, Falkenberg M, Spelbrink JN, Rovio AT, Bruder CE, Bohlooly YM, Gidlof S, Oldfors A, Wibom R, Tornell J, Jacobs HT, Larsson NG (2004) Premature ageing in mice expressing defective mitochondrial DNA polymerase. Nature 429:417–423

Turcani M, Rupp H (1997) Etomoxir improves left ventricular performance of pressure-overloaded rat heart. Circulation 96:3681–3686

Turrens JF (2003) Mitochondrial formation of reactive oxygen species. J Physiol 552:335–344

Tuunanen H, Knuuti J (2011) Metabolic remodelling in human heart failure. Cardiovasc Res 90:251–257

Tuunanen H, Engblom E, Naum A, Nagren K, Hesse B, Airaksinen KE, Nuutila P, Iozzo P, Ukkonen H, Opie LH, Knuuti J (2006) Free fatty acid depletion acutely decreases cardiac work and efficiency in cardiomyopathic heart failure. Circulation 114:2130–2137

Tuunanen H, Engblom E, Naum A, Nagren K, Scheinin M, Hesse B, Juhani Airaksinen KE, Nuutila P, Iozzo P, Ukkonen H, Opie LH, Knuuti J (2008) Trimetazidine, a metabolic modulator, has cardiac and extracardiac benefits in idiopathic dilated cardiomyopathy. Circulation 118:1250–1258

Ussher JR, Wang W, Gandhi M, Keung W, Samokhvalov V, Oka T, Wagg CS, Jaswal JS, Harris RA, Clanachan AS, Dyck JR, Lopaschuk GD (2012) Stimulation of glucose oxidation protects against acute myocardial infarction and reperfusion injury. Cardiovasc Res 94:359–369

Vadvalkar SS, Baily CN, Matsuzaki S, West M, Tesiram YA, Humphries KM (2013) Metabolic inflexibility and protein lysine acetylation in heart mitochondria of a chronic model of type 1 diabetes. Biochem J 449:253–261

van de Weijer T, Schrauwen-Hinderling VB, Schrauwen P (2011) Lipotoxicity in type 2 diabetic cardiomyopathy. Cardiovasc Res 92:10–18

van der Horst IC, Zijlstra F, Van 't Hof AW, Doggen CJ, de Boer MJ, Suryapranata H, Hoorntje JC, Dambrink JH, Gans RO, Bilo HJ, Zwolle Infarct Study G (2003) Glucose-insulin-potassium infusion in patients treated with primary angioplasty for acute myocardial infarction: the glucose-insulin-potassium study: a randomized trial. J Am Coll Cardiol 42:784–791

van Empel VP, Bertrand AT, van Oort RJ, van der Nagel R, Engelen M, van Rijen HV, Doevendans PA, Crijns HJ, Ackerman SL, Sluiter W, De Windt LJ (2006) EUK-8, a superoxide dismutase and catalase mimetic, reduces cardiac oxidative stress and ameliorates pressure overload-induced heart failure in the harlequin mouse mutant. J Am Coll Cardiol 48:824–832

Velloso LA, Folli F, Sun XJ, White MF, Saad MJ, Kahn CR (1996) Cross-talk between the insulin and angiotensin signaling systems. Proc Natl Acad Sci U S A 93:12490–12495

Wallhaus TR, Taylor M, DeGrado TR, Russell DC, Stanko P, Nickles RJ, Stone CK (2001) Myocardial free fatty acid and glucose use after carvedilol treatment in patients with congestive heart failure. Circulation 103:2441–2446

Wang XF, Zhang JY, Li L, Zhao XY, Tao HL, Zhang L (2011) Metformin improves cardiac function in rats via activation of AMP-activated protein kinase. Clin Exptl Pharmacol Physiol 38:94–101

Wargovich TJ, MacDonald RG, Hill JA, Feldman RL, Stacpoole PW, Pepine CJ (1988) Myocardial metabolic and hemodynamic effects of dichloroacetate in coronary artery disease. Am J Physiol 61:65–70

Weis BC, Cowan AT, Brown N, Foster DW, McGarry JD (1994) Use of a selective inhibitor of liver carnitine palmitoyltransferase I (CPT I) allows quantification of its contribution to total CPT I activity in rat heart. Evidence that the dominant cardiac CPT I isoform is identical to the skeletal muscle enzyme. J Biol Chem 269:26443–26448

Wells L, Vosseller K, Hart GW (2001) Glycosylation of nucleocytoplasmic proteins: signal transduction and O-GlcNAc. Science 291:2376–2378

Wentz AE, d'Avignon DA, Weber ML, Cotter DG, Doherty JM, Kerns R, Nagarajan R, Reddy N, Sambandam N, Crawford PA (2010) Adaptation of myocardial substrate metabolism to a ketogenic nutrient environment. J Biol Chem 285:24447–24456

White WB, Cannon CP, Heller SR, Nissen SE, Bergenstal RM, Bakris GL, Perez AT, Fleck PR, Mehta CR, Kupfer S, Wilson C, Cushman WC, Zannad F, Investigators E (2013) Alogliptin after acute coronary syndrome in patients with type 2 diabetes. New Engl J Med 369:1327–1335

Winter JL, Castro PF, Quintana JC, Altamirano R, Enriquez A, Verdejo HE, Jalil JE, Mellado R, Concepcion R, Sepulveda P, Rossel V, Sepulveda L, Chiong M, Garcia L, Lavandero S (2014) Effects of trimetazidine in nonischemic heart failure: a randomized study. J Cardiac Fail 20:149–154

Wong AK, AlZadjali MA, Choy AM, Lang CC (2008) Insulin resistance: a potential new target for therapy in patients with heart failure. Cardiovasc Ther 26:203–213

Wright JJ, Kim J, Buchanan J, Boudina S, Sena S, Bakirtzi K, Ilkun O, Theobald HA, Cooksey RC, Kandror KV, Abel ED (2009) Mechanisms for increased myocardial fatty acid utilization following short-term high-fat feeding. Cardiovasc Res 82:351–360

Wu P, Peters JM, Harris RA (2001) Adaptive increase in pyruvate dehydrogenase kinase 4 during starvation is mediated by peroxisome proliferator-activated receptor alpha. Biochem Biophys Res Comm 287:391–396

Yagyu H, Chen G, Yokoyama M, Hirata K, Augustus A, Kako Y, Seo T, Hu Y, Lutz EP, Merkel M, Bensadoun A, Homma S, Goldberg IJ (2003) Lipoprotein lipase (LpL) on the surface of cardiomyocytes increases lipid uptake and produces a cardiomyopathy. J Clin Invest 111:419–426

Yamauchi T, Kamon J, Minokoshi Y, Ito Y, Waki H, Uchida S, Yamashita S, Noda M, Kita S, Ueki K, Eto K, Akanuma Y, Froguel P, Foufelle F, Ferre P, Carling D, Kimura S, Nagai R, Kahn BB, Kadowaki T (2002) Adiponectin stimulates glucose utilization and fatty-acid oxidation by activating AMP-activated protein kinase. Nat Med 8:1288–1295

Young ME, Guthrie PH, Razeghi P, Leighton B, Abbasi S, Patil S, Youker KA, Taegtmeyer H (2002) Impaired long-chain fatty acid oxidation and contractile dysfunction in the obese Zucker rat heart. Diabetes 51:2587–2595

Yue TL, Bao W, Gu JL, Cui J, Tao L, Ma XL, Ohlstein EH, Jucker BM (2005) Rosiglitazone treatment in Zucker diabetic fatty rats is associated with ameliorated cardiac insulin resistance and protection from ischemia/reperfusion-induced myocardial injury. Diabetes 54:554–562

Zhang HX, Zang YM, Huo JH, Liang SJ, Zhang HF, Wang YM, Fan Q, Guo WY, Wang HC, Gao F (2006) Physiologically tolerable insulin reduces myocardial injury and improves cardiac functional recovery in myocardial ischemic/reperfused dogs. J Cardiovasc Pharmacol 48:306–313

Zhang L, Lu Y, Jiang H, Zhang L, Sun A, Zou Y, Ge J (2012) Additional use of trimetazidine in patients with chronic heart failure: a meta-analysis. J Am Coll Cardiol 59:913–922

Zhang L, Jaswal JS, Ussher JR, Sankaralingam S, Wagg C, Zaugg M, Lopaschuk GD (2013) Cardiac insulin-resistance and decreased mitochondrial energy production precede the development of systolic heart failure after pressure-overload hypertrophy. Circ Heart Fail 6:1039–1048

Zhao T, Parikh P, Bhashyam S, Bolukoglu H, Poornima I, Shen YT, Shannon RP (2006) Direct effects of glucagon-like peptide-1 on myocardial contractility and glucose uptake in normal and postischemic isolated rat hearts. J Pharmacol Exptl Therap 317:1106–1113

Zhou X, Chen J (2014) Is treatment with trimetazidine beneficial in patients with chronic heart failure? PLoS One 9:e94660

Zhu P, Lu L, Xu Y, Schwartz GG (2000) Troglitazone improves recovery of left ventricular function after regional ischemia in pigs. Circulation 101:1165–1171

Endogenous Mechanisms for Regulating Myocardial Contractility

5

Rolf Schreckenberg

Abstract

As the central organ in the circulatory system, the heart is fundamentally involved in the maintenance and stability of haemodynamics in the vascular system. The heart has to carry out this task 24 h a day without interruption. Over this period, it beats about 115,000 times, transporting about 8000 l of blood through both heart chambers. To be able to achieve this throughput, specific demands are placed on the heart: the contraction and relaxation of the myocardium is supposed to be rapid and autonomous while at the same time the heart rate, the ejection fraction and the pressure produced must always be adapted to the requirements at the periphery. Cardiomyocytes, which make up the majority of the working muscle (working myocardium) as well as the impulse generation and conduction system, are ultimately responsible for carrying out these tasks. The following chapter describes the range of endogenous mechanisms that can contribute to the regulation of the myocardial contractility and illustrates the various feedback loops and signalling pathways that can be induced via corresponding modulation of the cardiac function. A detailed description of the underlying cellular and molecular mechanisms will be preceded by a general overview of the composition and structure of the cardiomyocyte and a short summary of the specific characteristics and functions of their contractile apparatus.

R. Schreckenberg
Institute of Physiology, Justus-Liebig-Universität Gießen,
Aulweg 129, Giessen 35392, Germany
e-mail: Rolf.Schreckenberg@physiologie.med.uni-giessen.de

K.-D. Schlüter (ed.), *Cardiomyocytes – Active Players in Cardiac Disease*,
DOI 10.1007/978-3-319-31251-4_5

5.1 Composition and Structure of the Cardiomyocyte

With a length of 80–120 μm and a diameter of 15–25 μm, the cardiomyocytes of the right and left ventricles are considerably larger than those of the atria (length, 20–30 μm; diameter, 5–6 mm). Histologically, striated cardiac muscle tissue shares features with skeletal and smooth muscle. The serial arrangement of the sarcomeres, the organisation of the actin and myosin molecules and the parallel alignment of the Z-lines correspond largely to that of the skeletal muscle. On the other hand, the sometimes disordered and complex branching of the cardiomyocytes has significant parallels with smooth muscle.

Intercalated discs or dark bands that are characteristic of the myocardium form between the individual cardiomyocytes. These discs are made up of a disc- or band-shaped contact zone, the fascia adherens, which contains the desmosomes and forms stable contacts between the cells by anchoring the actin filament ends. The many gap junctions within the intercalated discs also ensure that the cardiomyocytes are electrically coupled, thus uniting the myocardium to form an electrical syncytium.

5.2 Specific Properties of the Contractile Apparatus

The structure of the contractile apparatus of cardiomyocytes largely corresponds to that of the cells of skeletal muscle. In both muscle types, actin, myosin, tropomyosin and troponin as well as titin make up the primary components of the sarcomere, although the cardiac muscle has specific cardiac isoforms that have different properties to the skeletal muscle isoforms. The reduced extensibility of cardiac sarcomeres compared to skeletal sarcomeres can be attributed to the lower elasticity of the titin present (Trombitás et al. 1995).

The structural homology of the three troponin subunits, troponin I (TnI), troponin T (TnT) and troponin C (TnC), is about 40–58 % between the cardiac isoforms and the skeletal isoforms. TnC isoforms of the skeletal and cardiac muscle have a total of four calcium-binding sites including two N-terminal sites which have considerably lower affinity for Ca^{2+} and are responsible for initiating contraction. A specific amino acid substitution inactivates the first N-terminal binding site in the cardiac isoform, however, and it is no longer available to trigger contractions as a result (Sweeney et al. 1990).

TnI has an additional structural difference that plays a functional role during relaxation. The cardiac isoform has an additional sequence on the N terminus made up of 32–33 amino acids which includes a phosphorylation site that downregulates the Ca^{2+} affinity of the cardiac TnC as a consequence of PKA activation, thus enabling the isoform to accelerate relaxation of the cardiomyocytes (Zhang et al. 1995).

5.3 Cellular Mechanisms of Cardiac Muscle Contraction

The contraction of the cardiac muscle starts with an electrical impulse which originates from the sinoatrial node and in turn generates an action potential on the membrane of the cardiomyocytes. The action potential first starts with the opening

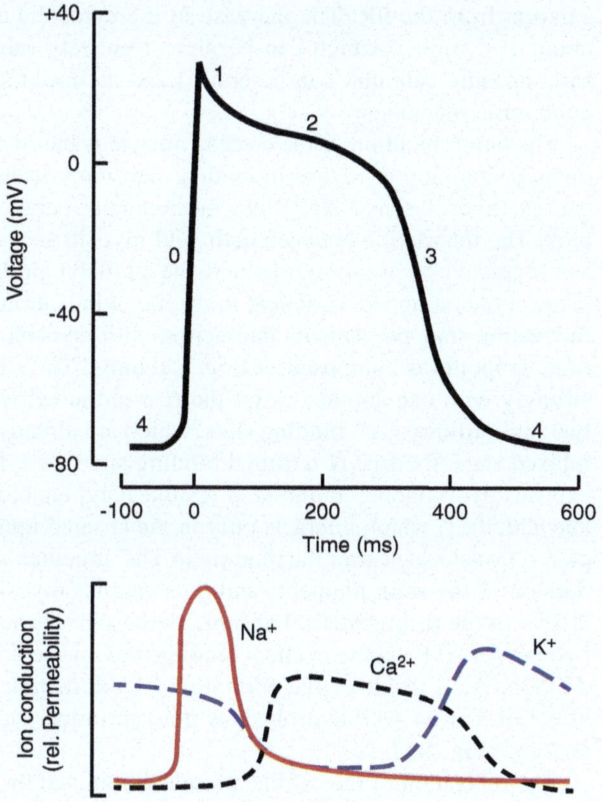

Fig. 5.1 Ventricular myocyte action potential. *Phase 0*: depolarisation, rapid influx of Na^+; *phase 1*: initial repolarisation, activation of transient outward I_{Kto} current; *phase 2*: plateau phase, inward Ca^{2+} movement through L-type voltage-gated Ca^{2+} channels; *phase 3*: repolarisation, L-type Ca^{2+} channels close, slow delayed rectifier K^+ channels open completely; *phase 4*: resting membrane potential, Na^+ and Ca^{2+} channels are closed, constant outward leak of K^+ through inward rectifier

of the fast sodium channels. Opening of sodium channels enables sodium ions to flow into the cell according to their chemical and electrostatic gradients. After a small delay, the potassium channels (K_{Ito}) open and thus already partially initiate repolarisation via the efflux of K^+ ions. These ion flows are superimposed by a prolonged, depolarising calcium influx through voltage-gated L-type calcium channels that protects the membrane against a progressing repolarisation and is responsible for the characteristic plateau phase (lasting up of 400 ms) of the action potential in the working myocardium. The repolarisation is accomplished by a massive efflux of potassium through iKr, iKs and iK1 channels. In this phase, the outward chemical driving forces for K^+ ions dominate over the inward electrostatic forces (Fig. 5.1).

The depolarisation is conducted down into the interior of the individual cell via the T-tubule system, which is made up of numerous, wide invaginations of the cell membrane that cross at the Z-lines. The membrane invaginations are located close to specialised calcium cisternae, which are part of the longitudinal system or the sarcoplasmic reticulum (SR). In the cardiomyocyte, a single cisterna contacts one T-tubule to form a diad. Around the diad there are numerous voltage-gated L-type calcium channels, which are responsible for the influx of the trigger calcium. This calcium influx activates ryanodine receptors (RyR2) in the membrane of the terminal cisternae, causing them to release large amounts of calcium from the SR. The increase in the cytosolic calcium from 10^{-7} mol/L to about 10^{-5} mol/L, which can be almost entirely (about 90 %) attributed to the sarcoplasmic calcium, can be considered as initial step in the activation of the contractile machinery.

The pump function of the cardiac muscle is based on a synchronised and coordinated contraction of the individual cardiomyocytes, which convert chemical energy (hydrolysis of ATP) into mechanical energy via the actin–myosin complex. The interaction between actin and myosin as well as the hydrolysis of ATP are regulated by tropomyosin and the globular protein troponin. Tropomyosin (Tm), a two-stranded, α-helical molecule, winds around the actin filament and in the resting state prevents its interaction with myosin and thus a possible contraction. Troponin is composed of three subunits, TnC, TnI and TnT, which interact strongly with one another. After the rise in the cytosolic calcium, as well as the two high-affinity Ca^{2+}-binding sites, which are already occupied by calcium in the relaxed state, the free N-terminal binding site in the TnC is also occupied, which exposes hydrophobic amino acid residues and enables TnC to interact with TnT and TnI. TnT, which interacts only in the striated musculature with Tm in a ratio of 1:1, translocates after the binding to TnC together with Tm away from the outer domain of the actin filaments and thus enables myosin to bind to actin (Lu et al. 2013). In the resting state, TnI inhibits the ATPase activity of actomyosin, which hydrolyses ATP via the cyclical binding and dissociation of actin with the myosin ATPase. As a result of the formation of the complex with TnC, the inhibitory effect of TnI on ATP hydrolysis is now curtailed and the cross-bridge cycle can proceed (Fig. 5.2).

The relaxation of the cardiac muscle is initiated by closure of the voltage-gated calcium channels, which prevents further release of sarcoplasmic calcium. The cytosolic calcium concentration must then be lowered, a task that is accomplished essentially by two mechanisms: calcium is either pumped back into the SR by primary active transport using the Ca–ATPase SERCA2a (sarcoplasmic–endoplasmic reticulum calcium ATPase, type 2a indicates the cardiac isoform) or transported from the cytosol into the extracellular space via a Na^{+}–Ca^{2+} exchanger (NCX). As calcium is taken up into the intracellular storage again, protons and other positively charged ions must leave the SR to maintain an electrically neutral state, because the formation of a positive SR potential would greatly impair the further uptake of Ca^{2+} ions. A large part of the free calcium within the SR is bound by the proteins calsequestrin and calreticulin. The activity of SERCA2a is determined by the regulatory protein phospholamban (PLB), which inhibits the activity of SERCA2a when bound to it in the non-phosphorylated state.

The passive Na^{+}–Ca^{2+} exchanger transports three Na^{+} ions from the extracellular space into the cytosol while moving a single Ca^{2+} ion out of the cell in the opposite direction. This transport process is thus highly dependent on the sodium gradient that must be continuously restored and maintained by the Na^{+}–K^{+}-ATPase. This active, ATP-consuming system transports sodium from inside the cell to the outside and potassium from outside the cell to the inside in a ratio of 3:2 (Fig. 5.3).

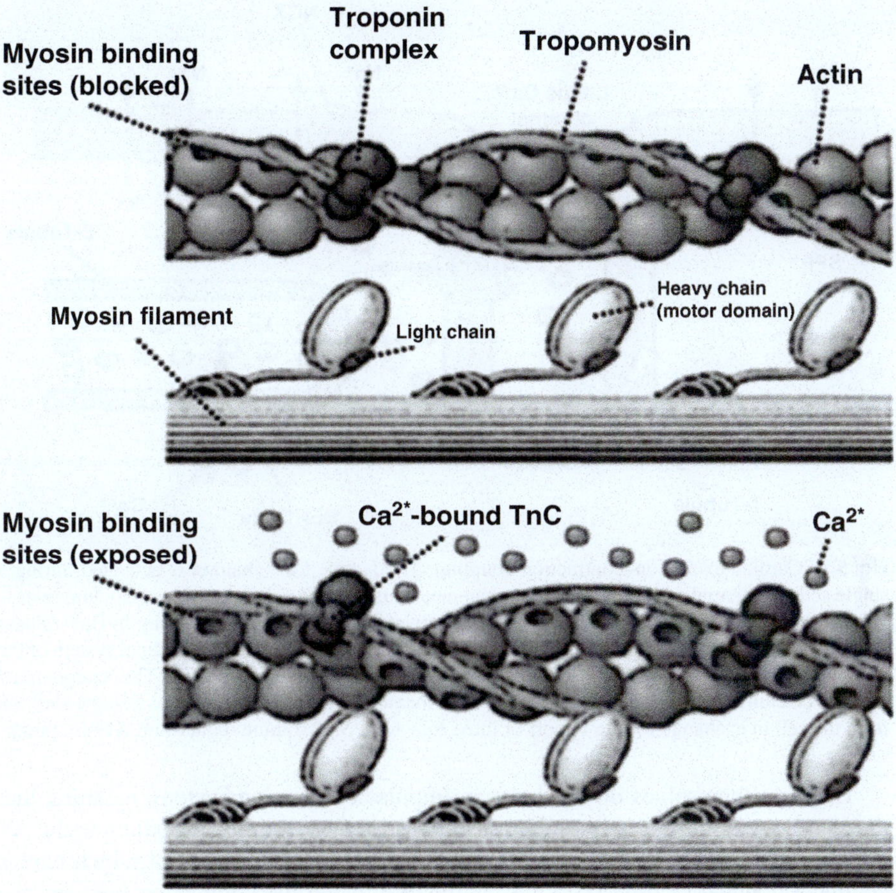

Fig. 5.2 Actin–myosin cross-bridge formation in cardiac myocytes. The highly ordered cardiac sarcomere consists of actin, myosin and a set of regulatory proteins. The interaction of actin and myosin is prevented in the absence of Ca^{2+}. Following an increase in free cytosolic Ca^{2+}, cardiac TnC and its associated protein tropomyosin undergo a conformational change and uncover myosin-binding sites on actin

5.4 The Principles of Cardiac Mechanics: The Operation of the Myocardium as a Functional Syncytium

All the cells in the myocardium are able to autonomously contract independently of nerve impulses. The electrical coupling of the cardiomyocytes and the specific impulse generation and conduction system of the heart are therefore the most important requirements for a coordinated contraction of the entire organ. Between the superior vena cava and the right cardiac auricle lie the sinoatrial node, the cells of which generate the most rapid internal rhythm and thus, as a pacemaker for all downstream structures, regulate the contraction of the entire organ.

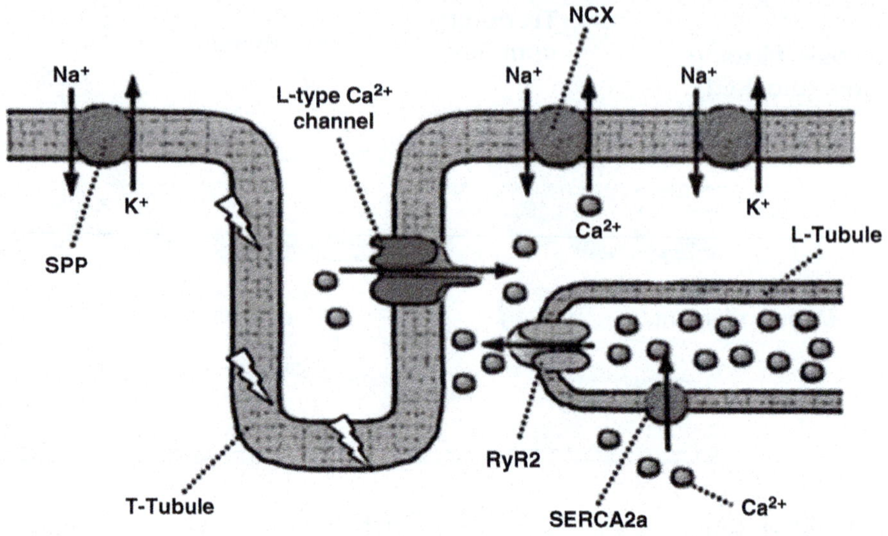

Fig. 5.3 Cardiac excitation–contraction coupling. The L-type Ca^{2+} channel is activated during a single action potential and represents the initial mechanism of the calcium-induced calcium release of cardiac myocytes. The Ca^{2+} entering the cell activates RyR2 channels resulting in Ca^{2+} release from the SR. The cardiac Ca^{2+}-ATPase SERCA2a is mainly located at the longitudinal system of the SR and transports Ca^{2+} from the cytoplasm into intracellular calcium stores. The sarcolemmal sodium–calcium exchanger (*NCX*) is an antiporter membrane protein that removes a single Ca^{2+} ion from the cell in exchange for the import of three *Na*+ ions. *SPP* sodium–potassium ATPase pump

The atria themselves do not have an impulse conduction system, meaning that the electrical impulses are directly transmitted via the atrium myocardium to the AV node (node of Aschoff and Tawara). The AV node, the thin fibres of which have a conduction speed of only 0.05 m/s, is the only electrical connection between the atrial and ventricular myocardium, which otherwise are electrically isolated from one another by a ring-shaped, fibrous band of connective tissue, the anulus fibrosus cordis. The delay in the conduction of the electrical impulse in the AV node ensures that the contraction of the atria and ventricles is temporally offset, which is an essential requirement for optimal filling of the two chambers. On the AV node, the bundle of His (0.8–2.0 m/s) and the two Tawara branches (left bundle branch and right bundle branch, 2.0–4.0 m/s) follow, which initially do not have any direct contact with the chamber myocardium because they are enclosed by a connective tissue sheath. The Tawara branches and their offshoots end at the Purkinje fibres which then extend uninterrupted into the ventricular working myocardium (Fig. 5.4).

Histologically, the specific myocytes of the impulse generation and conduction system differ from those of the working myocardium. The specialised cells of the impulse conduction system have few fibrils, are rich in sarcoplasm and need fewer mitochondria due to their lower metabolic output. Their T-tubule system is nowhere near as pronounced as that of the working muscle.

Unlike the skeletal muscle, in the myocardium, each electrical impulse affects every muscle fibre, that is, each heartbeat represents a ‘contraction maximum’ in a

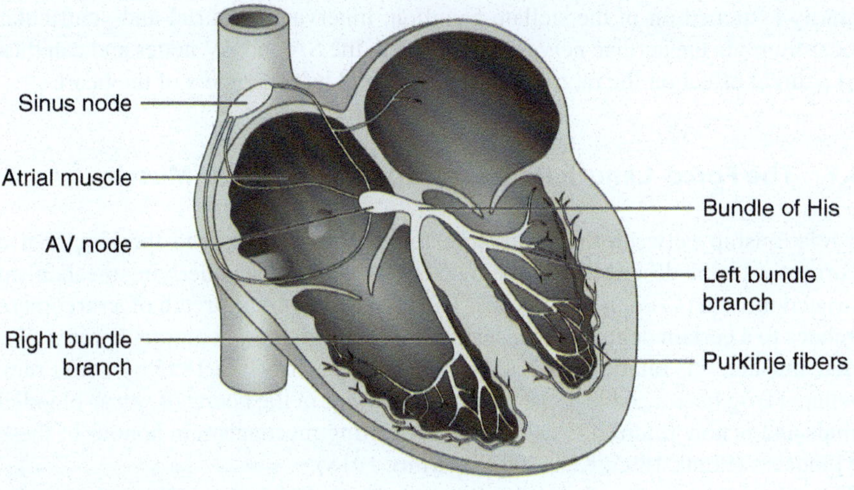

Fig. 5.4 The cardiac conduction system

manner of speaking. An increase in the contractile force therefore cannot be achieved by recruiting additional motor units. This property of the myocardium is ultimately the result of the extensive electrical coupling of all cardiomyocytes through the gap junctions within the intercalated discs. Another option to increase the contractile force in skeletal muscle involves being able to form superpositions, incomplete or complete tetanus. However, these types of contraction must not occur in the myocardium so that a rhythmical heart rate can be maintained. The contractions of the cardiac muscle are therefore made up solely of 'single twitches' that only last a little longer than the rather prolonged action potential of the myocardium (up to 400 ms, see above) and which thus ensure an adequately long refractory period for the cardiomyocytes.

The contractile force in the heart can therefore only be regulated by influencing downstream processes in the cells. These processes can be induced both by an appropriate prestretching of the ventricle (dependent on the particular end-diastolic filling, Frank–Starling mechanism, force–length relationship) and by the positive force–frequency relationship of the myocardium (Bowditch effect) as well as by neurohumoral activation.

Neurohumoral factors, which are partly controlled by the autonomous nervous system, usually reach the cardiomyocytes via the paracrine (synaptic transmission) or endocrine pathway and can affect the inotropy of the myocardium using second messengers in a receptor-mediated manner and appropriately adapt to external circumstances and requirements.

Despite a guaranteed autonomous cardiac activity, the heart has a rich autonomic innervation. The parasympathetic nervous system primarily affects the impulse generation and conduction system (sinoatrial (SA) and atrioventricular (AV) nodes) as well as the atrial musculature via the vagus nerve. Sympathetic fibres, which are primarily switched on in the stellate ganglion, innervate the atrial and ventricular myocardium via the cardiac nerves in addition to the SA and AV nodes and can thus exert a direct effect on the inotropy of the right and left chambers of the heart.

5.4.1 The Force–Length Relationship of the Cardiac Muscle

The relationship between the contractile force that develops and the length of a sarcomere was first described in 1966 by Gordon et al. as the underlying mechanism of striated muscle (Gordon et al. 1966). The increasing initial stretch of a sarcomere correlates to a certain degree with a steady rise in the resulting contractile force. For cardiac muscle, this relationship had already been described in 1895 by Otto Frank in isolated frog hearts and in 1918 by Ernest Starling in the hearts of warm-blooded animals and is now referred to as the Frank–Starling mechanism in honour of these two pioneers (Frank 1895; Katz 2002; Starling 1918).

This significant autonomic feedback loop creates a link between the end-diastolic ventricle filling and the resultant ejection fraction. Through the increase in the venous return, referred to as the preload, the end-diastolic volume consequently rises and also the end-diastolic pressure (LVEDP), which thus produces a greater initial stretching of the cardiomyocytes or sarcomeres. The resultant increase in the contractile force enables the cardiac muscle to achieve a correspondingly higher stroke volume (SV) without changing the heart rate.

The SV is calculated from the difference between the end-diastolic and the end-systolic volumes. With a constant afterload (= peripheral resistance), the increased stroke volume thus results from the additional end-diastolic blood volume that is available, with the end-systolic volume remaining unchanged (Fig. 5.5).

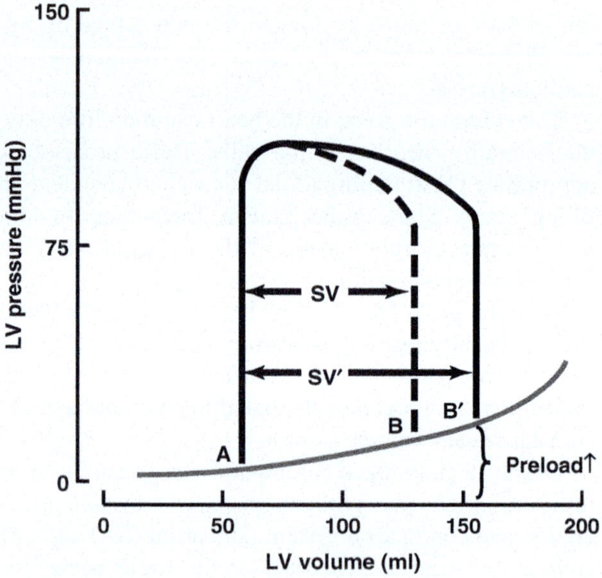

Fig. 5.5 Left ventricular pressure–volume (*PV*) relationship. The PV diagram illustrates how changes in preload affect the stroke volume (*SV*), the end-diastolic (*B*) and end-systolic volume (*A*)

The resultant Frank–Starling curve, which plots the SV relative to the LVEDP, can be affected by the peripheral resistance as well as by inotropic stimulation. A reduction in the afterload and/or positive inotropic influences shift the curve upwards and the SV rises. By increasing the afterload or negative inotropic stimulation, there is a corresponding fall in the SV (Fig. 5.6).

The heart can also respond indirectly to changes in the afterload via the Frank–Starling mechanism. By increasing the peripheral resistance, the heart is not at first able to maintain the ejection fraction and the end-systolic volume rises. In turn, the constant venous return results in an increase in the end-diastolic volume that can be ejected by means of the associated rise in the contractile force against the increased resistance.

The Frank–Starling mechanism enables the heart to autonomously accommodate short-term fluctuations in pressure and volume (change in the preload and/or the afterload) and to precisely adjust the pump power of the right and left ventricle relative to each other and in this way prevent volume shifts and subsequent pressure changes between the pulmonary and systemic circulation.

For a long time, the Frank–Starling mechanism was explained solely by the improved physical properties of the actin–myosin interaction because only with extension of the sarcomere from 1.9 μm to about 2.2–2.6 μm do actin and myosin filaments optimally overlap one another. The results of various experimental studies have, however, shown that other mechanisms must play a role in the force–length relationship of the cardiac muscle – as well as an increase in the calcium affinity of the troponin complex, a more efficient release of calcium from the SR has also been postulated (Moss and Fitzsimons 2002). Because new information has revealed that the Frank–Starling mechanism is not only based on an optimal actin–myosin overlap but instead seems to influence calcium homeostasis at the same time, it induces

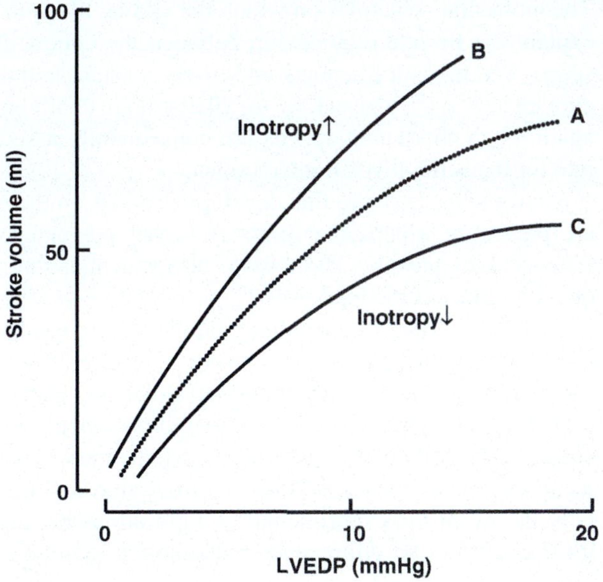

Fig. 5.6 Frank–Starling law of the heart. The Frank–Starling curve (*A*) is shifted upwards and to the left by increasing the inotropy (*B*). Decreasing the inotropic state shifts the curve down and to the right (*C*)

a positive inotropic effect that is based on similar mechanisms to those activated by inotropic influences that are independent on the prestretch.

In the initial stages of chronic heart failure, the cardiac minute volume is maintained by various adaptive mechanisms which can contribute in the short term to compensation but over the long term end up in a vicious cycle and can even contribute to progression and decompensation of the heart failure. The initial compensation mechanisms include an increase in the heart rate and permanent activation of the Frank–Starling mechanism.

A reduction in the SV inevitably leads to an elevated residual volume at the end of the diastole. By exploiting the Frank–Starling mechanism, the failing heart can initially ensure a largely normal SV again. However, the restricted pump function means that an elevated end-systolic volume remains in the ventricle, which with a constant preload is then responsible for the rise in the end-diastolic volume.

In the already failing myocardium, this situation leads directly to an increase in the oxygen requirement with a simultaneously reduced capillary perfusion because the subendocardial vessels in particular are compressed during the diastole by the high filling pressures. The initially functional dilation of the cardiac muscle can also shift to a 'structural dilation' that is characterised by structural modification of the myocardium with reduced elasticity and contractility.

The picture in terms of the efficiency of the Frank–Starling mechanism in advanced stages of heart failure is currently considered particularly controversial. The conclusions of various studies and experimental models range from a complete loss of the Frank–Starling mechanism and its underlying molecular mechanisms to complete functional preservation in NYHA stage IV (Holubarsch et al. 1996; Weil et al. 1998).

At this point, the relationship between the contractile force and the shortening speed of the myocardium, the force–velocity relationship, needs to be explained. The molecular principles on which the sliding filament theory is based can initially explain the inverse relationship between the contractile force and the shortening speed. The myosin filaments with their constant length can only slide against one another if cross-bridges along the filament are continuously produced and detached again. High shortening speeds are consequently associated with a high exchange rate for the actin–myosin interactions.

Conversely, the force that develops depends on the number of cross-bridges that are present at a particular time. A higher percentage of simultaneously formed cross-bridges prevents shortening, however, meaning that the speed must consequently reduce if the load rises.

In 1962, Edmund Sonnenblick published the results for the force–velocity relationship that he had obtained using a model of isolated cat papillary muscles (Sonnenblick 1962a). If the afterload exceeds a value at which the muscle can no longer shorten, it has reached its maximum isometric force and the shortening speed equals zero. Without the given afterload, the muscle would consequently shorten at its maximum speed (V_{max}). However, this value could not be determined experimentally in any of the experimental configurations because the weight of the muscle itself could not be eliminated. Sonnenblick therefore extrapolated the values for

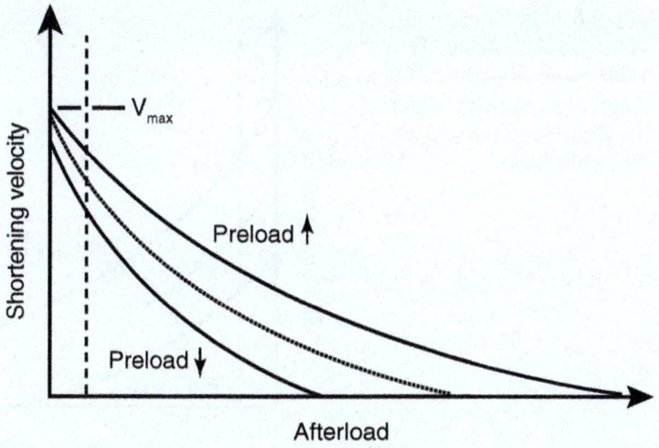

Fig. 5.7 The force–velocity relationship. The *solid curves* illustrate how changes in preload affect the shortening velocity at a given afterload

V_{max} that would theoretically correspond to a load of zero. The resulting curve for the force–velocity relationship for the individual muscle fibres can be influenced in the myocardium in a characteristic manner by the preload and inotropic stimuli (e.g. the sympathetic nervous system).

It has already been explained for the Frank–Starling mechanism that increasing the preload can improve the contractile force. Parallel to this, the shortening speed of the cardiomyocytes also increases, thus ensuring a more rapid pressure increase (dp/dt_{max}) during the isovolumetric contraction phase. During the ejection phase, the ventricle is thus able to eject more blood per unit of time – the SV increases.

Correspondingly, with a reduced preload, not only is the force lowered but also the speed with which the myocardium contracts. However, with the preload as the only variable, the theoretically achieved V_{max} of the myocardium cannot be influenced, or only to a slight degree, without an existing afterload (Sonnenblick 1962b) (Fig. 5.7).

Positive inotropic stimulation of the myocardium with a constant preload enables not only the maximum force or the speed of the pressure increase dp/dt_{max} to be increased but also the V_{max}. Unlike skeletal muscle, in which the V_{max} is a constant value that is primarily dependent on the available myosin isoform, the maximum shortening speed in cardiac muscle can be influenced by inotropic stimuli (Fig. 5.8).

Because the V_{max} in situ can be easily determined and its magnitude depends neither on the preload nor on the initial muscle length, it is often included as a criterion for assessing the myocardial contractility. Nevertheless, some authors doubt that the preload is unable to exert any effect on V_{max} of the myocytes. Furthermore, the extrapolation of the empirically obtained curve to V_{max} varies greatly, meaning that the use of V_{max} as a measure of contractility is not uncontroversial (Gülch 1974; Wolk et al. 1971).

5.4.2 The Force–Frequency Relationship of the Cardiac Muscle

Under physical stress, the heart minute volume (HMV) must be appropriately adjusted to meet the increased oxygen need. The increase in the HMV can in principle be achieved using two controlled variables: the volume ejected per beat is

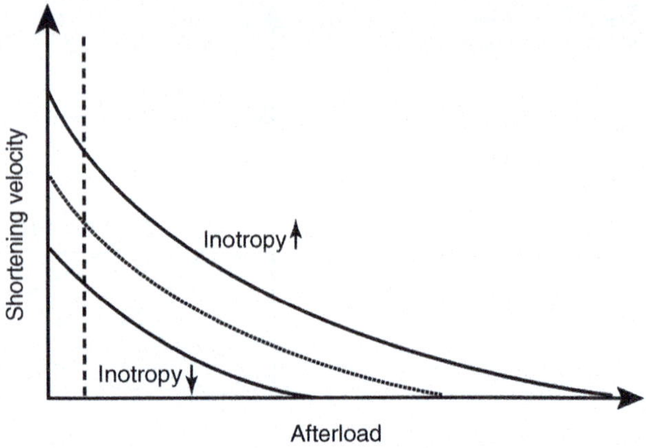

Fig. 5.8 The force–velocity relationship. The *solid curves* illustrate how changes in inotropy affect the shortening velocity at a given afterload

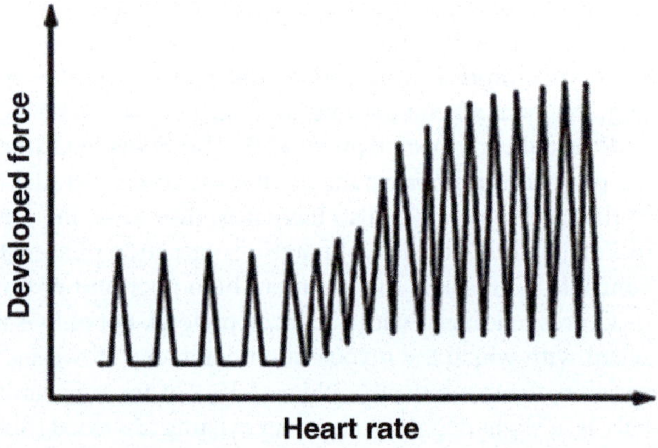

Fig. 5.9 The force–frequency relationship

increased and/or the increase in the heart rate is elevated. Independent of autonomic innervation or neurohumoral activation, the increase in the heart rate leads to improved contractility of the myocardium through autoregulation alone. This effect was described in 1871 by the US physiologist Henry Pickering Bowditch and is now referred to as the Bowditch or staircase effect (Bowditch 1871). An increase in the heart rate up to 120 beats per minute should normally be accompanied by a positive 'step', that is, a continuous increase in the contractility (Fig. 5.9).

Current explanations for the positive step assume that at higher heart rates, certain transport processes that are associated with the outward transport of calcium are overtaxed. Thus, at high heart rates, there is less time available for the NCX to transport the three Na^+ ions out of the extracellular space into the cytosol and the Ca^{2+} ion out of the cytosol into the extracellular space. The extra calcium that accumulates in the cytosol is alternatively taken up into the intracellular storage of the SR so that more calcium is released during the next depolarisation and thus the actin–myosin interaction can be improved.

An alternative theory to explain the Bowditch effect considers the capacity of the Na^+–K^+-ATPase to be the underlying reason for the positive step. Due to the relatively large reduction in the length of the diastole at higher heart rates, the Na^+–K^+-ATPase can no longer completely transport incoming sodium out of the cell during the depolarisation, leading to an increase in the submembrane Na^+ concentration. The rise in the internal sodium concentration reduces the height of the Na^+ gradient from outside to inside, however, and thus the driving force for the NCX. Consequently, more calcium is stored in the SR again (Vila Petroff et al. 2003).

The improvement in the electromechanical coupling must not come at the expense of the diastolic relaxation of the cardiac muscle, however. That is, the cytosolic calcium concentration must fall again to normal low levels during the diastole. This is assured by the high capacity of the SERCA2a that is able to balance out the functional deficit of the NCX with ease.

There are consistent study results available for the Bowditch effect that describe an increasingly negative effect of the positive force–frequency relationship and also a reversal of the positive step in the failing myocardium. The impaired calcium homeostasis in failing cardiomyocytes is considered responsible for the loss or formation of a negative force–frequency relationship. The underlying molecular and cellular mechanisms are still currently the subject of research. A reduced uptake capacity for cytosolic calcium in the SR and a reduced activity of the SERCA2a due to impaired interaction with its inhibitor, phospholamban, have been postulated as possible mechanisms. A reduction in the release of sarcoplasmic calcium and an increase in the activity of the NCX must also be considered as reasons behind the reduction in the contractility (Petretta et al. 2002).

5.4.3 Autonomic Innervation of the Heart

Cardiac function is influenced in a range of different ways by the autonomic nervous system. Via a dense network of vegetative innervation, both sympathetic and parasympathetic nerve fibres continuously adapt numerous cardiac functional parameters to the needs of the periphery and thus to current target values.

Like all vegetative systems, the superior centres of the cardiovascular system are arranged hierarchically but large parts of the precise cellular neuronal organisation have not yet been fully clarified. This is because vegetative control of an organ cannot often be attributed to a well-localised neuronal network, but instead series of nuclei and neurons that are immediately adjacent to each other interact in a complex manner before they exert an influence on an organ or organ system.

The most important control centres for the cardiovascular system are located in the bulbar or medullary cardiovascular centre, the nuclei of which are primarily located in the caudal sections of the brain stem, the medulla oblongata. The diencephalic cardiovascular centre is superior to these structures and its most important nuclei are located in the hypothalamus. The spinal centre is downstream of the medullary centre and its core areas are located in the lateral columns and the centre of the spinal cord. The medulla oblongata contains numerous structures that serve the autonomic nervous system. Four areas in particular have been identified that regulate the cardiovascular system and their connections are briefly described below.

The solitary nucleus (nucleus tractus solitarii (NTS)) is a brain nerve nucleus that contains special and general visceral afferent fibres from the three cranial nerves: the facial (VII), the glossopharyngeal (IX) and the vagus (X). Fibres from the glossopharyngeal and vagus nerves end in the medial and caudal nuclei to which they transmit information from the baroreceptors and chemoreceptors in the aortic arch and the carotid sinus. The NTS sends out excitatory glutamatergic fibres to the caudal ventrolateral medulla (CVLM) and the dorsal nucleus of the vagus nerve and activates them. The activated CVLM then sends inhibitory GABAergic fibres to the rostral ventrolateral medulla (RVLM) and inhibits it.

The RVLM is the primary nucleus that regulates the sympathetic nervous system. It sends excitatory fibres to the preganglionic neurons of the intermediolateral nucleus of the spinal cord via the reticulospinal tract. Consequently, when the blood pressure is high, baroreceptors activate the NTS which activates the CVLM, and the CVLM in turn inhibits the RVLM and thus inhibits the sympathetic nervous system. Conversely, low blood pressure leads to an increase in the sympathetic tone due to reduced inhibition of the RVLM. The CVLM is referred to as the depressor centre due to its inhibitory effects on the sympathetic nervous system, while the RVLM is correspondingly referred to as the pressor centre due to its stimulatory effect (Hsieh et al. 1998).

The second efferent trunk moves from the NTS to the dorsal nucleus of the vagus nerve, which transmits its information to the periphery via to the vagus nerve. Together with the nucleus ambiguus, which is also innervated by the NTS, this interconnection serves the parasympathetic nervous system and, when the blood pressure is high, it is activated by the excitatory fibres of the NTS or inhibited as a result of reduced activity of the NTS. The interconnections are modulated and regulated in the medulla oblongata by superior centres that are located primarily in the hypothalamus, partly in the reticular formation and in the limbic system and various parts of the cortex.

Even under resting conditions, the hypothalamus affects the tonic activity and the reflex connections of the brain stem. As the central regulatory point for the autonomic nervous system, the hypothalamus handles information coming from superior structures in complex processes and influences numerous parameters of the cardiovascular system by modulating the medullary neuron groups and the hormonal system. Parts of the posterior area of the hypothalamus have been identified which, when activated, cause an increase in the tone of the sympathetic nervous system, while stimulation of more anterior parts activates the parasympathetic nervous system.

Emotional responses to environmental and internal stimuli also affect downstream cardiovascular centres via the limbic system. Particularly in the motor and premotor fields of the neocortex, anticipatory and start responses are also initiated which adapt cardiovascular parameters under the control of the autonomic nervous system and somatic motor muscle power to one another even before physical activity starts.

The preganglionic sympathetic fibres originate in the thoracic segments Th1 to Th4. These fibres synapse with the superior cervical ganglion, the middle cervical ganglion and the cervicothoracic ganglion (= stellate ganglion), which is derived from the fusion of the inferior cervical ganglion with the first thoracic ganglion.

The superior and middle cervical cardiac nerves exit the cervical ganglia, and the thoracic cardiac nerves exit the stellate ganglion and the thoracic ganglia 2–4 and travel to the base of the heart where they form the cardiac plexus outside the pericardium. The cardiac plexus is divided into two parts, the superficial cardiac plexus (pars superficialis, also plexus cardiacus superficialis) and a larger, more posterior part, the deep cardiac plexus (pars profunda or plexus cardiacus profundus).

The preganglionic parasympathetic fibres originate predominantly in the dorsal nucleus of the vagus nerve. They reach the cardiac plexus via the superior and inferior cervical cardiac branches as well as the thoracic cardiac branches that synapse as lateral branches of the vagus nerve with minute groups of ganglia located in the subepicardial tissue of the atria. The sinoatrial node is primarily innervated by fibres of the right vagus nerve, while the AV node is predominantly under the control of the left vagus nerve. Postganglionic sympathetic and parasympathetic fibres, which can no longer be distinguished morphologically in the cardiac plexus, run from the coronary sulcus primarily along the path of the coronary vessels.

Also terminating in the cardiac plexus are visceral afferents that conduct information from the pressure receptors (aortic arch, atria and ventricles) and chemoreceptors (aortic body) towards the centre to the medulla oblongata as well as cardiac pain fibres that originate in the neuronal cell bodies of the spinal cord segments C3 to C4 and Th2 to Th7.

Sympathetic fibres initially travel primarily along the adventitia of the coronary vessels, forming numerous small, spindle-shaped enlargements (varicosities) where their transmitter noradrenaline is synthesised. The nerve fibres do not contact the individual cardiomyocytes using classical synapses, but noradrenaline is instead released from the numerous varicosities that are located close (distance of up to 500 nm) to the individual myocytes. The most important functional effects of the sympathetic nervous system in the heart are an increase in the heart rate (positive chronotropy), an increase in the contractility (positive inotropy), more rapid relaxation (positive lusitropy) and an acceleration of electrical conduction velocity (positive dromotropy).

Parasympathetic fibres also initially follow the path of the blood vessels. Again, they also do not form highly organised synaptic contacts with their target cells in the heart. In general, the parasympathetic nervous system induces functionally opposing effects to the sympathetic nervous system, even though its influence on the inotropy or lusitropy of the ventricle is nowhere near as pronounced.

5.4.3.1 The Influence of the Autonomic Nervous System on Heart Rate

In the resting state, the heart of an adult is subject to a vagotonic inhibition. A fully innervated heart beats with a frequency of 60–80 beats per minute. Denervation causes the heart rate to spontaneously increase to 100–120 beats per minute.

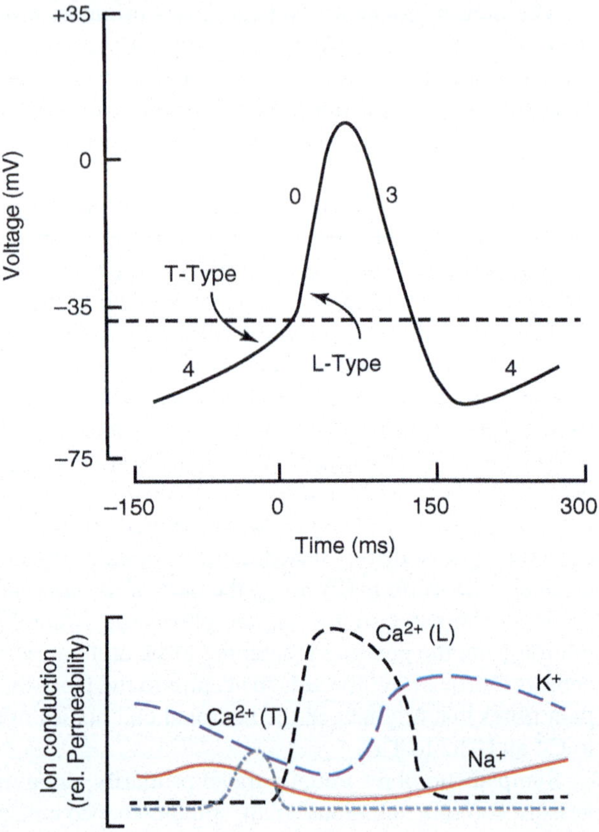

Fig. 5.10 Sinoatrial node action potential. *Phase 4*: diastolic depolarisation, inward 'funny' Na^+ current slowly depolarises cell membrane of the cell; *phase 0*: depolarisation phase of the action potential, inward *Ca2+* movement through T-type and L-type *Ca2+* channels; *phase 3*: repolarisation, K^+ channels open and cause a short-lived, hyperpolarizing outward K^+ current

The cellular network in the sinoatrial node takes on the role of the primary or physiological pacemaker in the cardiac muscle. The pacemaker cells do not have a stable resting potential and are therefore able to generate spontaneous and regular action potentials.

At the end of the repolarisation (start of phase 4), at a membrane potential of about −60 mV, HCN channels (pacemaker channel, HCN4) open which induce a slow inward flow of sodium ions (I_f). If the membrane potential exceeds −50 mV, T-type calcium channels open that contribute to further depolarisation (transition to phase 0). The calcium influx is supported by L-type calcium channels that open with a membrane potential of about −40 mV. While the HCN and T-type calcium channels in this phase are already inactivated, the L-type calcium channels maintain the slow calcium influx until a membrane potential of about +10 mV is reached. The repolarisation (phase 3) is initiated by the hyperpolarising efflux of potassium (I_{KR}, I_{KS}) that returns the membrane potential to −60 mV (Fig. 5.10).

An increase in the heart rate caused by the sympathetic nervous system presumes a simultaneous reduction in the parasympathetic tone. Activation of β_1-adrenergic receptors induced by the sympathetic nervous system via the transmitters noradrenaline or adrenaline leads to an increase in the cAMP level, resulting in activation of protein kinase A (PKA). It has been shown that HCN channels, which in the sinoatrial node are also referred to as pacemaker channels, can be modulated both by cAMP directly and by phosphorylation by PKA. PKA is also able to modulate the activity of RyR2 or SERCA2a and therefore to regulate the level and speed of the calcium cycles (Larsson 2010).

The cholinergic receptors in the heart are type M_2 receptors and they activate an inhibitory $G\alpha_i$ protein. As a result, K^+ channels of the IK_{Ach} type are opened and induce a potassium efflux that hyperpolarises the pacemaker cell even further. This makes it more difficult to trigger a pacemaker potential and the heart rate falls (Mesirca et al. 2013). As described previously in Sect. 5.4.2, the heart rate itself influences the contractility of the myocardium via the existing force–frequency relationship (Bowditch effect).

5.4.3.2 The Influence of the Autonomic Nervous System on the Conduction Velocity

The spread of the electrical stimulus via the specific impulse conduction system of the myocardium was described above. The autonomic nervous system can in principle modulate the conduction velocity in all sections of the myocardium. However, the conduction velocity in the pathway between the atria and the ventricles is primarily regulated by the AV node. Similar to the pacemaker potentials in the sinoatrial node, the potentials of the specialised cells of the AV node are generated not by fast sodium channels but by the opening of L-type calcium channels.

The sympathetic nervous system accelerates the conduction by activating the cAMP–PKA system which induces more rapid depolarisation and a steeper rise in the potential (phase 0) in the fibres of the AV node. The parasympathetic nervous system correspondingly reduces the conduction velocity by inhibiting the formation of cAMP.

5.4.3.3 The Influence of the Autonomic Nervous System on the Myocardial Inotropy and Lusitropy

At the level of the cardiomyocyte, modulating different signalling pathways can specifically induce or inhibit key mechanisms that can regulate and adapt the contractility (inotropy) and the relaxation (lusitropy) of the myocardium in response to a situation (Fig. 5.11).

The two opposing agents in the autonomic nervous system have the biggest effect on these regulatory mechanisms. The sympathetic nervous system exerts positive inotropic and lusitropic effects on the atria and on the right and left ventricle. In contrast, the inotropy can only be negatively affected in the two atria by the parasympathetic nervous system. In principle, it exerts an effect in the two ventricles but this is functionally of less importance. Although an independent negative lusitropic effect has been demonstrated for the parasympathetic nervous system in some animal experiments, its existence in humans is currently still disputed.

The inotropy and lusitropy are regulated at the cellular level primarily by modulating the intracellular calcium homeostasis. For this reason, the activity of specific

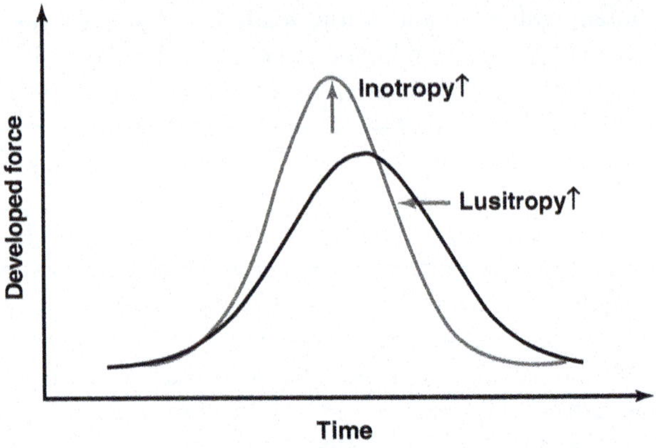

Fig. 5.11 Influence of the sympathetic nervous system on myocardial force generation (inotropy) and relaxation (lusitropy)

proteins that affect the myocardial calcium concentration during the excitation–contraction coupling and subsequent relaxation is under the control of the autonomic nervous system.

Thus, the concentration of the trigger calcium can be regulated during the plateau phase of a myocardial AP by modulating the L-type calcium channel. The subsequent release of sarcoplasmic calcium is largely under the control of RyR2, the activity of which can also be influenced by the vegetative nervous system. During relaxation, both the activity of SERCA2a (via the interaction with its specific inhibitor phospholamban) and the activity of the NCX can be actively regulated; this also applies to the calcium sensitivity of the troponin complex (Fearnley et al. 2011).

Signal transduction of the autonomic nervous system is assured in the myocardium by the stimulation of G protein–coupled receptors (GPCRs). Due to its numerous isoforms, the GPCR-type receptors are the largest and most diverse receptor superfamily. The number of GPCR isoforms that are formed by all the cell types in the cardiac muscle alone runs to about 200. As well as their significant influence on cardiac function, the GPCRs also play a critical role during differentiation and growth processes (Kang et al. 2007; see also Chaps. 2 and 6).

Any heptahelical receptor that is anchored in the cell membrane with seven helices and can bind and activate G proteins is referred to as a GPCR. Activation of the receptor results in the substitution of GDP for GTP (GTP exchange factor) at the α subunit of the G protein. The G protein complex becomes unstable due to the binding of GTP. Consequently, the conformation of the heterotrimeric G protein changes and the α and the βγ subunits are able to dissociate. The activated G proteins can now transmit the exogenous signals conveyed by the ligands to the interior of the cell. Depending on the α subunit, other proteins are either activated or deactivated. Thus, for example, the α subunits of the G_s proteins modulate the activity of adenylyl cyclase while the α subunits of the $G_{q/11}$ proteins activate phospholipase C. Shortly after, the intrinsic GTPase activity of the α subunit of the G protein cleaves the bound GTP into $GDP+P_i$. The α subunit can thus reassociate with the βγ subunit and again bind to a receptor.

5.4.3.4 Effects of Sympathetic Stimulation on Cardiomyocytes

Noradrenaline, which is released from the varicosities of postganglionic sympathetic neurons, and adrenaline, which is secreted from the adrenal medulla into the bloodstream following sympathetic stimulation, are the two most important transmitters in the sympathetic nervous system. Their signal transduction is mediated by adrenergic α and β receptors. The ratio of β_1 to β_2 receptors in the heart is about 70:30. Functionally, their effect is primarily synergistic, but the β_1 receptor has a greater effect on the myocardium, while the β_2 receptor has a greater effect on the smooth musculature of the vessels. Expression of β_3 receptors has also been demonstrated in the myocardium, but their importance and their effect on cardiac function are currently unclear.

While adrenergic α_1 receptors in the heart are predominantly expressed in cardiomyocytes, α_2 receptors are primarily located at the synapses of postganglionic neurons where they regulate the further release of transmitters using negative feedback. Activation of the $G\alpha_s$ protein by stimulation of β-adrenergic receptors leads to the activation of adenylyl cyclase, which converts ATP to cAMP and in turn (cAMP mediated) activates protein kinase A (PKA).

A positive inotropic effect by stimulating the β_1 and/or β_2 receptor can already be achieved by lengthening the AP and the resultant increase in the influx of trigger calcium (Bouchard et al. 1995). A PKA-mediated phosphorylation of the L-type calcium channel at Ser1928 of the α_{1c} subunit or at Ser478/Ser479 of the β_{2c} subunit is the underlying cause (Kamp and Hell 2000). However, the cytosolic concentration of the sarcoplasmic calcium is increased under the influence of the catecholamines adrenaline and noradrenaline not only by the improvement in the calcium-induced release but also by the increase in the activity of RyR2 due to phosphorylation at Ser2030 by PKA, leading to an independent positive inotropic effect (Shan et al. 2010; Xiao et al. 2006).

Another protein whose exact function has not yet been conclusively elucidated but which current information suggests an influence on the contractility of the myocardium is myosin-binding protein C (MyBP-C). MyBP-C in its non-phosphorylated state has a certain braking function on cross-bridge cycling that is reversed by PKA-dependent phosphorylation, and it can therefore have a positive inotropic effect (Flashman et al. 2004; Sadayappan et al. 2011).

Although β_1 and β_2 receptors both activate PKA via $G\alpha_s$ proteins, there are nevertheless some qualitative and quantitative differences in terms of their influences on the mechanisms described, the reasons for which are not yet known in detail (Xiao and Lakatta 1993).

The ability of the myocardium to relax, known as the lusitropy, is also improved by β-adrenergic stimulation. The most effective measure to accelerate diastolic relaxation is to lower the cytosolic calcium concentration as rapidly as possible. The PKA is thus able to increase the activity of both SERCA2a and the NCX.

For each ATP molecule, the calcium ATPase SERCA transports two Ca^{2+} ions out of the cytosol into the SR. An important mechanism to control SERCA2a activity is its affinity for calcium. The ability to bind calcium is regulated by the small phosphoprotein phospholamban (PLN, 52 amino acids, MW: 6 kDa). In the cell,

PLN is present as a monomer, pentamer and SERCA2a–PLN complex in a dynamic equilibrium. Only the monomer interacts with SERCA, however, stabilising the calcium ATPase in the E2 conformation. Dissociation of the PLN allows the transition to the E1 form, which empties two high-affinity Ca^{2+}-binding sites that can be sequentially occupied by calcium. Once the first Ca^{2+}-binding site is occupied, the binding of the second Ca^{2+} ion is facilitated by a cooperative effect. At the same time, the first binding site that was occupied prevents reassociation with PLN (Chen et al. 2007). The most important mechanism for PLN dissociation from SERCA2a is the change in its conformation which can be induced by phosphorylation at Ser10 (PKC), Ser16 (PKA) or Thr17 (CaMKII). In vivo studies have consistently demonstrated that the restoration of cytosolic calcium by phosphorylation of PLN at Ser16 and Thr17 and subsequent disinhibition of SERCA2a can be significantly accelerated by β-adrenergic stimulation (Ablorh et al. 2014).

The transport capacity of the NCX can also be increased by sympathetic stimulation. Indications of accelerated calcium transport have been identified in animal experiments which can probably be attributed to PKA-dependent phosphorylation of NCX. A definitive phosphorylation site within the protein has not yet been identified, however (Wei et al. 2007).

The contractility of the myocardium can also be controlled independently of modulation of the Ca^{2+} transients via the Ca^{2+} sensitivity of the myofilaments. The particular properties of the cardiac TnI (cTnI) in this regard are the focus of a number of studies. Only the cardiac isoform of the TnI can be phosphorylated by β-adrenergic stimulation and subsequent PKA activation at the N-terminal Ser23/Ser24. This leads to a significant reduction of the Ca^{2+} sensitivity of the myofilaments and particularly of the Ca^{2+} affinity of the cTnC. In various animal models, the phosphorylation of cTnI led to a significant improvement in the lusitropy. This effect was accompanied by an increase in the rate of cross-bridge cycling and an increase in the unencumbered shortening speed. In combination, these influences on the interaction between actin and myosin lead not only to an accelerated relaxation, but the inotropy is also improved by phosphorylation of the cTnI, particularly in the auxotonic contraction phase (Kögler and Rüegg 1997; Layland et al. 2005).

Along with PKA, the formation of cAMP leads to direct activation of the exchange protein directly activated by cAMP (EPAC). The precise significance of EPAC during the excitation–contraction process in cardiomyocytes is not currently known, but initial results indicate that there is a PKA-independent influence on RyR2 and SERCA2a (Pereira et al. 2007; Ruiz-Hurtado et al. 2013; Okumura et al. 2014).

Another protein that may have a direct effect on the inotropy and the lusitropy is the cardiac isoform of Ca^{2+}–calmodulin-dependent protein kinase (CaMKII). Activation of CaMKII leads to phosphorylation of the L-type calcium channel, RyR2, PLN and various cation channels. Primarily activated by calcium or calmodulin, it is postulated that there is a direct β-adrenergic influence and indirect activation of CaMKII by EPAC (Greenstein et al. 2014; Grimm and Brown 2010; Oestreich et al. 2009).

The complex interaction between β-adrenoceptor stimulation and inotropy caused by cellular effects in cardiomyocytes is summarised in Table 5.1.

Table 5.1 Potential targets of β-adrenoceptor-dependent effects in cardiomyocytes

Target protein	Trigger	Phosphorylation site
Improvement of contraction		
L-type Ca^{2+} channel (α_{1c})	PKA	Ser1928
L-type Ca^{2+} channel (β_{2a})	PKA	Ser478, Ser479
Ryanodine receptor type 2 (RyR2)	PKA	Ser2030 (Ser2809)
Ryanodine receptor type 2 (RyR2)	cAMP–EPAC	–
Myosin-binding protein C (MyBP-C)	PKA	Ser282 (Ser273, Ser302)
Improvement of relaxation		
Phospholamban (PLN)	PKA	Ser16
Phospholamban (PLN)	cAMP–EPAC	(Ser16)
Na^{+}–Ca^{2+} exchanger (NCX)	PKA	Not defined
Troponin I (TnI)	PKA	Ser23/Ser24
Improvement of calcium-dependent processes		
Phospholamban (PLN)	CaMK	Thr17
L-Type Ca^{2+} channel	CaMK	Ser1512/Ser1570
Ryanodine receptor type 2 (RyR2)	CaMK	Ser2815 (Ser2809)
Ca^{2+}–calmodulin kinase (CaMK)	cAMP–EPAC	Thr286

Phosphorylation sites provided in brackets are of uncertain functional significance

α_1-Adrenoceptors are $G\alpha_{q/11}$ protein–coupled receptors which, after stimulation by adrenaline or noradrenaline, are able to form the second messengers inositol trisphosphate (IP_3) and diacylglycerol (DAG) via activation of phospholipase C. The phospholipid IP_3 can release Ca^{2+} from intracellular depots as a result of binding to its specific receptor. IP_3 receptors are non-specific cation channels that on opening can cause a rapid release of Ca^{2+} ions from the SR and thus an increase in the cytoplasmic Ca^{2+} concentration.

The availability of DAG, which remains localised in the plasma membrane, creates an important prerequisite for activation of the protein kinase C (PKC). Currently, at least 12 isoforms of PKC have been described that differ in their amino acid sequence and their regulatory mechanisms. On the basis of these differences, they are divided into four subfamilies: classical, novel, eccentric and atypical. For the classical PKC isoforms, calcium is an important cofactor that facilitates the translocation of the enzyme at the plasma membrane and thus the activation of the enzyme with the help of DAG. Stimulation of the α_1-adrenergic receptors can also achieve positive inotropic and lusitropic effects, but these are inferior to β-adrenergic stimulation in terms of their efficiency.

Molecular biology analyses have shown that both the α_{1c} and the β_{2a} subunits of the L-type calcium channel have potential phosphorylation sites for PKC. However, the studies carried out to date do not provide a coherent picture of the α-adrenergic-mediated effects on the AP duration or the initial calcium influx. Depending on the model, stimulation of PKC leads to no change, an increase, a reduction or complex biphasic curves for the Ca^{2+} influx (Bouchard et al. 1995; Kamp and Hell 2000).

Table 5.2 Potential targets of α-adrenoceptor-dependent effects in cardiomyocytes

Target protein	Trigger	Phosphorylation site
Improvement of contraction		
L-type Ca^{2+} channel (α_{1c})	DAG–PKC	Not defined
L-type Ca^{2+} channel (β_{2a})	DAG–PKC	Thr27, Thr31
Myosin-binding protein C (MyBP-C)	PKA	(Ser273, Ser302)
IP_3-receptors	IP_3	–
Improvement of relaxation		
Phospholamban (PLN)	PKC	Ser10
Troponin I (TnI)	PKC	Ser23/Ser24, Ser43/Ser45, Thr144

Phosphorylation sites provided in brackets are of uncertain functional significance

Another mechanism to achieve positive inotropic effects via $G\alpha_{q/11}$-protein–coupled receptors is the ability to mobilise additional calcium by means of IP_3 or activation of sarcoplasmic IP_3 receptors. The IP_3-induced release of sarcoplasmic calcium increases the RyR2-mediated cytoplasmic Ca^{2+} concentration and thus induces a positive inotropic effect (Roderick and Bootman 2007).

By activating PKC, positive lusitropic effects can also be achieved. The PLN has a specific PKC phosphorylation site at Ser10, which also leads to a change in conformation and subsequent SERCA2a dissociation. There is already in vitro evidence and an indication of the relevance of a PKC-induced increase in activity of SERCA2a. Confirmation of these results in vivo is not yet available (Colyer 1998; Mattiazzi et al. 2005).

Within the cTnI, three positions have been identified so far that can be phosphorylated dependent on PKC. Phosphorylation at Ser23/Ser24, which can be carried out both by PKC and by PKA, leads to the positive lusitropic and inotropic effects previously described. Specific phosphorylation at Ser43/Ser45, on the other hand, reduces the rate of cross-bridge cycling and lowers the inotropy. Both effects result in a reduction in the energy requirement and economises the cardiac workload. PKC-dependent phosphorylation at Thr144 achieves primarily a reduction in the shortening speed of the sarcomeres. The complex interactions of PKA- and PKC-dependent phosphorylation sites and the influence of various PKC isoforms on the function of cTnI are the subject of ongoing studies (Layland et al. 2005). Potential targets of α-adrenoceptor-dependent activation are summarised in Table 5.2.

As a classic second messenger, calcium plays an important role in many intracellular signalling steps. By interacting with the calcium-sensing receptor (CaSR), calcium also acts as a first messenger, however, and regulates cell functions that have a direct effect on calcium homeostasis. An isolated ventricular myocyte model has already shown that as a $G\alpha_{q/11}$-protein–coupled receptor, CaSR exerts positive inotropic effects by producing IP_3 and positive lusitropic effects by increasing the phosphorylation of PLN (Schreckenberg et al. 2015).

5.4.3.5 Effects of Parasympathetic Stimulation on Cardiomyocytes

The parasympathetic nervous system, which mainly innervates the heart via the vagus nerve, predominantly uses the neurotransmitter acetylcholine for signal transduction to the heart. The receptors available are divided into nicotinic and muscarinic acetylcholine receptors. The nicotinic-type receptors are ligand-gated ion channels that are present mostly in skeletal musculature, autonomic ganglia and the CNS. The muscarinic receptors are part of the large superfamily of metabotropic receptors that accomplish their signal transduction with the help of the G proteins. The members M_1, M_3 and M_5 are coupled to $G\alpha_{q/11}$ proteins and can induce the formation of the second messengers inositol trisphosphate (IP_3) and diacylglycerol (DAG) via the activation of phospholipase C. Muscarinic receptors of type M_2 and M_4, on the other hand, are coupled to $G\alpha_i$ proteins which contribute to the reduction of the cAMP level by inhibiting adenylyl cyclase. The precise mechanism by which $G\alpha_i$ proteins inhibit adenylyl cyclase has not yet been completely deciphered. Direct inhibition of the adenylyl cyclase by the α_i subunit, competitive mechanism of the α_i and α_s subunits, direct inhibition of adenylyl cyclase by the $\beta\gamma$ subunits and inhibition of the α_s subunit by the $\beta\gamma$ subunits of the $G\alpha_i$ protein have been postulated as possible mechanisms. The M_2-type receptor is also involved in the opening of the K^+ channels of IK_{Ach} type, and this induces a potassium efflux that hyperpolarises the cell further.

Cardiac muscle primarily expresses the M_2 receptor which, when stimulated, induces negative chronotropic effects at the sinoatrial node and the negative dromotropic effects at the AV node (see above). Negative inotropic effects mediated by the M_2 receptor are induced primarily in the atria. In contrast, the influence of the parasympathetic nervous system on the inotropy in the ventricle is considered rather low. The expression of M_1- and M_3-type muscarinic receptors has been shown in principle in the myocardium and their functional relevance demonstrated in vitro, but there are no current data that suggest an effect on the inotropy or the lusitropy in vivo (Tsuchida et al. 2013).

Most of the negative inotropic effects are explained by the M_2-dependent inhibition of adenylyl cyclase with a subsequent reduction in the cAMP level. The regulation of the calcium-handling proteins and its consequences for calcium homeostasis also has priority here. Under basal conditions, the parasympathetic nervous system has only a negligible effect on the L-type calcium channel; however, functionally relevant effects on the calcium influx under β-adrenergic stimulation have been demonstrated, but these can only be partly attributed to a reduction in the cAMP–PKA system. Activation of certain phosphatases and NO-mediated stimulation of cGMP-dependent phosphodiesterases must be considered as alternatives.

The activity of the NCX is also inhibited primarily under β-adrenergic stimulation dependent on M_2, meaning that the reduction in the cytosolic calcium concentration is delayed and the diastolic relaxation under these conditions is curbed. It is postulated that there is a dephosphorylation of the PKA-dependent phosphorylation that occurred previously by activating various phosphatases (Wei et al. 2007).

Similar mechanisms that have negative effects on the lusitropy are assumed for the M_2-dependent influence of the cTnI and the PLN. It is highly likely that the serine–threonine phosphatase PP2A (type 2A protein phosphatase) is critically involved in the dephosphorylation of cTnI and PLN and thus inhibits the previously described sympathetic modulation of both proteins and their functional effects (Gupta et al. 1998; Wijnker et al. 2011).

The need for a balance between the sympathetic and parasympathetic influences cannot be reconciled with the very sparse parasympathetic innervation that is present, particularly in the ventricular myocardium. Cardiomyocytes themselves are likely to be the most important cell population in terms of acting as a non-neuronal source for the production and secretion of the neurotransmitter acetylcholine (Roy et al. 2013).

5.4.3.6 The Influence of Humoral Factors on Cardiac Function

As well as the classical transmitters of the autonomic nervous system, cardiac function is also modulated by a number of humoral factors that can also activate many of the previously described G protein–coupled receptors and influence the heart function depending on the subunit coupled ($G\alpha_s$, $G\alpha_{q/11}$, $G\alpha_i$).

Activation of the $G\alpha_s$ subunit is induced in the myocardium not only by β-adrenergic receptors – the histamine H2 receptor, the serotonin 5-HT_4 receptor and the VIP receptor are also coupled to the $G\alpha_s$ proteins and can induce positive inotropic effects in a similar manner. However, their efficiency in stimulating adenylyl cyclase and thus having a relevant influence on the contractility of the myocardium does not approach the level of β-adrenergic stimulation (Brodde et al. 1995).

Along with angiotensin II, endothelin-1, a peptide hormone that is principally produced and secreted by endothelial cells, is one of the most important factors with a vasoconstrictor effect. Its effects are mediated primarily by the ET_A or ET_B receptor. Both receptor isoforms are expressed in the myocardium, but only the ET_A receptor induces positive inotropic effects in the ventricular myocardium by means of activation of its $G\alpha_{q/11}$ subunit. In the atria, the ET_A receptor can also couple with $G\alpha_i$ proteins, but stimulation also leads to a positive inotropic response in the atrial myocardium. The ET_B receptor is of less importance for cardiac function (Pönicke et al. 1998).

The octapeptide angiotensin II, the many functions of which throughout the cardiovascular system have been the subject of study for many years, binds to the AT_1 receptor in the atrial myocardium and causes positive inotropic effects that are predominantly mediated via its coupling to the $G\alpha_{q/11}$ protein. The effects of angiotensin II on the function of the ventricle and the localisation and distribution of the two receptors (AT_1 and AT_2) have not yielded clear results depending on the experimental approach (Feolde et al. 1993; Moravec et al. 1990).

$G\alpha_i$ proteins are activated by stimulation of the adenosine A_1 receptor and the neuropeptide Y receptor in addition to the muscarinic M_2 receptor. Both adenosine and neuropeptide Y (NPY) induce negative inotropic effects in the atria by inhibiting adenylyl cyclase. For NPY, which exerts a strong vasoconstrictive effect in the vascular system, a negative inotropic effect has also been described in the ventricular myocardium. The influence of NPY on the contractility of the ventricle is observed mostly under sympathetic stimulation, meaning that mostly inhibitory effects on α/β-adrenergic signalling pathways are postulated and less the induction of independent negative inotropic mechanisms (Piper et al. 1989; Woo and Ganguly 1995).

5.4.4 The Consequences of a Neurohumoral Imbalance on Cardiac Function: Pathophysiology, Symptoms and Treatment

From a pathophysiological perspective, heart failure is defined as a complex syndrome with the underlying cause being a functional disorder of the myocardium. A direct consequence of this is the inability of the cardiac muscle to provide the tissue with sufficient blood, and therefore oxygen, to ensure localised metabolism at rest or under a load. The consequences of this unsatisfactory perfusion of the body or organs lead to the characteristic but non-specific symptoms such as dyspnoea, reduced performance, fatigue and oedema (clinical definition of heart failure).

A diverse aetiology often underlies the structural or functional damage to the myocardium. In descending order, coronary heart disease, which is accompanied by arterial hypertension in about 50 % of patients, as well as cardiomyopathy, defects, pericardial disease and inflammatory processes are the most common triggers of heart failure in Western countries.

As a response to the declining pump function of the heart, compensation mechanisms are initiated by the body that are intended to maintain the blood pressure and ensure the blood supply to organs. The renin–angiotensin–aldosterone system is activated in addition to the sympathetic nervous system. The synthesis and secretion of humoral factors are also induced, which, in terms of their effects, include vasoconstrictors, vasodilators and cytokines. As heart failure progresses, those mediators that facilitate vasoconstrictor as well as positive chronotropic and inotropic effects generally predominate (Spinarova and Vitovec 2007).

The disproportionate and sustained increase in activity of the sympathetic nervous system is intended to counteract the declining pump function of the myocardium. Although the sympathetic activation can contribute to the compensation in the short term, over the long term it induces functional and structural remodelling that further accelerates the progression of heart failure (Weil and Schunkert 2006).

Sustained sympathetic stimulation not only increases the energy requirements of the already failing myocardium, it also reduces the diastolic filling time of the coronary arteries due to the increase in the heart rate and thus lowers the available energy while exerting an arrhythmogenic effect.

The ratio of β_1 receptors to β_2 receptors is only 50:50 in the failing heart (usually 70:30). This change can be attributed to a reduction in the expression of the β_1-adrenergic receptor but the β_2 receptor is only negligibly affected. However, not only the receptor density but also the coupled subunits undergo a change in expression in the failing myocardium. A massive upregulation of the $G\alpha_i$ subunit is often accompanied by a reduction in the stimulating $G\alpha_s$ subunit. There can also be a change from the $G\alpha_s$ to the $G\alpha_i$ subunit as a result of PKA-dependent phosphorylation, but this only affects the β_2 receptor (Lohse et al. 2003).

Together with β-arrestin, the expression of the β-adrenergic receptor kinase GRK2 is massively induced as heart failure progresses. Both proteins are directly involved in the decoupling and internalisation of G protein–coupled receptors. It was originally postulated that elevation of the GRK2 level as heart failure progressed may be a sensible protective mechanism that protects the heart against excessive exposure to catecholamines. Sustained and long-term upregulation of GRK2 with the resultant desensitisation or decoupling of cardiac receptors leads inevitably to impaired cardiac function and the loss of the positive inotropic reserve (Dzimiri 1999).

The mechanisms described (reduction in the β_1 receptor, induction of the $G\alpha_i$ subunit and GRK2) should consequently lead to a reduction in the phosphorylation of specific target proteins, but RyR2, for example, in the failing myocardium, is present in a hyperphosphorylated state and contributes to the functional disorder by releasing sarcoplasmic calcium during the diastole (Lohse et al. 2003).

The precise mechanisms by which the β_1-specific receptor blockers mediate their effects have not yet been fully clarified. Some of the well-documented cardioprotective actions include their negative chronotropic and anti-arrhythmogenic properties. However, beta blockers do not simply inhibit all β-adrenergic-mediated signalling pathways, but there is a partial resensitisation of the system that takes place which can also contribute to an improvement in the clinical symptoms (Lymperopoulos 2013; Madamanchi 2007).

Baroreceptor activation therapy (BAT) is a device-based therapeutic procedure that could make a useful addition to pharmacotherapy for heart failure. During BAT procedures, the baroreceptors in the carotid sinus are stimulated by electrical impulses. Afferent and efferent paths in the autonomic nervous system subsequently lead by counter-regulatory means to inhibition of the sympathetic nervous system and activation of the parasympathetic nervous system (Olshansky et al. 2008; Zile et al. 2015).

5.5 Closing Summary

To maintain the required perfusion of the entire periphery, cardiac function is continuously adapted to the particular situation. Along with autonomous regulatory mechanisms (Frank–Starling, Bowditch), cardiac function is also continuously modified by dense autonomic innervation and numerous humoral factors. A thorough understanding of the cellular and molecular regulatory mechanisms underlying cardiac contraction is in turn a basic requirement for identifying the specific pathomechanisms and subsequently those opportunities for intervention that are essential for effective pharmacotherapy of heart failure.

References

Ablorh NA, Dong X, James ZM, Xiong Q, Zhang J, Thomas DD, Karim CB (2014) Synthetic phosphopeptides enable quantitation of the content and function of the four phosphorylation states of phospholamban in cardiac muscle. J Biol Chem 289:29397–29405

Bouchard RA, Clark RB, Giles WR (1995) Effects of action potential duration on excitation-contraction coupling in rat ventricular myocytes. Action potential voltage-clamp measurements. Circ Res 76:790–801

Bowditch HP (1871) Über die Eigenthumlichkeiten der Reizbarkeit, welche die Muskelfasern des Herzens zeigen. Arbeitsphysiologie 6:139–176

Brodde OE, Michel MC, Zerkowski HR (1995) Signal transduction mechanisms controlling cardiac contractility and their alterations in chronic heart failure. Cardiovasc Res 30:570–584

Chen Z, Akin BL, Jones LR (2007) Mechanism of reversal of phospholamban inhibition of the cardiac Ca^{2+}-ATPase by protein kinase A and by anti-phospholamban monoclonal antibody 2D12. J Biol Chem 282:20968–20976

Colyer J (1998) Phosphorylation states of phospholamban. Ann N Y Acad Sci 853:79–91

Dzimiri N (1999) Regulation of beta-adrenoceptor signalling in cardiac function and disease. Pharmacol Rev 51:465–501

Fearnley CJ, Roderick HL, Bootman MD (2011) Calcium signaling in cardiac myocytes. Cold Spring Harb Perspect Biol 3:a004242. doi:10.1101/cshperspect.a004242

Feolde E, Vigne P, Frelin C (1993) Angiotensin AT1 receptors mediate a positive inotropic effect of angiotensin II in guinea pig atria. Eur J Pharmacol 245:63–66

Flashman E, Redwood C, Moolman-Smook J, Watkins H (2004) Cardiac myosin binding protein C: its role in physiology and disease. Circ Res 94:1279–1289

Frank O (1895) Zur Dynamik des Herzmuskels. Z Biol 32:370–447

Gordon AM, Huxley AF, Julian FJ (1966) The variation in isometric tension with sarcomere length in vertebrate muscle fibres. J Physiol 184:170–192

Greenstein JL, Foteinou PT, Hashambhoy-Ramsay YL, Winslow RL (2014) Modeling CaMKII-mediated regulation of L-type Ca^{2+} channels and ryanodine receptors in the heart. Front Pharmacol 5:60. doi:10.3389/fphar.2014.00060.eCollection2014

Grimm M, Brown JH (2010) β-adrenergic receptor signaling in the heart: role of CaMKII. J Mol Cell Cardiol 48:322–330

Gülch RW (1974) A critical analysis of myocardial force-velocity relations obtained from damped quick-release experiments. Basic Res Cardiol 69:32–46

Gupta RC, Neumann J, Watanabe AM, Sabbah HN (1998) Muscarinic-cholinoceptor mediated attenuation of phospholamban phosphorylation induced by inhibition of phosphodiesterase in ventricular cardiomyocytes: evidence against a cAMP-dependent effect. Mol Cell Biochem 187:155–161

Holubarsch C, Ruf T, Goldstein DJ, Ashton RC, Nickl W, Pieske B, Pioch K, Lüdemann J, Wiesner S, Hasenfuss G, Posival H, Just H, Burkhoff D (1996) Existence of the Frank-Starling mechanism in the failing human heart. Investigations on the organ, tissue, and sarcomere levels. Circulation 94:683–689

Hsieh JH, Wu JJ, Yen CT, Chai CY (1998) The depressor caudal ventrolateral medulla: its correlation with the pressor dorsomedial and ventrolateral medulla and the depressor paramedian reticular nucleus. J Auton Nerv Syst 70:103–114

Kamp TJ, Hell JW (2000) Regulation of cardiac L-type calcium channels by protein kinase A and protein kinase C. Circ Res 87:1095–1102

Kang M, Chung KY, Walker JW (2007) G-protein coupled receptor signaling in myocardium: not for the faint of heart. Physiology (Bethesda) 22:174–184

Katz AM (2002) Ernest Henry Starling, his predecessors, and the "Law of the Heart". Circulation 106:2986–2992

Kögler H, Rüegg JC (1997) Cardiac contractility: modulation of myofibrillar calcium sensitivity by beta-adrenergic stimulation. Isr J Med Sci 33:1–7

Larsson HP (2010) How is the heart rate regulated in the sinoatrial node? Another piece to the puzzle. J Gen Physiol 136:237–241

Layland J, Solaro RJ, Shah AM (2005) Regulation of cardiac contractile function by troponin I phosphorylation. Cardiovasc Res 66:12–21

Lohse MJ, Engelhardt S, Eschenhagen T (2003) What is the role of β-adrenergic signaling in heart failure? Circ Res 93:896–906

Lu QW, Wu XY, Morimoto S (2013) Inherited cardiomyopathies caused by troponin mutations. J Geriatr Cardiol 10:91–101

Lymperopoulos A (2013) Physiology and pharmacology of the cardiovascular adrenergic system. Front Physiol 4:240. doi:10.3389/fphys.2013.00240

Madamanchi A (2007) β-adrenergic receptor signalling in cardiac function and heart failure. J Med 10:99–104

Mattiazzi A, Mundiña-Weilenmann C, Guoxiang C, Vittone L, Kranias E (2005) Role of phospholamban phosphorylation on Thr17 in cardiac physiological and pathological conditions. Cardiovasc Res 68:366–375

Mesirca P, Marger L, Toyoda F, Rizzetto R, Audoubert M, Dubel S, Torrente AG, Difrancesco ML, Muller JC, Leoni AL, Couette B, Nargeot J, Clapham DE, Wickman K, Mangoni ME (2013) The G-protein-gated K+ channel, IKACh, is required for regulation of pacemaker activity and recovery of resting heart rate after sympathetic stimulation. J Gen Physiol 142:113–126

Moravec CS, Schluchter MD, Paranandi L, Czerska B, Stewart RW, Rosenkranz E, Bond M (1990) Inotropic effects of angiotensin II on human cardiac muscle in vitro. Circulation 82:1973–1984

Moss RL, Fitzsimons DP (2002) Frank-Starling relationship: long on importance, short on mechanism. Circ Res 90:11–13

Oestreich EA, Malik S, Goonasekera SA, Blaxall BC, Kelley GG, Dirksen RT, Smrcka AV (2009) Epac and phospholipase Cepsilon regulate Ca2+ release in the heart by activation of protein kinase Cepsilon and calcium-calmodulin kinase II. J Biol Chem 284:1514–1522

Okumura S, Fujita T, Cai W, Jin M, Namekata I, Mototani Y, Jin H, Ohnuki Y, Tsuneoka Y, Kurotani R, Suita K, Kawakami Y, Hamaguchi S, Abe T, Kiyonari H, Tsunematsu T, Bai Y, Suzuki S, Hidaka Y, Umemura M, Ichikawa Y, Yokoyama U, Sato M, Ishikawa F, Izumi-Nakaseko H, Adachi-Akahane S, Tanaka H, Ishikawa Y (2014) Epac1-dependent phospholamban phosphorylation mediates the cardiac response to stresses. J Clin Invest 124:2785–2801

Olshansky B, Sabbah HN, Hauptman PJ, Colucci WS (2008) Parasympathetic nervous system and heart failure: pathophysiology and potential implications for therapy. Circulation 118:863–871

Pereira L, Métrich M, Fernández-Velasco M, Lucas A, Leroy J, Perrier R, Morel E, Fischmeister R, Richard S, Bénitah JP, Lezoualc'h F, Gómez AM (2007) The cAMP binding protein Epac modulates Ca^{2+} sparks by a Ca^{2+}/calmodulin kinase signalling pathway in rat cardiac myocytes. J Physiol 583:685–694

Petretta M, Vicario ML, Spinelli L, Ferro A, Cuocolo A, Condorelli M, Bonaduce D (2002) Combined effect of the force-frequency and length-tension mechanisms on left ventricular function in patients with dilated cardiomyopathy. Eur J Heart Fail 4:727–735

Piper HM, Millar BC, McDermott BJ (1989) The negative inotropic effect of neuropeptide Y on the ventricular cardiomyocyte. Naunyn Schmiedebergs Arch Pharmacol 340:333–337

Pönicke K, Vogelsang M, Heinroth M, Becker K, Zolk O, Böhm M, Zerkowski HR, Brodde OE (1998) Endothelin receptors in the failing and nonfailing human heart. Circulation 97:744–751

Roderick HL, Bootman MD (2007) Pacemaking, arrhythmias, inotropy and hypertrophy: the many possible facets of IP3 signalling in cardiacmyocytes. J Physiol 581:883–884

Roy A, Fields WC, Rocha-Resende C, Resende RR, Guatimosim S, Prado VF, Gros R, Prado MA (2013) Cardiomyocyte-secreted acetylcholine is required for maintenance of homeostasis in the heart. FASEB J 27:5072–5082

Ruiz-Hurtado G, Morel E, Domínguez-Rodríguez A, Llach A, Lezoualc'h F, Benitah JP, Gomez AM (2013) Epac in cardiac calcium signaling. J Mol Cell Cardiol 58:162–171

Sadayappan S, Gulick J, Osinska H, Barefield D, Cuello F, Avkiran M, Lasko VM, Lorenz JN, Maillet M, Martin JL, Brown JH, Bers DM, Molkentin JD, James J, Robbins J (2011) A critical function for Ser-282 in cardiac myosin binding protein-C phosphorylation and cardiac function. Circ Res 109:141–150

Schreckenberg R, Dyukova E, Sitdikova G, Abdallah Y, Schlüter KD (2015) Mechanisms by which calcium receptor stimulation modifies electromechanical coupling in isolated ventricular cardiomyocytes. Pflugers Arch 467:379–388

Shan J, Kushnir A, Betzenhauser MJ, Reiken S, Li J, Lehnart SE, Lindegger N, Mongillo M, Mohler PJ, Marks AR (2010) Phosphorylation of the ryanodine receptor mediates the cardiac fight or flight response in mice. J Clin Invest 120:4388–4398

Sonnenblick EH (1962a) Force-velocity relations in mammalian heart muscle. Am J Physiol 202:931–939

Sonnenblick EH (1962b) Implications of muscle mechanics in the heart. Fed Proc 21:975–990

Spinarova L, Vitovec J (2007) Neurohumoral changes in chronic heart failure. Biomed Pap Med Fac Univ Palacky Olomouc Czech Repub 151:201–207

Starling EH (1918) The Linacre Lecture on the Law of the Heart. Longmans/Green and Co., London

Sweeney HL, Brito RM, Rosevear PR, Putkey JA (1990) The low-affinity Ca^{2+}-binding sites in cardiac/slow skeletal muscle troponin C perform distinct functions: site I alone cannot trigger contraction. Proc Natl Acad Sci U S A 87:9538–9542

Trombitás K, Jin JP, Granzier H (1995) The mechanically active domain of titin in cardiac muscle. Circ Res 77:856–861

Tsuchida K, Mizukawa Y, Urushidani T, Tachibana S, Naito Y (2013) An inotropic action caused by muscarinic receptor subtype 3 in canine cardiac purkinje fibers. ISRN Pharmacol 2013:207671. doi:10.1155/2013/207671.eCollection2013

Vila Petroff MG, Palomeque J, Mattiazzi AR (2003) Na^{+}-Ca^{2+} exchange function underlying contraction frequency inotropy in the cat myocardium. J Physiol 550:801–817

Wei SK, Ruknudin AM, Shou M, McCurley JM, Hanlon SU, Elgin E, Schulze DH, Haigney MC (2007) Muscarinic modulation of the sodium-calcium exchanger in heart failure. Circulation 115:1225–1233

Weil J, Schunkert H (2006) Pathophysiology of chronic heart failure. Clin Res Cardiol 95:1–15

Weil J, Eschenhagen T, Hirt S, Magnussen O, Mittmann C, Remmers U, Scholz H (1998) Preserved Frank-Starling mechanism in human end stage heart failure. Cardiovasc Res 37:541–548

Wijnker PJ, Boknik P, Gergs U, Müller FU, Neumann J, dos Remedios C, Schmitz W, Sindermann JR, Stienen GJ, van der Velden J, Kirchhefer U (2011) Protein phosphatase 2A affects myofilament contractility in non-failing but not in failing human myocardium. J Muscle Res Cell Motil 32:221–233

Wolk MJ, Keefe JF, Bing OH, Finkelstein LJ, Levine HJ (1971) Estimation of Vmax in auxotonic systoles from the rate of relative increase of isovolumic pressure: (dP-dt)kP. J Clin Invest 50:1276–1285

Woo ND, Ganguly PK (1995) Neuropeptide Y prevents agonist-stimulated increases in contractility. Hypertension 26:480–484

Xiao RP, Lakatta EG (1993) β_1-adrenoceptor stimulation and β_2-adrenoceptor stimulation differ in their effects on contraction, cytosolic Ca^{2+}, and Ca^{2+} current in single rat ventricular cells. Circ Res 73:286–300

Xiao B, Zhong G, Obayashi M, Yang D, Chen K, Walsh MP, Shimoni Y, Cheng H, Ter Keurs H, Chen SR (2006) Ser-2030, but not Ser-2808, is the major phosphorylation site in cardiac ryanodine receptors responding to protein kinase A activation upon β-adrenergic stimulation in normal and failing hearts. Biochem J 396:7–16

Zhang R, Zhao J, Mandveno A, Potter JD (1995) Cardiac troponin I phosphorylation increases the rate of cardiac muscle relaxation. Circ Res 76:1028–1035

Zile MR, Abraham WT, Weaver FA, Butter C, Ducharme A, Halbach M, Klug D, Lovett EG, Müller-Ehmsen J, Schafer JE, Senni M, Swarup V, Wachter R, Little WC (2015) Baroreflex activation therapy for the treatment of heart failure with a reduced ejection fraction: safety and efficacy in patients with and without cardiac resynchronization therapy. Eur J Heart Fail 17:1066–1074

Part II

Cardiomyocytes in Ageing, Disease, and Protection

Growth Regulation of Cardiomyocytes: Control of Cell Size and Its Role in Cardiac Hypertrophy

6

Klaus-Dieter Schlüter

Abstract

Cardiac hypertrophy in its pure sense describes the ability of cardiomyocytes to increase their cell shape in the absence of cell proliferation. As the large majority of mammalian cardiomyocytes lose the ability to cytokinesis shortly after birth, this is the most important process by which the heart muscle mass can be adapted to increased mechanical stress. Cardiac hypertrophy is a physiological process during adolescence in which the total heart size is increased nearly eightfold from birth to adolescence, but there is only a limited ability of the adult heart to further increase heart size without getting dysfunctional. As cardiac hypertrophy is an established risk factor for heart failure, a lot of attention has been dropped to identify key pathways that trigger cardiac hypertrophy and the transition from physiological hypertrophy to pathophysiological hypertrophy. A large number of hormones, cytokines and neurotransmitters have been identified that trigger cardiac hypertrophy and specifically those related to the sympathetic nervous system, and the renin-angiotensin system is of importance for the transition to heart failure. An increase in the translational capacity is the basis of cardiac growth, but changes in the expression, mainly linked to the sarcomere but also linked to cardiac metabolism, characterise pathophysiological hypertrophy. Although initially not considered as a realistic perspective, it is clear meanwhile that even pathophysiological hypertrophy is a reversible process. The main challenge for the future is to indentify key pathways that normalise heart sizes and normalise the cardiac phenotypes. This chapter reviews our current understanding of key processes leading to cardiac hypertrophy.

K.-D. Schlüter
Institute of Physiology, Justus-Liebig-Universität Gießen,
Aulweg 129, 35392 Giessen, Germany
e-mail: Klaus-Dieter.Schlueter@physiologie.med.uni-giessen.de

K.-D. Schlüter (ed.), *Cardiomyocytes – Active Players in Cardiac Disease*,
DOI 10.1007/978-3-319-31251-4_6

6.1 About the Definition of the Term "Hypertrophy"

In its pure sense the term "hypertrophy" simply describes the ability of an organ to increase its size by growth of cells without cell proliferation. The term "cardiac hypertrophy", however, is used in a slightly different way. Mammalian cardiomyocytes lose their ability to replicate early after birth. However, the size of the heart further increases during adolescence. In man, the heart volume increases within the first 20 years by approximately 100–800 ml. This is accompanied by increasing the size of individual cardiomyocytes. Nevertheless, even in the adult phase, the heart is able to increase its size further under conditions of volume overload as it occurs in case of high physical activity or pregnancy. This type of cardiac hypertrophy is therefore named "physiological hypertrophy", and it is believed that it is, at least in principle, reversible. In many cases however the term "hypertrophy" is used to describe a situation of "pathophysiological hypertrophy". In this case the increase in cardiac mass is associated with a drop in cardiac function. In general, cardiac hypertrophy is associated with a high risk of development of heart failure. Triggers that favour pathophysiological hypertrophy are chronic pressure overload, myocardial infarction (ischaemia and reperfusion), chronic ischaemia, genetic disorders or myocardial inflammation just to mention a few. In general, the assumption is that even under these conditions, a hypertrophic response is initially a compensatory mechanism because an increase in wall thickness normalises wall stress. However, the same process initiates a transition from hypertrophy to heart failure. The reason for a transition of physiological, compensatory hypertrophy to pathophysiological, dysfunctional hypertrophy is not the increase in cell and tissue size but a corresponding cellular remodelling process. Therefore, the phrase "cardiac remodelling" has been used to describe the transition of cardiac tissue from an adaptive organ into a maladaptive organ. As this process is maladaptive, it is also termed "adverse remodelling". In a complex in vivo system, it is rather impossible to decide which signal transduction pathways are leading to hypertrophy and/or adverse remodelling. As outlined later in detail, some receptor systems are linked to protein synthesis while others to phenotypic changes caused by translational activation. Therefore, different signalling pathways interact to create hypertrophy and remodelling.

Although the term "hypertrophy" is mainly used to simply describe the increase in heart size, it must be noted that within the heart, only cardiomyocytes grow by hypertrophy, but cardiac remodelling includes proliferation of cardiac fibroblasts and potentially cardiac endothelial cells. Therefore, the term "cardiac hypertrophy" is a little bit misleading in a pure sense. It should be noted that inflammatory cells, cardiac fibroblasts and cardiomyocytes influence each other in a complex way. However, in this chapter the focus is on cardiomyocytes. Therefore, the term "hypertrophy" will be used here to describe the adaptation of cardiomyocytes to external growth stimuli. These stimuli can be a hyperactivation of neurotransmitter release, mechanical stress to cells or pro-inflammatory activation of cardiomyocytes. Under most pathophysiological situations that cause cardiac hypertrophy, a battery of these stimuli activates the cardiomyocytes to grow and change their phenotype. Consequently, current therapies against heart failure and thereby treatment regimes against viac hypertrophy combine various pharmacological inhibitors directed against catecholamines, angiotensin II, aldosterone and load reduction. It should be mentioned clearly that the complex regulation of cardiac growth has not yet been fully understood and that locally produced cytokines and growth factors are known. It is a challenge for future research to determine whether additional inhibition of such factors will optimise heart failure treatment. It is another challenge for future research to investigate whether intracellular key pathways are a more suitable target for optimised treatment. Given the still unacceptable high rate of mortality for patients with heart failure, future development in this field are certainly required.

6.2 Hormones, Cytokines and Neurotransmitters That Regulate Cardiac Growth

The identification of hormones and neurotransmitters that induce myocardial hypertrophy has led to the successful development of treatment strategies that attenuate cardiac growth. Mainly they are directed against transmitters of the sympathetic nervous system (SNS) or the renin-angiotensin-aldosterone system (RAAS). Both pathways are the best characterised pathways that are associated with cardiac hypertrophy and heart failure. The SNS releases catecholamines from its terminal nerve endings, like norepinephrine (=noradrenaline, NOR) and epinephrine (=adrenaline, ADR). NOR and ADR bind to a couple of distinct receptors on cardiomyocytes, namely, α_1-, β_1-, β_2- and β_3-adrenoceptors. From the pharmacological profile, NOR has a preference to β-adrenoceptors, while ADR preferentially binds to α-adrenoceptors. Whereas NOR represents mainly a neurotransmitter from the terminal nerve endings of the ventricles, ADR represents a co-transmitter that is additionally released into the blood from the adrenals and acts as a neurotransmitter and hormone. Neuronal regulation offers normally an acute but short signal. Hormonal stimulation is delayed but longer lasting. NOR, preferentially activating β-adrenoceptors, acts mostly as a neurotransmitter. Stimulation of β-adrenoceptors is linked to quick changes in cardiac performance (positive inotropy) but not strongly connected to growth regulation (long-term adaptation). ADR, preferentially activating α-adrenoceptors, acts more as a hormone. Stimulation of α-adrenoceptors is strongly connected to cell growth (long-term activation) and moderately coupled to inotropy. Historically, there was a long debate about the type of adrenergic receptors that contribute to cardiac hypertrophy among the various adrenoceptors. At least for the rat, it is clear that selective stimulation of α_{1A}-adrenoceptors increases the rate of protein synthesis in cardiomyocytes and thereby induces hypertrophy (Pönicke et al. 2001). Of note, cardiomyocytes also express functionally coupled α_{1B}-adrenoceptors, but these receptors do not activate pathways that increase protein synthesis. The stimulating effect of α_{1A}-adrenoceptor activation on protein synthesis does not tell us anything about the role of these receptors for the transition process of cardiac hypertrophy to heart failure. In principle, moderate stimulation of α-adrenoceptors induces protein expression that maintains proper cardiac contractility. This pathway seems to trigger physiological hypertrophy (Anwar et al. 2005). In line with this, no α-adrenoceptor antagonists have been successfully implicated into clinical practice as an anti-hypertrophic therapy. In contrast, selective stimulation of β-adrenoceptors does not increase protein synthesis (Schlüter and Piper 1992). In line with this in vitro observation, pressure overload induces cardiac hypertrophy in β_1/β_2-adrenergic knockout mice (Palazzesi et al. 2006). The situation is clearly established for adult and terminally differentiated ventricular rat cardiomyocytes but may not represent a correct view of receptor coupling to growth control in all mammalians. Cardiomyocytes from mice respond in a different way (Sabri et al. 2000). However, many clinical observations are in line with findings initially found in rat cardiomyocytes. Catecholamines, the natural agonists of adrenergic receptors, are not specific for one of the different adrenoceptors. Therefore, it is important to notice that α- and β-adrenoceptor-activated pathways interfere with each other. Stimulation of β-adrenoceptors has an inhibitory effect on the α-adrenoceptor-activated protein synthesis. Therefore, natural and non-selective catecholamines like NOR stimulate cardiac growth to a lesser extent than selective pharmacological agonists (Schäfer et al. 2001). This inhibitory effect is caused by β_1- or β_2-adrenoceptor stimulation increasing intracellular cAMP levels but probably also via stimulation of β_3-adrenoceptors leading to increased activation of cGMP pathways. Interestingly, lowering blood pressure in male moderate hypertensive patients with a selective β_1-adrenoceptor antagonist (atenolol) increased heart size further suggesting that either blockade of β_1-adrenoceptors improves α-adrenoceptor-dependent hypertrophy (see above) or that blockade of β_1-adrenoceptors directs catecholamines to cardiac β_2-adrenoceptors that may be linked to protein synthesis in hypertensive patients (see below) (Mehlsen et al. 1993).

Are β-adrenoceptor-dependent pathways in adult terminally differentiated cardiomyocytes always anti-hypertrophic? The answer to this important question is no, because receptor coupling can be modified by transforming growth factor (TGF)-β_1. The modification of receptor coupling with respect to growth regulation in cardiomyocytes may play a prominent role in the transition of hypertrophy to heart failure. In vitro studies indicated that TGF-β_1 couples β_2-adrenoceptors to cardiac growth regulation, and this is of particular interest because β_2-adrenocpeptors are mostly insensitive to β-adrenoceptor downregulation as it occurs in heart failure patients (Schlüter et al. 1995b). In vivo, β-adrenoceptor coupling was improved in transgenic TGF-β_1 expressing mice accompanied by upregulation of uncoupling protein (UCP)-3 (Rosenkranz et al. 2002). Finally, β-blockers attenuated the upregulation of UCP-3 in heart failure patients without affecting TGF-β_1 (Huntgeburth et al. 2011). All these data underline the relevance of findings with cultured adult ventricular rat cardiomyocytes for general aspects.

In contrast to α-adrenoceptor stimulation, stimulation of β-adrenoceptors may favour maladaptive hypertrophy because it potentially activates stress kinases and induces apoptosis. Apoptosis triggers cardiac hypertrophy indirectly because cell death of individual myocytes has to be compensated by hypertrophic growth of remaining cardiomyocytes. This would also explain the lack of pro-hypertrophic effects of β-adrenoceptor stimulation in vitro but pro-hypertrophic effects in vivo.

Although some reports are given in the literature for all of these examples, a mechanistic basis for the use of β-adrenoceptor antagonists (β-blocker) is not obvious because a stressed heart will be overloaded by catecholamines, and this desensitises β-adrenoceptors. In fact, β-blockers can reintroduce a better coupling of β-adrenoceptors, a finding not in line with a prominent role for β-adrenoceptor coupling in initiating cardiac hypertrophy and a transition to heart failure. Inhibition of G protein-coupled receptor kinase-2 (β-GRK-2) has similar effects as β-blockers indicating that reintroduction of physiological responsiveness of β-adrenoceptors to ligand-dependent stimulation is required for maintaining cardiac function. It may be that the intracellular coupling of β-adrenoceptor-dependent signalling is optimised in the presence of β-blockade into a direction of a more adaptive type of hypertrophy. Although the role of β-adrenoceptor stimulation for the progression of heart failure is not well understood, the use of β-blockers in heart failure is established.

The second major maladaptive pathway involved in cardiac hypertrophy is the renin-angiotensin-aldosterone system (RAAS). At first, it must be stated that SNS and RAAS are not working independent from each other. The SNS induces the release of enzymatic active renin from juxtaglomerular cells of the kidney. Renin acts as a protease that cleaves circulating angiotensinogen in angiotensin I. Angiotensin I is finally processed by angiotensin-converting enzyme-1 (ACE-1) into angiotensin II (Ang-II). Angiotensin II is the active component of the RAAS and a main hormone involved in blood pressure control. Hyperactivation of the RAAS increases blood pressure. This mechanical load on cardiomyocytes may trigger a cardiac growth. In contrast to foetal and neonatal cardiomyocytes, adult and terminally differentiated cardiomyocytes only weakly respond to Ang-II by increasing the rate of protein synthesis (Ruf et al. 2002). However, Ang-II induces the expression of proteins in cardiomyocytes that affect contractility, and thereby, Ang-II directly contributes to heart failure, whereas its effect on cardiac growth seems to be limited. Systemically, Ang-II contributes to cardiac hypertrophy and heart failure also by an activation of aldosterone release from the adrenals that then act as a pro-fibrotic hormone. Ang-II-dependent stimulation of cardiac fibroblasts increases the local concentration of TGF-β that is responsible for a shift of β-adrenoceptor coupling in cardiomyocytes (see above). All these processes contribute to cardiac remodelling, and thereby, Ang-II contributes to cardiac hypertrophy in vivo without being a strong activator of protein synthesis in cardiomyocytes. Inhibition of the RAAS is possible at multiple points: direct inhibition of renin, inhibition of ACE activity or blockade of angiotensin type 1 receptors (AR_1B). Further support to beneficial effects of RAAS inhibition is given by antagonising aldosterone receptors on top of ACE inhibitors. However, the beneficial effects of RAAS inhibition are not simply the consequence of the inhibition of the Ang-II-dependent effects. Besides the various effects of Ang-II on cardiac tissue, it induces a negative feedback on renin release from the kidney. Therefore, an inhibition of RAAS leads to an increase of plasma renin levels. Renin is not simply a protease-converting angiotensinogen into angiotensin I. It binds to two distinct receptors, the pro-renin receptor and M6P/IGFII receptor, both of which are expressed on cardiomyocytes (Heger and Schlüter 2013). Specifically activation of M6P/IGFII receptors has shown to modify the remodelling process of cardiomyocytes, to depress hypertrophic growth responses and to antagonise further β-adrenoceptor stimulation (Hinrichs et al. 2011). Inhibition of ACE leads to an activation of the alternative ACE molecule, ACE-2. Unlike ACE-1, ACE-2 cleaves angiotensin I into Ang-(1–7) that exhibits anti-hypertrophic effects on cardiomyocytes via binding to sarcolemmal Mas receptors. Furthermore, inhibition of ACE-1 increases plasma levels of bradykinin, a hormone that increases cGMP levels in cardiomyocytes again reducing growth stimulation. Finally, selective inhibition of angiotensin receptor type 1 (AT_1R) directs Ang-II to angiotensin receptor type 2 (AT_2R) that antagonises the effect of AT_1R stimulation. However, it is unclear whether this is of clinical relevance because adult and terminally differentiated cardiomyocytes do only barely express AT_2R, and recent clinical trials did not display any additive effect of AT_1R on ACE inhibition as would be expected if cardiomyocytes express both receptor subtypes (reviewed in Schlüter and Wenzel 2008). In conclusion, there is an intense cross reactivity between RAAS and SNS and a very complex cross reactivity between the different receptor systems that are not well understood on the cellular level. However, in principle β-blockers and ACE inhibitors can be combined and lead to at least a delay in the progression of heart failure.

Despite the obvious role of SNS and RAAS in pathophysiological hypertrophy, various hormones have been identified that directly stimulate cardiac growth. Among them are endothelin-1 that increases protein synthesis and promotes eccentric growth of cardiomyocytes via activation of endothelin receptor-A (ET_A); thyroid hormone that does not induce a foetal-type phenotype of cardiomyocytes and thereby seems to induce selectively a physiological type of hypertrophy; insulin and insulin-like growth factors (IGF), specifically IGF-I; parathyroid hormone and parathyroid hormone-related protein; coupling systemic calcium regulation to growth regulation of cardiomyocytes; and finally neuropeptide Y, a co-transmitter of the SNS (reviewed in Schlüter et al. 1995a, b). In most cases these hormones share their ability to increase protein synthesis with that of α-adrenoceptors because they all belong to a subclass of G protein-coupled receptors that are linked to $G\alpha_q$, and therefore, they all induce similar pathways. As inhibition of α-adrenoceptors did not improve the situation under clinical conditions, it is evident that inhibitors of some of these similar receptor agonists are not developed either. This does, however, not mean that they do not play a physiological role in cardiac biology rather than that they are not main players in the transition from hypertrophy to heart failure.

6.3 Mechanical Stress Triggering Cardiac Growth

For a long time, mechanical stretch was regarded as an important trigger of cardiac growth adaptation. The concept is based on the principle understanding that increasing wall thickness, eccentric hypertrophy, normalises wall stress as it occurs under conditions of chronic pressure overload, such as hypertension. However, it turned out to be difficult to show on a cellular level that mechanical load rather than neurohumoral factors stimulates cardiac growth. Adult, terminally differentiated cardiomyocytes do not attach properly to flexible culture material and could not be used to study the effect of mechanical load on protein synthesis. Therefore, neonatal cardiomyocytes were used and stretched, and it was found that cardiomyocytes can release Ang-II that then triggers the increase in protein synthesis (Komuro et al. 1990). However, as mentioned above, Ang-II is a poor direct activator of protein synthesis in adult, terminally differentiated cardiomyocytes. Therefore, alternative mechanisms have been suggested to transfer mechanical load into cardiac growth. As such, transient receptor potential channels (TRPC) that act as stretch-activated channels are likely candidates. The cellular mechanism is not established, but at least in a mouse model of pressure overload, TRPC inhibitors reduced hypertrophy (Kiyonaka et al. 2009). Nevertheless, from such in vivo experiments, it cannot be concluded that activation of stretch-activated TRPCs trigger hypertrophy directly as they may simply affect neurohumoral stimulation. Arguments for a direct role of mechanical stretch on cardiac hypertrophy comes further from the observation that patients with familiar hypertrophic cardiomyopathy (FHC) have often mutations in sarcomere proteins suggesting that reduced mechanical force can cause a compensatory increase of protein synthesis. Altogether, it is clear that mechanical stress somehow modifies the rate of protein synthesis and degradation in cardiomyocytes and that it favours a situation of hypertrophy and leads to transition of hypertrophy into heart failure, but whether these are indeed direct effects or indirect effects is not clear. Moreover, inhibition of RAAS with an ACE inhibitor like ramipril reduced cardiac hypertrophy without lowering blood pressure indicating that mechanical stretch performed by chronic pressure overload itself is not sufficient to induce or maintain cardiac hypertrophy (Mathew et al. 2001). Similarly, etanercept, a compound affecting circulating TNF-α, reduced cardiac mass without lowering blood pressure (Daien et al. 2013). In contrast to mechanical stretch, at least one publication has clearly demonstrated that an increase in beating frequency of load-free-contracting cardiomyocytes increases protein synthesis (Ivester et al. 1993). In this case it is possible that higher beating frequencies elevate diastolic cell calcium to trigger increased protein synthesis although this has not been analysed.

6.4 Ischaemic Events Affecting Growth Regulation of Cardiomyocytes

Ischaemia and reperfusion, also described as myocardial infarction, cause an irreversible loss of cardiac tissue that has to be replaced by surviving cardiomyocytes from the border zone between infarcted areas with irreversibly damaged cardiomyocytes and healthy myocardium or from the remote area of the cardiac tissue. The situation has similarities and differences to other triggers of myocardial hypertrophy. Similar to the situation in chronic pressure overload, mechanical stress is exposed to cardiomyocytes because these cells have to compensate the loss of function of damaged cardiomyocytes. Similarly to chronic pressure overload, the heart is exposed to excessive catecholamines at least during the onset of reperfusion. On the other side, there are also differences between both stimuli. Differences occur because damaged cardiomyocytes release cytosolic molecules with intrinsic biological activity into the extracellular fluid and because inflammatory processes are much stronger as one would expect during chronic pressure overload. Extracellular events that potentially modify the rate of protein synthesis and thereby initiate a hypertrophic response are, for example, the acidification of the tissue due to anaerobic metabolism during ischaemia and the accumulation of extracellular RNA (eRNA) released by damaged cardiomyocytes. eRNA binds to a TNF-α-converting enzyme (TACE) that then gets activated and releases TNF-α. Furthermore, infiltrating macrophages release pro-inflammatory cytokines that activate corresponding sarcolemmal chemokine receptors (Heymans et al. 2013). Differences between hypertrophy caused by pressure overload and that caused by ischaemia and reperfusion also occur in the role of intracellular calcium and reactive oxygen species (ROS). Intracellular calcium, mainly diastolic calcium, and ROS are directly involved in signalling processes defining the phenotype of hypertrophied cardiomyocytes (see below). At the time of reperfusion, intracellular calcium concentration is much higher than even under unfavourable conditions of chronic heart failure, and ROS generation is much stronger than under conditions of pressure overload. In conclusion, ischaemia and reperfusion injury leads to cardiac hypertrophy that differs slightly from known pathways of cardiac hypertrophy caused by pressure overload.

6.5 Intracellular Signalling Triggering Cardiac Growth

As mentioned above, the basis of myocardial hypertrophy is an increase of protein synthesis in adult terminally differentiated cardiomyocytes. This requires an increase in translational activity. At least in principle, this can be evoked by two different ways: an increase in translational capacity, in other words an increase in the number of ribosomes per cell, and an increase in translational efficiency, that is, an increase in turnover rates of existing ribosomes. These two pathways are the basis of hypertrophic growth of cardiomyocytes. Increases in protein synthesis are linked to stimulation of distinct receptors. Most receptors that are proven to increase the rate of protein synthesis whenever stimulated belong to the family of G protein-coupled receptors (GPCR). GPCR are associated with G proteins. These are trimers built up of α, β and γ subunits. The principle activation pathway of GPCRs is as follows: a ligand binds to the receptor at an extracellular receptor-binding domain. This activates the receptor complex leading to a release of the Gα subunit that translocates the signal into intracellular compartments. The remaining β and γ dimer remains bound at the sarcolemmal membrane but may trigger additional signalling processes although such signals have not directly linked to increased protein synthesis. GPCRs are subdivided into three different subgroups depending on the type of Gα subunits. These three subgroups are α_s, α_i and α_q. The different Gα isoforms determine the type of intracellular signal that is generated in the cell. $G\alpha_s$-coupled receptors activate adenylyl cyclase that forms cAMP as a second messenger from ATP. Examples for such receptors that are expressed in cardiomyocytes are the β_1- and β_2-adrencoceptors. $G\alpha_i$-coupled receptors inhibit adenylyl cyclase. An example of such a receptor expressed in cardiomyocytes is the muscarinic receptor M_2. $G\alpha_q$-coupled receptors activate phospholipase C, an enzyme that forms diacylglycerol (DAG) and inositol triphosphate (IP_3). An example for this type of receptor expressed in cardiomyocytes is the α-adrenoceptor that can selectively be activated by phenylephrine (PE). This receptor system is indeed the best characterised model of a GPCR that directly increases translational activity. Its ability to increase translational activity is first linked to the activity of DAG. DAG activates protein kinase C (PKC). As all protein kinases, PKC is able to phosphorylate target proteins and thereby changes their activity. There are three different families of PKCs known, and members of each of these families are expressed in cardiomyocytes. These are classic calcium-activated serine-threonine kinase PKC isoforms (α, β and γ) that are sensitive to calcium and DAG, novel PKCs (δ, ε, η, θ and μ) that are insensitive to calcium but sensitive to DAG and atypical PKC isoforms (ξ and λ) that are insensitive to calcium and DAG. The heart expresses the PKC isoforms α, β, δ, ε and ξ. PKC isoforms activated by GPCRs have attracted a lot of attention. Experiments in which specific isoforms in adult terminally differentiated cardiomyocytes were selectively downregulated by antisense oligonucleotides showed that calcium-insensitive isoforms δ and ε are required for proper PE-dependent linkage to increased protein synthesis (Schreckenberg et al. 2004). Using different transgenic mouse models, PKCε has been linked to physiological hypertrophy, and PKCδ has been linked to maladaptive hypertrophy (Sabri and Steinberg 2003). A potential downstream target of PKC that links stimulation of GPCRs with translational capacity is phosphatidyl inositol 3-kinase (PI3K). PI3K possess regulatory (p85, p101) and catalytic subunits (p110). The catalytic subunit can be subdivided into different isoforms. PI3K can be activated by GPCR signalling in a PKC-dependent and PKC-independent way. Specific downregulation of p110α abrogated the coupling of PE to protein synthesis (Wenzel et al. 2006). However, downregulation of p110β abrogated Ang-II-dependent stress activation in cardiomyocytes but did not affect protein synthesis. PI3K p110α is also implicated in protective pathways activated by growth factors via tyrosine kinase receptors. Collectively, the data suggests that α-adrenoceptor signalling is linked to PI3K/p110α activation, thereby favouring cardiac growth but not remodelling, while AT_1R signalling is linked to PI3K/p110β, thereby favouring adverse remodelling rather than hypertrophy in its pure sense. PI3K activation leads to the activation of another kinase, namely, Akt1, that then activates the mammalian target of rapamycin (mTOR)/p70S6K complex. The 40S subunit of ribosomes includes ribosomal protein S6 that is finally phosphorylated by $p70^{S6K}$ and subsequently increases the translational efficiency of ribosomes. Via this signalling cascade, protein synthesis is increased. Furthermore, Akt-1 inhibits glycogen synthase kinase 3β (GSK3β) which is a suppressor of cell growth. These signals have been linked to α-adrenoceptors in cardiomyocytes, and selective stimulation of such receptors induces an adaptive type of hypertrophy rather than heart failure. As expected selective activation of PI3K pathways improves inotropy and protects from cardiac injury (Bisping et al. 2014). Moreover, inhibition of mTOR leads to feedback activation of its upstream kinases PI3K and Akt-1, and accumulating evidences show that mTOR inhibition is cardioprotective. Activation of Akt-1 pathways can also be achieved by selective activation

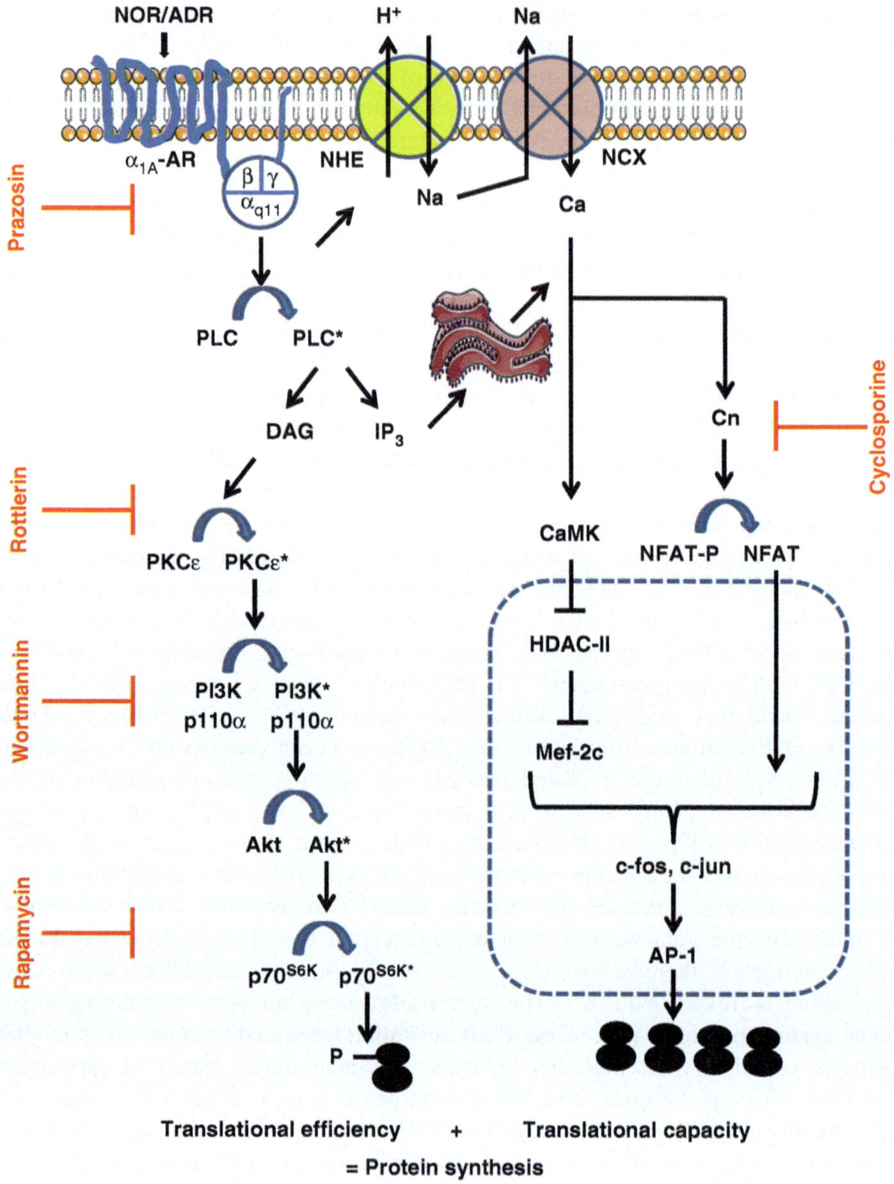

Fig. 6.1 Intracellular signalling of α1A-adrenoceptors leading to cardiac hypertrophy via increasing translational capacity and efficiency. Potential inhibitions are indicated in red

of tyrosine kinase receptors that are normally linked to cytokines. This might be the basis of the growth-promoting effects of insulin and insulin-like growth factors. The underlying pathway coupling of $\alpha 1_A$-adrenoceptors to improved protein synthesis is shown in Fig. 6.1 in detail.

As mentioned above the process of increasing protein synthesis in cardiomyocytes is based on two different steps, namely, an increase in translational efficiency, as described above, and an increase in translational capacity. The latter was clearly shown in adult terminally differentiated cardiomyocytes exposed to PE in which growth stimulation did not only increase phenylalanine labelling of proteins but also uridine labelling of RNA (Pinson et al. 1993). The vast majority of cellular RNA, nearly 90 % of total RNA, is rRNA, and therefore, it is likely to assume that the translational capacity of cardiomyocytes is increased during hypertrophy as in proliferating cells. Of note, increase in translational efficiency and capacity is tightly connected to each via the mTOR/p70S6K pathway. However, increasing translational capacity requires additional activation of 5′-terminal oligopyrimidine (5′-TOP) tract mRNAs that are required for ribosomal protein translation and activation of RNA polymerase I by RNA polymerase I transcription factor UBF. This leads to the question which signals allow to modify transcription factor activity. Such examples are known for nuclear factor of activated T cells (NFAT) and transcription activator protein 1 (AP-1) although none of the two transcription factors have linked to ribosomal formation. NFAT is activated by calcineurin that dephosphorylates NFAT and allows translocation into the nucleus. Calcineurin is activated by increased diastolic calcium concentration. α-Adrenoceptor stimulation increases intracellular calcium indirectly by activating the sodium-proton exchanger (NHE) in a PKC-independent way (Schäfer et al. 2002). Accumulating sodium can be removed by the reverse mode of the sodium-calcium exchanger (NCX) increasing the cellular calcium concentration. Furthermore, AP-1 binding to DNA is required to increase protein synthesis in cardiomyocytes although the connection between both events (activation of AP-1-dependent transcription and increasing protein synthesis) is not clear (Taimor et al. 2004). Collectively, these data show that increasing protein synthesis requires signalling within the cytoplasm but also to the nucleus.

Not all $G\alpha_q$-coupled receptors increase protein synthesis in a similar way. AT_1R for instance are linked to $G\alpha_{q12/13}$, whereas α-adrenoceptors are linked to $G\alpha_{q11}$. $G\alpha_{q12/13}$ links receptor activation to Rho-kinase pathways that activate NADPH oxidases and therefore increases ROS levels in cells. Oxidative stress activates stress-activated kinase pathways; the best characterised is the p38 MAP kinase pathway. It also causes post-translational modification of proteins but has minor effects on protein synthesis. AT_1R stimulation activates p110β kinase isoforms that are not linked to pathways that accelerate protein synthesis (Wenzel et al. 2006). These molecular differences are the basis for the aforementioned different responsiveness of cardiomyocytes to selective α-adrenoceptors and Ang-II with respect to cardiac hypertrophy.

In contrast to $G\alpha_q$-coupled receptors, $G\alpha_s$-coupled receptors use cAMP as a second messenger and activate protein kinase A (PKA). PKA-dependent pathways inhibit the aforementioned effect of selective stimulation of $G\alpha_q$-coupled receptors on translational efficiency (Schäfer et al. 2001). This leads to a significant inhibition of the physiological part of cardiac hypertrophy. To what extent the success of β-blockers is linked to an inhibition of pathways linked to physiological hypertrophy in vivo is not completely clear because they also reduce adverse remodelling caused by β-adrenoceptor stimulation.

6.6 Phenotypic Adaptation During Hypertrophic Growth Regulation

An increase in cell size is the basis of myocardial hypertrophy. However, the increase of protein synthesis efficiency and capacity is supported by a switch of proteins that are expressed in cells with fixed cell sizes and those that grow. The first findings about specific proteins expressed in hypertrophic vs. normotrophic cardiomyocytes already revealed a reactivation of a translational programme that is characteristic for foetal and neonatal cardiomyocytes. The term "foetal shift" had been formed to describe this aspect of myocardial hypertrophy. An overview of proteins that are reexpressed during hypertrophy is given in Table 6.1. As found for hypertrophic growth, also the reprogramming of cardiomyocytes is independent of the blood pressure (Saadane et al. 1999). It should be mentioned that the phenotype of hypertrophic cardiomyocytes is somehow unique as it combines the size and activity of adult cardiomyocytes (lack of spontaneously beating) with proteins expressed during foetal or neonatal development (rearrangement of sarcomeres). Other aspects of the so-called foetal shift are related to a metabolic switch that favours the use of carbohydrate sources (glucose, lactate) versus lipids as it does in the adult heart (Kolwicz and Tian 2011).

The reactivation of a translational machinery that is responsible for the foetal shift requires the transition of the hypertrophic signal into the nucleus and the activation of specific transcription factors. Therefore, additional signal transduction pathways must be taken into account that may trigger this phenotypic shift of cardiomyocytes. In general five transcription factors are mainly responsible for the phenotypic change although others may also participate. These are NFAT, GATA4, NFκB, serum response factor (SRF) and myocyte enhancer factor (MEF)-2. The activation of these transcription factors is based on the following signalling pathways. GPCRs using $G\alpha_{q11}$, examples are the aforementioned α-adrenoceptors or endothelin receptors, activate PKC. PKCε, known to be associated with physiological hypertrophy, downregulates cardiac-specific transcription factors such as Nkx.2.5 and GATA4 (Galli et al. 2013). PKC mainly activates mitogen-activated protein kinase (MAPK) pathways and exerts its effect indirectly. MAPK pathways consist of a series of kinases that are activated by proteolytic cleavage. Members of this cascade are Ras, Raf and MAPK kinase (MEK). In general, three different MAPK pathways must be distinguished: the extracellular responsive kinase (ERK) pathway with an activation of p42/p44 MAPK as a final mediator, the stress-activated p38-MAPK pathway with p38-MAPK as the terminal mediator and c-jun N-terminal kinase (JNK). Each of these three pathways activates distinct pathways. GPCRs linked to $G\alpha_{q11}$ have a preference to the ERK pathway. They activate PKCs that activate a cascade of protein kinase, namely, Raf, MEK and finally ERK

Table 6.1 Examples for proteins that are involved in the foetal shift as phenotypic adaptation of cardiomyocytes

Foetal isoform	Adult isoform
β-Myosin heavy chain	α-Myosin heavy chain
Atrial natriuretic peptide	–
Brain natriuretic peptide	–
Creatine kinase B	Creatine kinase M
Glucose transporter-1, glucose transporter-4	–
–	Trifunctional protein

– not expressed or expressed in small amounts

(p42/44 MAP kinase). Initially it was speculated that these pathways are also coupled to the regulation of protein synthesis, but this seems not to be the case in adult terminally differentiated cardiomyocytes. ERK activation leads to a translocation into the nucleus where it phosphorylates the transcription factor Elk-1. Phosphorylated Elk-1 forms a dimer with SRF and binds to the serum response element of the *c-fos* promoter. Newly synthesised c-Fos can then combine with Jun to form AP-1 complexes that trigger cardiac transcription (Neary 1997). By this pathway, ERK activation can modify a couple of cardiac-specific genes. PKC activation also activates protein kinase D (PKD), a rather new type of protein kinases that translocates into the nucleus where it phosphorylates histone deacetylase (HDAC)-II. HDAC-II constitutively represses the activity of the transcription factor Mef-2c that is required for cardiac gene transcription. The initial second messenger of GPCRs linked to $G\alpha_{q11}$, DAG, is responsible for the activation of PKC and a product of activated PLC with IP_3 being another product of this enzyme. IP_3 increases the level of cytosolic calcium via binding to and activation of IP_3 receptors located on the SR. Calcium activates the CaM-kinase II and calcineurin. Calcineurin dephosphorylates NFAT in the cytosol causing its translocation into the nucleus where it interacts with GATA4 and activating hypertrophic foetal genes (Zhang 2002). NFAT can also be directly activated by NHE suggesting another cross point of pathways linked to regulation of protein synthesis and transcriptional control (Hasamitsu et al. 2012). In summary, GPCRs linked to $G\alpha_{q11}$ can activate or inhibit transcription factors via direct phosphorylation (PKC), via activation of MAPK pathways and by activation of CaM-kinase II. The aforementioned Akt-1 pathway that is responsible for the link to the translational machinery is also able to modify transcription factor activity by inhibition of GSK3β. As mentioned before, GSK3β inhibits translational activity, but it also represses the activity of NFAT. Therefore, $G\alpha_{q11}$-dependent signalling indirectly increases NFAT activity. A summary of $G\alpha_{q11}$-dependent signalling that affects transcriptional regulation is given in Fig. 6.2.

GPCRs linked to $G\alpha_{q12/13}$, such as Ang-II receptors, activate a different signal pathway, namely, the Rho-kinase pathway, which is required to activate NADPH oxidase (NOX). Activation of NOX requires also the translocation of p47/p67-phox that is mediated by PKC-δ. Activated NOX increases the cytosolic concentration of reactive oxygen species (ROS). As expected, PKC-δ activation and oxidative stress

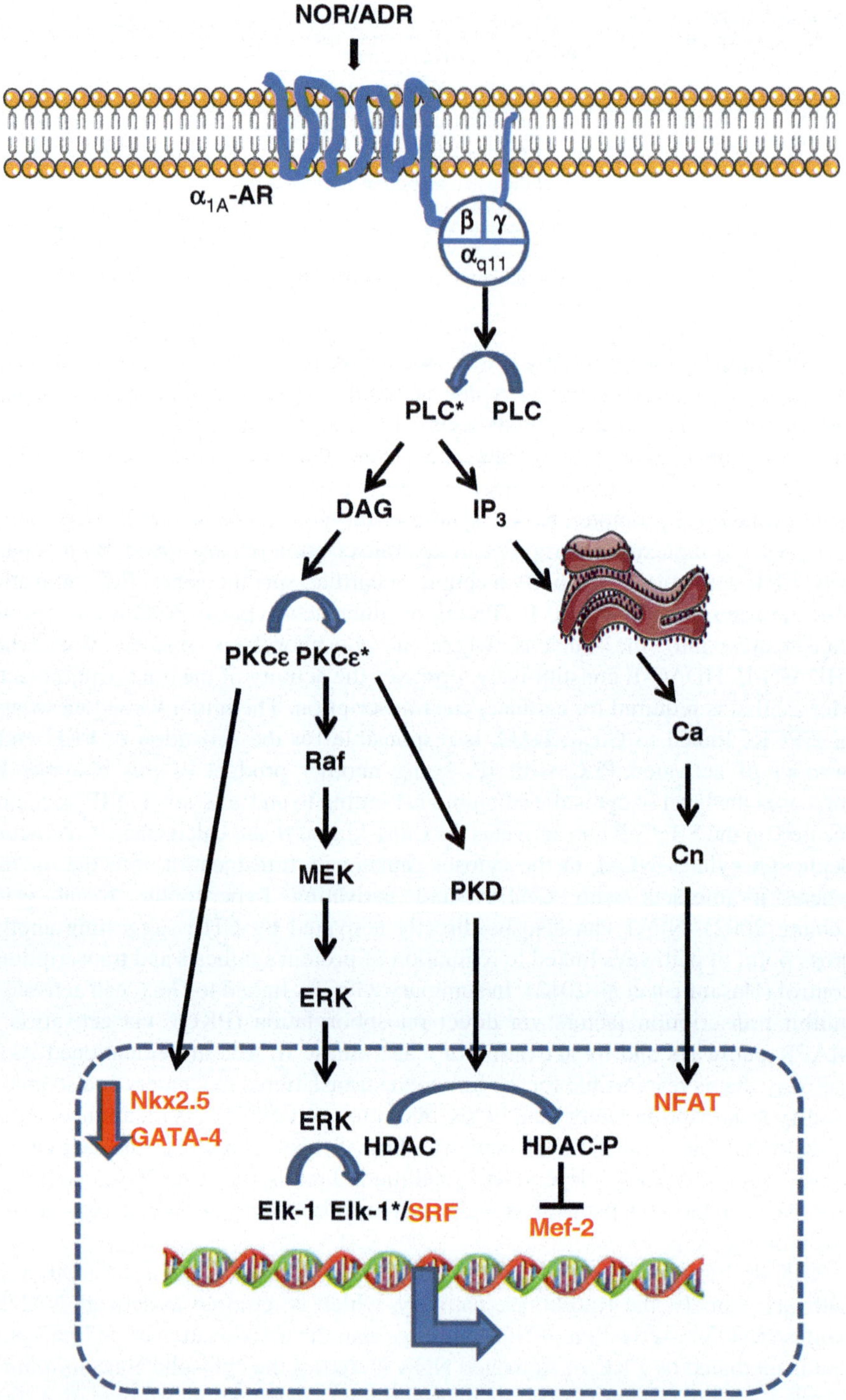

Fig. 6.2 Intracellular signalling of α1A-adrenoceptors leading to cardiac phenotypic adaptation of myocardial hypertrophy via its effect on transcription

are associated with pathophysiological hypertrophy. The aforementioned ROS interferes again with CaMKII. CaMKII can activate histone deacetylase class II (HADAC II) that inhibits the activity of Mef-2. As stimulation of angiotensin receptors is linked to pathophysiological hypertrophy and stimulation of α-adrenoceptors is linked to physiological hypertrophy, Fig. 6.3 shows main signal transduction pathways for Ang-II and indicates differences between the two receptor systems.

GPCRs linked to $G\alpha_s$ activate adenylyl cyclase that increases the cytosolic concentration of the second messenger cAMP. In the heart two isoforms of adenylyl cyclase are dominant: adenylyl cyclase V and VI (Ishikawa and Homcy 1997). In adult hearts type V is the main isoform, whereas in neonatal hearts, VI is the dominant isoform. Interestingly, there is a recent clinical trial ongoing evaluating the effect of cardiac administering an adenovirus-5 encoding human adenylyl cyclase type VI, the neonatal isoform of adenylyl cyclase, to patients with heart failure (NCT-00787059). Initial reports in a pig model of heart failure sound promising (Lai et al. 2004). Similarly, normalised expression of adenylyl cyclase V in a mouse model restored normal β-adrenoceptor-dependent coupling without affecting hypertrophy (Tepe and Liggett 1999). cAMP activates PKA that inhibits GSK3β and thereby activates NFAT activity. PKA can also activate the cAMP response element-binding protein (CREB) that is required for proper transcription factor activation. Finally, cAMP activates exchange proteins activated by cAMP (Epac), but their role in cardiac hypertrophy is less understood. Genetic deletion of Epac-1 protected hearts from the development of heart failure but not against hypertrophy (Okumura et al. 2014). PKA phosphorylates L-type calcium channels at the sarcolemmal membrane and thereby increases intracellular calcium and activates CaMKII. GPCRs linked to $G\alpha_s$ indirectly modify transcriptional activation by inhibition of Raf activation that is required for ERK-dependent transcriptional activation. A summary of $G\alpha_s$-dependent pathways is given in Fig. 6.4.

Stretch-activated TRPCs increase intracellular calcium concentration that is again coupled to the activation of CaMKII and calcineurin.

Collectively, the activation of transcription factors changes significantly the phenotype of a hypertrophic cardiomyocyte in comparison with non-hypertrophic cardiomyocytes. Whether this is specific for pathophysiological hypertrophy or not remains elusive. However, during adolescence, it is clear that cardiomyocytes significantly increase their cell shape but replace foetal-type proteins by adult isoforms. Therefore, one may consider the foetal shift mainly as associated with pathophysiological hypertrophy or even directly linked to heart failure.

6.7 Phenotypic Changes in Cardiomyocytes Linked to Heart Failure

In the chapter above signalling pathways have been reviewed that potentially affect transcriptional regulation. But which specific proteins are linked to heart failure and should therefore be targeted in therapy? In general, three different groups of proteins have to be mentioned. First, proteins have to be mentioned that are part of the foetal shift during hypertrophic adaptation such as reexpression of atrial natriuretic peptide (ANP). Second, downregulation of proteins that are required for proper

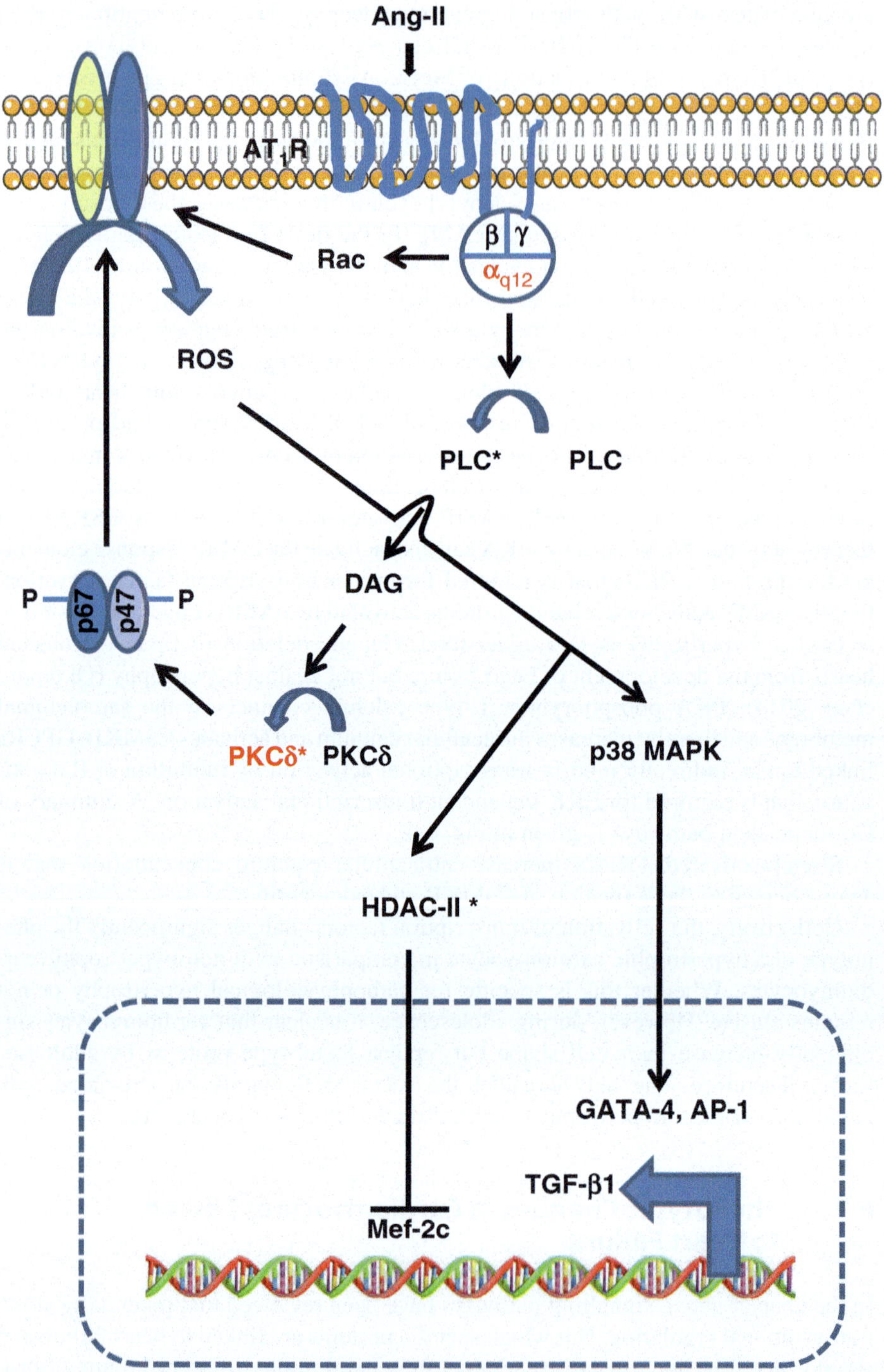

Fig. 6.3 Intracellular signalling pathways of angiotensin receptors affecting stress-activated pathways leading to oxidative stress and functional impairment

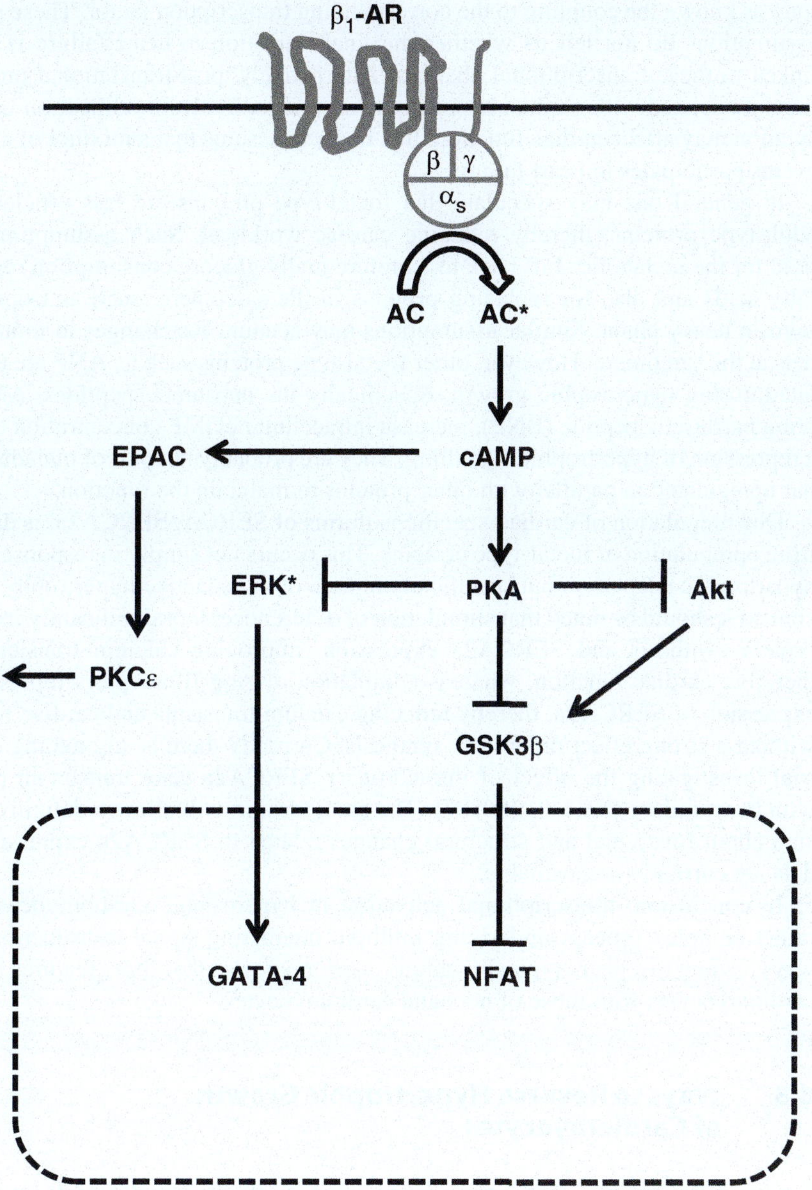

Fig. 6.4 Effect of β-adrenoceptor-dependent signalling on hypertrophic pathways affecting translational capacity and phenotypic adaptation

function of adult cardiac myocytes, such as SR-calcium-ATPase (SERCA), has to be noted. Third, proteins that are normally expressed in adult cardiomyocytes but that are increased in their expression thereafter, such as sodium-calcium exchanger (NCX), are forming another aspect of transcriptional regulation of cardiac phenotypes during hypertrophy. A summary of such proteins is given in Table 6.1 and as long as known the coupling to the corresponding transcription factor. These general assumptions do not tell us whether the final transition to heart failure is indeed linked to these transcriptional changes. Alternatively, post-translational modifications associated with cardiac hypertrophy such as oxidative modification or nitrosylation may affect cardiac function. It is likely to assume that a mixture of all these events participates in heart failure.

In general one may speculate that foetal-type proteins are less efficient than adult-type proteins, thereby reducing cardiac workload. Such assumptions may hold for the metabolic shift back to a preferentially glucose consumption replacing fatty acids and also for replacing proteins in the sarcomere, such as troponin or myosin heavy chain. Similar assumptions may account for changes in ionic channels at the sarcomere. However, other foetal-type proteins such as ANP are upregulated during hypertrophic growth. Specifically the natriuretic peptides, ANP and brain natriuretic peptide (BNP), increase intracellular cGMP concentration causing a repression of hypertrophic signalling. They are probably not part of maladaptation but upregulated in parallel with other proteins to maintain the function.

Downregulation of cardiac-specific isoforms of SERCA, SERCA2a, is different from upregulation of foetal-type proteins. This occurs not simply in response to pro-hypertrophic adaptation but specifically in case of maladaptive hypertrophy. In this context it should be noted that stimulation of α-adrenoceptors significantly increases protein synthesis and SERCA2a expression, improving calcium transients and therefore cardiac function, whereas stimulation of Ang-II receptors represses the expression of SERCA2a, thereby reducing calcium transients and cardiac function without a strong effect on protein synthesis. Currently there is an ongoing clinical trial investigating the effect of intracoronary SERCA2a gene transfer in patients with heart failure (NCT-01966887). The study design will allow a definitive decision about functional and structural changes related to SERCA2a expression, but data are currently not available.

In conclusion, transcriptional activation in hypertrophic cardiomyocytes is a rather complex process underlying multiple interacting signal transduction pathways leading to a phenotypically unique cardiomyocytes that share features of adult cardiomyocytes with those of neonatal cardiomyocytes.

6.8 Ways to Reverse Hypertrophic Growth of Cardiomyocytes

Cardiac hypertrophy is an adaptive process that eventually translates adaptation into heart failure. This translation is mainly based on an unfavourable phenotype of cardiomyocytes, and the term "adverse remodelling" has been chosen for this process. When evaluating ways to reverse hypertrophic growth, it is mainly expected that this unfavourable phenotype is converted into a phenotype allowing the cell to maintain its function in a highly efficient way. For this process the term "reverse remodelling" has been chosen. Whether this requires also a reduction in cell size is not yet clear. However, data for successful reverse remodelling suggest a coupling of this process.

A couple of well-known pathways are able to interrupt the hypertrophic adaptation of cardiomyocytes due to mechanical or neurohumoral stimulation. Outside-inside signalling of growth factors, such as IGF-1, is linked to receptor tyrosine kinases that activate PI3K (p110α) and subsequently the Akt pathway. As described above this pathway increases protein synthesis but is not linked to phenotypic changes. However, PI3K activation suppresses the expression of miRNA 34 which increases hypertrophy and attributes to adverse remodelling. Based on these findings stimulation of such pathways may not be sufficient to normalise heart size but probably adverse remodelling. Four distinct pathways are known to activate the NO/cGMP-dependent pathways. At first, increased workload in the presence of adverse remodelling will reduce ATP levels leading to increased concentrations of AMP that activate AMP kinase (AMPK). AMPK phosphorylates and activates constitutively expressed NO synthases (NOS) in cardiomyocytes. Second, activation of cardiac-specific β_3-adrenoceptors activates NOS in a mechanistically different way because β_3-adrenoceptors are $G\alpha_i$-coupled GPCRs and activate NOS via this pathway. Third, angiotensin I can be converted into Ang-(1–7) by cardiac-specific ACE2. Ang-(1–7) binds to Mas receptors that again activate NOS. Fourth, reexpression of foetal-type proteins in cardiomyocytes increases the expression of natriuretic peptides (ANP, BNP) that can act in an autocrine way on cardiomyocytes via binding to and activation of receptors coupled to particular guanylyl cyclase, thereby increasing the intracellular concentration of cGMP. cGMP activates protein kinase G (PKG) that reduces workload of cardiomyocytes and represses the opening probability of TRPCs, thereby reducing intracellular calcium concentrations. All these NO/cGMP-dependent events attenuate the development of heart failure from hypertrophied hearts.

Although enough pathways have been identified to stop the progression of the transition from hypertrophy to heart failure, interrupting these pathways does not necessarily lead to a reduction of hypertrophy in a sense of heart size. Whether this is indeed necessary or whether it is sufficient to interrupt adverse remodelling is not clear. However, if hypertrophy in its pure sense of organ enlargement should indeed be reversed, this requires a shift of the relationship between protein synthesis and protein degradation into the direction of degradation, and this can obviously not be done by simply interrupting pathways that increase protein synthesis but also requires an activation of pathways to stimulate protein degradation. From this assumption it is clear that this must be performed by activation of pathways distinct from those involved in pro-hypertrophic signalling.

Myocardial hypertrophy is not simply the reaction of the heart to external stimuli, most prominent hyperactivation of receptors linked to SNS or RAAS. Locally derived factors such as intermedin, adrenomedullin, vasopressin and others contribute and modify locally the general activation of neurohumoral-activated pathways. An unexpected mechanistic insight in the ability of the heart to revert from a hypertrophic stage into a non-hypertrophic stage came from clinical observations with left ventricular assist devices (LVAD). LVAD has been used to mechanistically unload the heart. LVAD was initially intended as a bridging therapy for patients with end-stage heart disease requiring transplantation to improve survival. The idea behind this concept was simply to bridge the time lap between diagnosis and waiting for a donor heart. However, it turned out that mechanical unloading substantially reduced the size of cardiomyocytes and improved cardiac function. Mechanistically, vascular endothelial growth factor (VEGF) was identified as a major factor responsible for these favourable effects of LVAD. VEGF induces angiogenesis in hypertrophic hearts, and this seems to be required for physiological hypertrophy. Furthermore, VEGF improves cardiac function directly in cardiomyocytes (own unpublished data). In pathophysiological hypertrophy, however, cardiac expression of VEGF is low. VEGF is regulated by hypoxia-induced transcription factor (HIF-1α) that requires copper as a cofactor. Copper depletion of hypertrophied hearts reduces VEGF expression. VEGF activity has to be translated into intracellular signals by binding to and activation of VEGF receptors. Two distinct VEGF receptors have been described. VEGF receptor 1 (VEGF-1R) activates cGMP pathways that interfere with protein synthesis. Therefore, cardiac hypertrophy is reduced. VEGF receptor 2 (VEGF-2R), however, activates MAP kinase pathways and contributes to phenotypic changes of cardiomyocytes. Interestingly, HIF-1α increases the expression of VEGF-1R and VEGF. Therefore, VEGF and VEGF-dependent pathways are unique in the ability to reverse hypertrophy.

6.9 Therapeutic Options

The most obvious option to attenuate a further progress of myocardial hypertrophy into heart failure is already in clinical practice. This consists of a combined blockade of the SNS (i.e. by β-blockers) or by blockade of the RAAS (i.e. by ACE inhibitors, AT1R blocker, aliskiren, a direct renin antagonist, and spironolactone, an aldosterone antagonist). Each of these treatment procedures has been positively confirmed to delay the progression of hypertrophy into heart failure, either individually or in combination with others. Moreover, the treatments improve life quality and improve longevity. Additive therapeutic options in case of heart failure are diuretics, calcium antagonists, heart rate reduction and statins. As long as these therapeutic options reduce afterload, it remains elusive whether this really affects hypertrophy. From experimental studies that directly couple mechanical load to hypertrophy, there is no clear linkage. From studies with LVAD, it may be possible that reverse pathways are distinct from pro-hypertrophic stimulation and dependent indeed on mechanical unloading. However, patients exposed to LVAD are normally patients out-treated with standard therapeutics including reduction in afterload, and still LVAD has an additive effect. Therefore, LVAD may be linked to other specific pathways not targeted by diuretics or calcium antagonists. Nevertheless, the current treatment options are still not convincing, and the risk to develop heart failure if hypertrophy is diagnosed is still high, and the prognosis for patients with heart failure is unacceptably bad. Therefore, there is a great interest to further develop therapeutic options that target hypertrophic pathways. Currently, a couple of new drugs are in clinical testing. These are related to the signal transduction pathways mentioned above. Recombinant human neuregulin (rhNRG)-1β activates tyrosine kinase receptors that potentially convert pathophysiological hypertrophy in physiological hypertrophy by downregulation of miRNA 34 and activation of mTOR pathways. There is currently a phase III trial with 1,600 heart failure patients ongoing in which rhNRG-1 is added to standard therapy. BRL-37433 selectively activates β_3-adrenoceptors and increases the activity of nNOS in cardiomyocytes (Watts et al. 2013). Despite promising in vitro data, there are currently no clinical trials in progress. A recent progress in RAAS treatment was observed in the PARADIGM-HF trial which showed a significant improvement in strong clinical outcomes such as survival due to a combined blockade of AT_1R and neprilysin, resulting in improvement of natriuretic peptide function in cardiomyocytes (McMurray et al. 2014). The study underlines the concept that a combination of inhibiting pathological hypertrophic pathways and increasing protective pathways is indeed superior to selective inhibition of pathophysiological hypertrophy pathways. However, data on cardiac hypertrophy are warranted. There are a couple of further drugs in development that are not yet specifically intended to be used in hypertrophic models but depending on safety may potentially become attractive in the field. In this direction one should mention direct activators of soluble guanylyl cyclase, activators of AMPL targets directed against the Rho/Rock-pathway and inhibitors of GRK2.

6.10 Conclusive Remarks

Cardiac hypertrophy is associated with a high risk to develop heart failure. This is most likely linked to the phenotypic differentiation of hypertrophied cardiomyocytes. Whether mechanical loading is indeed a major component of pro-hypertrophic stimulation is not yet clear. However, mechanical unloading activates pathways that are linked to reversal of cardiac hypertrophy. It remains unclear at the moment whether anti-hypertrophic effects are additive to reverse remodelling, in other words whether normalisation of phenotypic changes is sufficient or whether reduction in cell size is of further benefit. Nevertheless, the current progress in the development of treatment regimes affecting hypertrophied cardiomyocytes, specifically those obtained with LVAD, is indicating that even pathological hypertrophy is a reversible process, and this is at least in contrast to the long-term dogma that only physiological hypertrophy is a reversible process.

References

Anwar A, Taimor G, Korkususz H, Schreckenberg R, Berndt T, Abdallah Y, Piper HM, Schlüter KD (2005) PKC-independent signal transduction pathways increase SERCA expression in adult rat cardiomyocytes. J Mol Cell Cardiol 39:911–919

Bisping E, Wakula P, Poteser M, Heinzel FR (2014) Targeting cardiac hypertrophy: toward a causal heart failure therapy. J Cardiovasc Pharmacol 64:293–305

Daien CI, Fesler P, du Cailar G, Daien V, Mura T, Dupuy AM, Cristol JP, Ribstein J, Combe B, Morel J (2013) Etanercept normalises left ventricular mass in patients with rheumatoid arthritis. Ann Rheum Dis 72:881–887

Galli D, Gobbi G, Carrubbi C, di Marcantonio D, Benedetti L, de Angelis MGC, Meschi T, Vaccarezza M, Sampaolesi M, Mirandola P, Vitale M (2013) The role of PKCε-dependent signaling for cardiac differentiation. Histochem Cell Biol 139:35–46

Hasamitsu T, Nakamura TY, Wakabayashi S (2012) Na(+)/H(+) exchanger 1 directly binds to calcineurin A and activates downstream NFAT signaling, leading to cardiomyocyte hypertrophy. Mol Cell Biol 32:3265–3280

Heger J, Schlüter KD (2013) Renin and TGFII/M6P receptor system in cardiac biology. Sci World J. doi:10.1155/2013/260298

Heymans S, Corsten MF, Verhesen W, Carai P, van Leeuwen REW, Custers K, Peters T, Hazebroek M, Stöger L, Wijnands E, Janssen BJ, Creemers EE, Pinto YM, Grimm D, Schürmann N, Vigorito E, Thum T, Stassen F, Yin X, Mayr M, de Windt LJ, Lutgens E, Wouters K, de Winther MPJ, Zacchinga S, Giacca M, van Bilsen M, Papageorgiou AP, Schroen B (2013) Macrophage microRNA-155 promotes cardiac hypertrophy and failure. Circulation 128:1420–1432

Hinrichs S, Heger J, Schreckenberg R, Wenzel S, Euler G, Arens C, Bader M, Rosenkranz S, Caglayan E, Schlüter KD (2011) Controlling cardiomyocytes length: the role of renin and PPARγ. Cardiovasc Res 89:344–352

Huntgeburth M, Tiemann K, Shahverdyan R, Schlüter KD, Schreckenberg R, Gross ML, Mödersheim S, Caglayan E, Müller-Ehmsen J, Ghanem A, Vantler M, Zimmermann WH, Böhm M, Rosenkranz S (2011) Transforming growth factor β1 oppositely regulates the hypertrophic and contractile response to β-adrenergic stimulation in the heart. PLoS One 6, e26628

Ishikawa Y, Homcy CJ (1997) The adenyly cyclase as integrators of transmembrane signal transduction. Circ Res 80:297–304

Ivester CT, Kent RL, Tagawa H, Tsutsui H, Imamura T, Cooper G, McDermott PJ (1993) Electrically stimulated contraction accelerates protein synthesis rates in adult feline cardiomyocytes. Am J Physiol 265:H666–H674

Kiyonaka S, Kato K, Nishida M, Mio K, Numaga T, Sawaguchi Y, Yoshida T, Wakamori M, Mori E, Numata T, Ishii M, Takemoto H, Ojida A, Watanabe K, Uemura A, Kurose H, Morii T, Kobayashi T, Sato Y, Sato C, Hamachi I, Morio Y (2009) Selective and direct inhibition of TRPC3 channels underlies biological activities of a pyrazole compound. Proc Natl Acad Sci U S A 106:5400–5405

Kolwicz SC Jr, Tian R (2011) Glucose metabolism and cardiac hypertrophy. Cardiovasc Res 90:194–201

Komuro I, Kaida T, Shibazaki Y, Kurabayashi M, Katoh Y, Hoh E, Takaku F, Yazaki Y (1990) Stretching cardiac stimulates protooncogene expression. J Biol Chem 265:3595–3598

Lai NC, Roth DM, Gao MH, Tang T, Dalton N, Lai YY, Spellman M, Clopton P, Hammond HK (2004) Intracoronary adenovirus encoding adenylyl cyclase VI increases left ventricular function in heart failure. Circulation 110:330–336

Mathew J, Sleight P, Lonn E, Johnstone D, Pogue J, Yi Q, Bosch J, Sussex B, Probstfield J, Yusuf S, HOPE Investigators (2001) Reduction in cardiovascular risk by regression of electrocardiographic markers of left ventricular hypertrophy by the angiotensin-converting enzyme inhibitor ramipril. Circulation 104:1615–1621

McMurray JJ, Packer M, Desai AS, Gong J, Lefkowitz MP, Rizkala AR, Rouleau JL, Shi VC, Solomon SD, Swedberg K, Zile MR, PARADIGM-HF Investigators and Committees (2014) Angiotensin-neprilysin inhibition versus enalapril in heart failure. N Eng J Med 371:993–1004

Mehlsen J, Gleerup G, Haedersdal C, Winther K (1993) Beneficial effects of isradipine on the development of left ventricular hypertrophy in mild hypertension. Am J Hypertens 6:95S–97S

Neary JT (1997) MAPK cascades in cell growth and death. News Physiol Sci 12:286–293

Okumura S, Fujita T, Cai W, Jin M, Namekata I, Mototani Y, Jin H, Ohnuki Y, Tsuneoka Y, Kurotani H, Tsunematsu T, Bai Y, Suzuki S, Hidaka Y, Umemura M, Ichikawa Y, Yokoyama U, Sato M, Ishikawa F, Izumi-Nakaseko H, Adachi-Akahane S, Tanaka H, Ishikawa Y (2014) Epac1-dependent phospholamban phosphorylation mediates the cardiac response to stresses. J Clin Invest 124:2785–2801

Palazzesi S, Musumeci M, Catalano L, Patrizio M, Stati T, Michienzi S, di Certo MG, Mattei E, Vitelli L, Marano G (2006) Pressure overload causes cardiac hypertrophy in beta1-adrenergic and beta2-adrenergic receptor double knockout mice. J Hypertension 24:563–571

Pinson A, Schlüter K-D, Zhou XJ, Schwartz P, Kessler-Icekson G, Piper HM (1993) Alpha- and Beta-adrenergic stimulation of protein synthesis in cultured adult ventricular cardiomyocyted. J Mol Cell Cardiol 25:477–490

Pönicke K, Schlüter KD, Heinroth-Hoffmann I, Seyfarth T, Goldberg M, Osten B, Piper HM, Brodde OE (2001) Noradrenaline-induced increase in protein synthesis in adult rat cardiomyocytes: involvement of only α_{1A}-adrenoceptors. Naunyn-Schmiedeberg's Arch Pharmacol 364: 444–453

Rosenkranz S, Flesch M, Amann K, Häuseler C, Kilter H, Seeland U, Schlüter KD, Böhm M (2002) Alterations of β-adrenergic signaling and cardiac hypertrophy in transgenic mice overexpressing TGF-β1. Am J Physiol 283:H1253–H1262

Ruf S, Piper HM, Schlüter KD (2002) Specific role for the extracellular signal-regulated kinase pathway in angiotensin II-but not phenylephrine-induced cardiac hypertrophy in vitro. Pflügers Arch–Eur J Physiol 443:483–490

Saadane N, Alpert S, Chalifour LE (1999) Expression of immediate early genes, GATA-4, and Nkx-2.5 in adrenergic induced cardiac hypertrophy and during regression in adult mice. Br J Pharamcol 127:1165–1176

Sabri A, Steinberg SF (2003) Protein kinase C isoform-selective signals that lead to cardiac hypertrophy and the progression of heart failure. Mol Cell Biochem 251:97–101

Sabri A, Pak E, Alcott SA, Wilson BA, Steinberg SF (2000) Coupling function of endogenous alpha(1)- and beta-adrenergic receptors in mouse cardiomyocytes. Circ Res 86:1047–1053

Schäfer M, Pönicke K, Heinroth-Hoffmann I, Brodde OE, Piper HM, Schlüter KD (2001) Beta-adrenoceptor stimulation attenuates the hypertrophic effect of alpha-adrenoceptor stimulation in adult rat ventricular cardiomyocytes. J Am Coll Cardiol 37:300–307

Schäfer M, Schäfer C, Piper HM, Schlüter KD (2002) Hypertrophic responsiveness of cardiomyocytes to α- or β-adrenoceptor stimulation requires sodium-proton-exchanger-1 (NHE-1) activation but not cellular alkalization. Eur J Heart Fail 4:249–254

Schlüter KD, Piper HM (1992) Trophic effects of catecholamines and parathyroid hormone on adult ventricular cardiomyocytes. Am J Physiol 263:H1739–H1746

Schlüter KD, Wenzel S (2008) Angiotensin II: a hormone involved in and contributing to pro-hypertrophic cardiac networks and target of anti-hypertrophic cross-talks. Pharmacol Therapeut 119:311–325

Schlüter KD, Millar BC, McDermott BJ, Piper HM (1995a) Regulation of protein synthesis and degradation in adult ventricular cardiomyocytes. Am J Physiol 269:C1347–C1355

Schlüter KD, Zhou XJ, Piper HM (1995b) Induction of hypertrophic responsiveness to isoproterenol by TGF-β in adult rat cardiomyocytes. Am J Physiol 269:C1311–C1316

Schreckenberg R, Taimor G, Piper HM, Schlüter K-D (2004) Inhibition of Ca^{2+}-dependent PKC isoforms unmasks ERK-dependent hypertrophic growth evoked by phenylephrine in adult ventricular cardiomyocytes. Cardiovasc Res 63:553–560

Taimor G, Schlüter KD, Helmig S, Piper HM (2004) Transcription activator protein 1 mediates α- but not β-adrenergic hypertrophic growth responses in adult cardiomyocytes. Am J Physiol 286:H2369–H2375

Tepe NM, Liggett SB (1999) Transgenic replacement of type V adenylyl cyclase identifies a critical mechanism of beta-adrenergic receptor dysfunction in the Gαq overexpressing mouse. FEBS Lett 458:236–240

Watts VL, Sepulveda FM, Cingolani OH, Ho AS, Niu X, Kim R, Miller KL, Vandegaer K, Bedja D, Gabrielson KL, Rameau G, O'Rourke B, Kass DA, Barouch LA (2013) Anti-hypertrophic and anti-oxidant effect of beta3-adrenergic stimulation in myocytes requires differential neuronal NOS phosphorylation. J Mol Cell Cardiol 62:8–17

Wenzel S, Abdallah Y, Helmig S, Schäfer C, Piper HM, Schlüter KD (2006) Contribution of PI 3-kinase isoforms to angiotensin II- and α-adrenoceptor signalling pathways in cardiomyocytes. Cardiovasc Res 71:352–362

Zhang W (2002) Old and new tools to dissect calcineurin's role in pressure-overload cardiac hypertrophy. Cardiovasc Res 53:294–303

Protein Degradation in Cardiomyocytes: Target Proteins and Clinical Consequences

7

Oliver Drews

Abstract

Protein degradation or proteolysis is casually considered an uneventful process of decomposition and hence its role in health and disease underestimated. Current understanding though describes proteolysis as part of versatile and dynamic signalling networks, encompassing hundreds, potentially more than a thousand, of proteins. Proteolytic events are adjusted in response to physiological stimuli and in turn have the capacity to alter cellular as well as systemic function. Similarly, controlled protein degradation influences the pathophysiology with dysregulated or insufficient proteolysis considered driving disease progression, or even being part of early pathophysiological development. Therefore, a rapidly increasing number of studies incorporate involvement of proteolytic regulation in their hypothesis. Our view of proteolytic regulation via autophagy and the ubiquitin-proteasome pathway in particular evolved to the level that they are considered decisive in disease development. Indeed, their manipulation in vitro as well as in vivo influence cardiomyocyte function and cardiac disease outcome. Substrates subject to degradation range from single proteins (ubiquitin-proteasome pathway) to complete organelles (autophagy). In this chapter, protein degradation via the UPS and autophagy are discussed within the context of physiological function and pathophysiological impact.

O. Drews
Division of Cardiovascular Physiology, Institute of Physiology and Pathophysiology, Heidelberg University, Im Neuenheimer Feld 326, Heidelberg 69120, Germany
e-mail: drews@physiologie.uni-heidelberg.de

K.-D. Schlüter (ed.), *Cardiomyocytes – Active Players in Cardiac Disease*,
DOI 10.1007/978-3-319-31251-4_7

7.1 Key Function of the Ubiquitin-Proteasome System

All eukaryotic cells rely on the presence and basic function of the ubiquitin-proteasome system (UPS), including cardiomyocytes. Ubiquitin-proteasome-dependent degradation is the major non-lysosomal pathway to control the abundance of intracellular proteins (Rock et al. 1994), including structural, contractile, and signaling proteins (see Sect. 7.1.13). Proteins with direct participation in the human UPS are estimated to exceed 2 % of the genome (Drews and Taegtmeyer 2014), but a much lower number seems to be essential, such as proteasome subunits. Protein degradation via the UPS is a particularly evolved pathway that identifies its substrates via a series of highly specific events (Fig. 7.1) (Glickman and Ciechanover 2002). Thus, the UPS has the capacity to modulate cardiac structure and function precisely or at large. In the canonical pathway for protein degradation via the UPS, substrate proteins are labeled by poly-ubiquitination (Thrower et al. 2000) (synonymously called ubiquitination). For this, ubiquitin requires activation, conjugation, and ligation by E1, E2, and E3 enzymes, respectively. Subsequently, poly-ubiquitinated proteins are recognized by 26S proteasome complexes and degraded into peptides. Ubiquitin itself is a small protein that is recycled during the process and not subject to proteasome-dependent degradation (see Sect. 7.1.3). Removal of ubiquitin is performed by deubiquitinases, which may counteract targeting substrates for proteasome degradation. Multiple steps in the pathway require energy in the form of ATP. However, proteolysis itself is a catalytic process not requiring ATP. According to experiments performed in vitro, peptides generated by proteasomes have a length of 3–22 amino acids with about two thirds being smaller than eight residues (Kisselev et al. 1999). Their terminal amino acids depend on the active site that cleaves the polypeptide chain (see Sect. 7.1.4). In general, peptides are considered to be further degraded and the amino acids recycled in the cell. On the other hand, the peptides seem to play an important role in the immune response in the form of epitopes in antigen presentation (see Sect. 7.1.6). Thus, if an autoimmune response or pathogens are contributing factors in cardiac disease, the UPS may play a role in facilitating inflammation and pathogen elimination. Targeting the UPS in human diseases is a rapidly growing field of research and clinical application. So far, targeting is mostly directed against proteasome complexes and clinically approved for specific anticancer chemotherapy against multiple myeloma (Drews and Taegtmeyer 2014; Kisselev et al. 2012). Comprehensive understanding of the UPS in the physiology and pathophysiology of the heart may be rewarded with novel therapies for heart disease and minimizing adverse effects in anticancer therapy.

7.1.1 Ubiquitin Accumulation and Reduced Proteasome Function in Heart Disease

Several disease phenotypes are associated with accumulation of ubiquitinated proteins and less proteasome activity compared to control groups. Studies of human heart tissue include dilated and hypertrophic cardiomyopathy, ischaemic heart

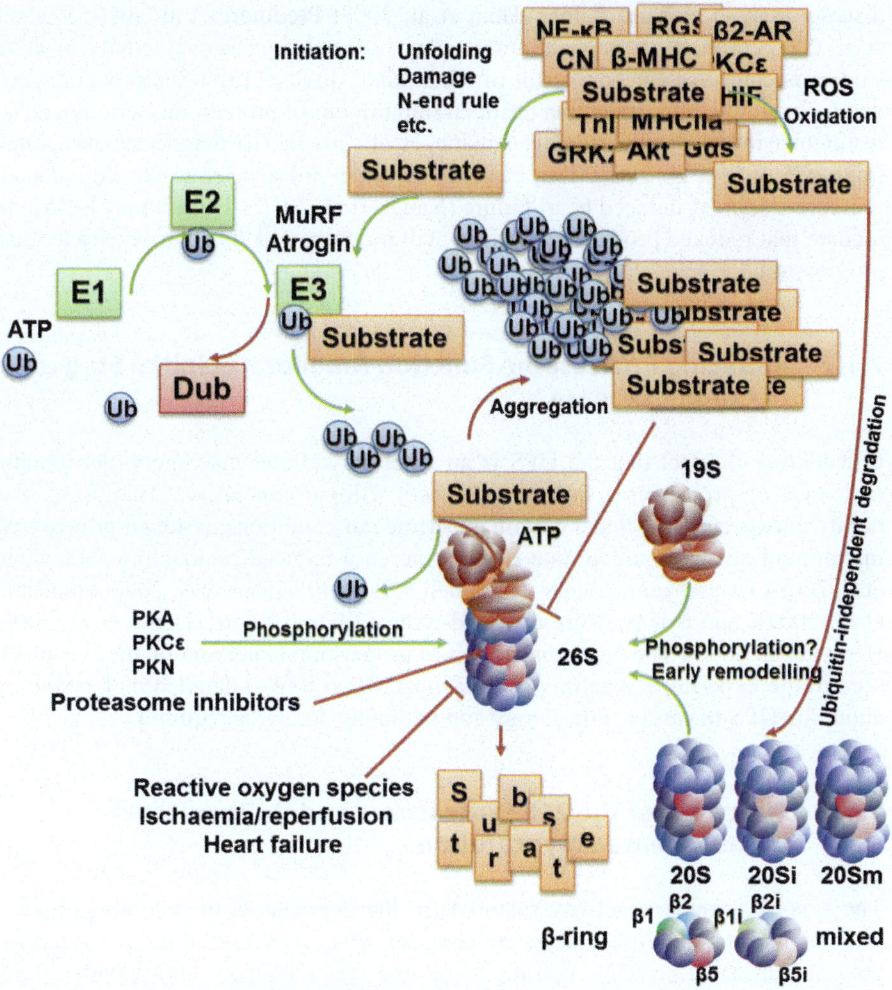

Fig. 7.1 Regulation of the ubiquitin-proteasome system in cardiac physiology and pathophysiology. Individual or multiple substrate proteins are poly-ubiquitinated subsequently to an initiation signal. Ubiquitinated substrates are degraded by 26S proteasomes. Alternatively, substrates may be degraded by 20S proteasomes without prior ubiquitination upon oxidative damage. Subpopulations of 20S proteasome exist in hearts due to alternative incorporation of proteolytic β-subunits in β-rings. During compensated hypertrophic remodeling, the composition of subpopulations is altered in favor of inducible (i) subunits. In addition, 26S activity is increased due to enhanced 19S with 20S proteasome assembly. In contrast, proteasome activity is impaired after ischaemia/reperfusion injury and in failing hearts. Impairment may be caused by dephosphorylation and oxidative damage. Excessive accumulation of substrate proteins associated with heart failure and familial cardiomyopathies may aggregate and contribute to proteasome impairment. Recovery or enhancement of proteasome activity can be achieved via kinases, such as PKA

disease, as well as heart failure (Hein et al. 2003; Predmore et al. 2010; Weekes et al. 2003). To date, it is uncertain whether reduced proteasome activity in these studies is causative for or a result of the studied disease. From reciprocal experimental settings, it is certain that artificial impairment of proteasome activity can (i) result in cardiac dysfunction (Herrmann et al. 2013), (ii) deteriorate ischaemic injury (Powell et al. 2005; Tian et al. 2012), and (iii) accelerate development of pressure-overload-induced heart failure (Ranek et al. 2015). Thus, it may be safe to assume that reduced proteasome activity will more than likely contribute to disease progression.

7.1.2 Reducing Proteasome Function Ameliorates Initial Stages of Cardiac Disease

Considering the fact that the UPS is an essential cellular machinery, detrimental outcomes of proteasomal intervention seem a logic conclusion. Therefore, it is highly unexpected that under certain experimental conditions, artificial proteasome impairment seems to ameliorate (i) isoproterenol-induced remodeling (Stansfield et al. 2008), (ii) ischaemic injury (Campbell et al. 1999; Huang et al. 2008; Stansfield et al. 2007), and (iii) pressure-overload-induced heart failure (Depre et al. 2006; Hedhli et al. 2008). The beneficial effects of proteasomal intervention may be attributed to the experimental setups and readouts. Therefore, a detailed understanding about the UPS in cardiac physiology and pathophysiology is required.

7.1.3 Functions of the 26S Proteasome and Its Regulation in the Diseased Myocardium

The canonical enzyme activity required for the degradation of poly-ubiquitinated proteins harbors the 26S proteasome complex (Fig. 7.1). The complex recognizes poly-ubiquitinated proteins, unfolds them, and degrades them into peptides (Lam et al. 2002). Ubiquitin is not degraded, but rather recycled after deubiquitination (Verma et al. 2002). All of the mentioned functions of the 26S proteasome complex are attributed to the 19S regulatory particle (synonymous with 19S proteasome complex), except degradation. The latter is carried out by the 20S core particle (synonymous with 20S proteasome complex). Thus, each 26S proteasome complex is actually a megaprotein complex of 19S and 20S proteasome complexes, ranging at about 2000 kDa (Drews et al. 2007b). Up to two 19S proteasome complexes can dock to one single 20S proteasome complex (Peters et al. 1993), which is also called 30S proteasome complex. Assembly of 26S proteasomes and unfolding of proteins targeted for degradation are at least in part dependent on ATP (Orino et al. 1991; Peters et al. 1994). Therefore, 26S proteasome activity is measured in the presence of ATP. A biochemical study on cardiac tissue emphasizes the utilization of optimized protocols to facilitate the detection of differences in a comparative analysis and provides specific suggestions including optimal ATP concentration (Powell et al. 2007).

As a consequence, cardiac disease phenotypes associated with energy depletion may cause 26S proteasome dissociation and loss of 26S proteasome activity, which in turn may further disease progression. Indeed, acute global cardiac ischaemia of isolated hearts results in reduced 26S proteasome activity within 30 minutes (Powell et al. 2005). However, similar reduction in activity was observed for 20S proteasome complexes as well. Therefore, additional regulatory events seem to be involved and potentially directed at 20S proteasome complexes (discussed in Sect. 7.1.8). Further reduction of proteasome activities by pharmaceutical intervention in the aforementioned study was associated with deteriorated cardiac function.

In contrast to the activity loss induced by acute ischaemia, acute or early cardiac remodeling induced by isoproterenol or transaortic constriction is associated with increased 26S proteasome activity (Depre et al. 2006; Drews et al. 2010; Ranek et al. 2015). The additional activity is paralleled by higher quantities of assembled 26S proteasome complexes, while 20S proteasome quantities remain stable, indicating stimulated 19S with 20S proteasome assembly (Drews et al. 2010). Reduction of proteasome activities by pharmaceutical intervention under the aforementioned conditions reduced cardiac remodeling and improved cardiac function (Depre et al. 2006; Hedhli et al. 2008; Stansfield et al. 2008). The severity of stimulus for remodeling and the duration likely impact the level of 26S proteasome the regulation and eventually lead to reduced 26S proteasome activity in failing hearts (Ranek et al. 2015; Tsukamoto et al. 2006). In such animal models of heart failure, observations converge with studies on humans, reporting reduced 26S proteasome activity and accumulation of ubiquitinated proteins (Hein et al. 2003; Predmore et al. 2010; Weekes et al. 2003). Notably, mechanical unloading of failing hearts by ventricular assist devices improves cardiac function concomitantly with recovery of proteasome expression and activity (Kassiotis et al. 2009; Wohlschlaeger et al. 2010), suggesting UPS and cardiac function are interdependent.

7.1.4 Structure and Function of the 20S Proteasome: The Proteolytic Core Complex

The dynamic regulation of 26S and 20S proteasome complexes can be concerted as, previously mentioned for acute ischaemia or divergent (Drews et al. 2010). The individual analysis of both complexes is elaborate, and often assumed to be congruent. Therefore, only a limited number of manuscripts precisely identify whether 26S or 20S proteasome activity was measured and even less actually provide both measurements. Access to the proteolytic chamber of the 20S proteasome complex is limited by the two entry points to the complex, a gated pore of only a few angstrom in diameter (Groll et al. 2000). Regulatory particles, such as the 19S proteasome, enable opening of that pore and facilitate substrate protein sequestration into the proteolytic chamber. Access to the pore and measurement of 20S proteasome activity can be facilitated by low detergent levels, such as sodium dodecyl sulfate (Dahlmann et al. 1993; Tanaka et al. 1986). Differentiating 26S and 20S proteasome activities based on ATP dependence and pore opening revealed individual regulation of these proteasome complexes during isoproterenol-induced cardiac remodeling (Drews et al. 2010). While 26S proteasome activity is stimulated via increased 26S assembly, 20S proteasome function is at least in part reduced.

Inside the proteolytic chamber of 20S proteasome complexes, there exist six locations that catalyze hydrolysis of a protein (Unno et al. 2002). Structurally, these six locations are formed by three distinct β-subunits, which all exist twice in the complex, as each 20S proteasome complex consists of two sets of 7 α- and 7 β-subunits (Fig. 7.1). The three β-subunits conferring proteolytic activity have different preferences for the residue at which a polypeptide chain is cleaved and are mostly distinguished by their chymotrypsin- (subunit β5, hydrophobic preference), trypsin- (β2, alkaline), and caspase-like (β1, acidic) activity (Groll et al. 1999; Rivett 1989). Assaying all proteasome activities requires detailed elaboration of protocols as particularly 26S proteasome complexes are unstable (Drews et al. 2007b). In contrast to 26S vs. 20S proteasome activities though, it is long known that chymotrypsin-, trypsin-, and caspase-like activities can be differentially altered and ratios vary. In isoproterenol-induced cardiac remodeling, all 26S proteasome activities are increased uniformly, but only the trypsin- and caspase-like 20S proteasome activities are reduced (Drews et al. 2010). The chymotrypsin-like 20S proteasome activity remained unaffected.

7.1.5 Regulation of Proteasome Activity by Inducible Subunits: Structural Basis for Cardiac 20S Proteasome Subpopulations

Only a few studies with focus on the myocardium include a comprehensive and differential analysis of proteasome activities. One reason for excluding analyses on different types of peptidolytic preference may be that the corresponding predominant mechanism for regulation was poorly investigated in cardiac tissue: subunits β1, β2, and β5 can be replaced during proteasome assembly by β1i, β2i, and β5i, respectively (Fig. 7.1) (Huber et al. 2012). The switch in subunits affects activities due to the different preferences at which they cleave a polypeptide chain. Briefly, caspase-like activity is reduced in favor of trypsin- and chymotrypsin-like activities (Gaczynska et al. 1994, 1996). Changes in activities are considered to originate from allosteric effects as well (Gaczynska and Osmulski 2014). In cardiac tissue, protein expression of β1i, β2i, and β5i is much lower than their counterparts under unstimulated conditions. Therefore, they may not have been detected in early studies. Their expression is much higher in cells of the immune system, and they are further inducible by, e.g., cytokines, such as interferon-γ (Strehl et al. 2005). Therefore, they are also called the inducible or immuno-subunits. Comprehensive characterization of cardiac proteasome complexes by proteomics identified their presence under baseline conditions (Gomes et al. 2006; Zong et al. 2006). We could further demonstrate that incorporation of inducible subunits at different ratios contributes to the existence of 20S proteasome subpopulations with distinct activity profiles (Fig. 7.1) (Drews et al. 2007a). Interestingly, proteasome subpopulations vary between organs, such as heart and liver, potentially signifying specific requirements for cleaved peptides generated by proteasomes (Drews et al. 2007a). In addition, differences in proteasomes between organs or cell types may support preferential pharmaceutical targeting.

7.1.6 Role of Inducible 20S Proteasome Subunits in Inflammatory Heart Disease and Cytoprotection

So far, the stimulated incorporation of inducible subunits is thought to serve two particular purposes: the immune response and cytoprotection (Angeles et al. 2012; Basler et al. 2013; Kruger and Kloetzel 2012). Increased cleavage after hydrophobic residues mediated by incorporation of inducible subunits along with the aid of the proteasome activator complex PA28 and additional aminopeptidases seem to improve antigen presentation via major histocompatibility I complexes (Goldberg and Rock 1992; Strehl et al. 2005). For example, mice lacking β1i generate fewer influenza nucleoprotein-specific cytotoxic T lymphocyte precursors (Van Kaer et al. 1994). Inflammation of cardiac tissue is associated with matrix remodeling and fibrosis (Braunwald 2008), which in turn may negatively impact cardiac contractility, relaxation, and action potential propagation among other factors. Thus, via its participation in antigen presentation, the UPS has the potential to influence inflammatory cardiac diseases positively by facilitating pathogen elimination and negatively by enabling inflammation. In a murine model of coxsackievirus-induced myocarditis, the expression of inducible proteasome subunits is associated with effective virus elimination and resistance against chronic myocarditis (Szalay et al. 2006). Besides of the role in antigen processing, a more immediate cytoprotective role has been ascribed to inducible subunits. They are thought to improve removal of proteins, which are damaged by interferon γ-induced oxidative stress (Seifert et al. 2010). The cytoprotective role of inducible subunits via enhancement of proteasome activities is controversially regarded, because substrate recognition and entrance to the proteolytic chamber of proteasomes via their pores are considered key rate limiting steps in degradation of poly-ubiquitinated proteins (Nathan et al. 2013). In murine cardiomyocytes lacking β5i and subjected to interferon-γ treatment, poly-ubiquitinated proteins and oxidatively damaged proteins accumulate (Opitz et al. 2011). Therefore, increased proteotoxic stress was suggested to be a mechanism contributing to exacerbated coxsackievirus-induced myocarditis in β5i knockout mice (Opitz et al. 2011). Whether pro-inflammatory or cytoprotective qualities of inducible proteasome subunits modulate the physiology and pathophysiology of cardiomyocytes in noninflammatory models remains to be shown.

7.1.7 Cardioprotection Against Oxidative Stress by Proteasome Degradation

With regard to oxidative stress, proteasomes seem to assume specialized roles (Drews and Taegtmeyer 2014). On one hand, they have been ascribed the ability to degrade oxidatively damaged proteins even without prior poly-ubiquitination

(Fig. 7.1) (Grune et al. 1997; Shringarpure et al. 2003). Thus, maintaining or even enhancing proteasome activity may be particularly important for cardiac disease phenotypes with reportedly increased oxidative stress, such as aging (Dai et al. 2012), ischaemia/reperfusion injury (Steinberg 2013), or heart failure (Koitabashi and Kass 2011). In support of this notion are investigations showing increased accumulation of oxidatively damaged proteins and deteriorating cardiac function after artificial reduction of proteasome activity during ischaemia/reperfusion injury (Divald et al. 2010; Divald and Powell 2006). It could be shown that PA28α (a subunit of the proteasome activator complex PA28) overexpression in cultured neonatal cardiomyocytes enhances proteasome function and reduces the accumulation of hydrogen peroxide-induced accumulation of oxidatively damaged proteins (Li et al. 2011b). Furthermore, overexpression of PA28α in mice subjected to ischaemia/reperfusion injury limited infarct size and prevented postreperfusion cardiac dysfunction (Li et al. 2011a). Interestingly, proteasome activity can be stimulated by ischaemic preconditioning (Asai et al. 2009). Thus, enhanced proteasome activity by ischaemic preconditioning may be a contributing factor for conferring cardioprotection by the procedure. This hypothesis is corroborated by investigations showing increased proteasome activity after ischaemia/reperfusion injury in animals subjected to ischaemic preconditioning (Churchill et al. 2010; Divald et al. 2010). Pharmaceutical intervention with proteasome activity interferes with the cardioprotective effect of ischaemic precondition, leading to less preservation of cardiac function (Asai et al. 2009; Churchill et al. 2010; Divald et al. 2010).

7.1.8 Impact of Oxidative Damage Associated with Cardiac Disease on Proteasomes

Although proteasome degradation mitigates the accumulation of oxidatively damaged proteins and hence contributes to cytoprotection against reactive oxygen species, proteasome complexes are also susceptible to oxidative modification (Fig. 7.1) (Drews and Taegtmeyer 2014). Oxidative modification of proteasome subunits in the human myocardium has been implicated as a mechanism for reduced proteasome activity in failing hearts (Predmore et al. 2010). In mice, increased carbonylation and 4-hydroxy-2-nonenal modification of cardiac 20S proteasome subunits concomitant with reduced proteasome activity were reported within treatments of 30 minutes occlusion of the left anterior descending coronary artery followed by 60 minutes reperfusion (Bulteau et al. 2001). Similarly, ischaemia-/reperfusion-induced carbonylation of 19S proteasome subunits is associated with reduced proteasome activity in murine hearts (Powell and Divald 2010). Thus, previous episodes of myocardial infarction may reduce proteasome activities in humans, leaving hearts more prone to additional stress at cellular and organ level. Natural molecular aging of proteasome complexes induced by oxidative stress accumulated over a lifetime may also contribute to vulnerable cardiac conditions with senescence (Farout and Friguet 2006).

7.1.9 Reciprocal Effect of Proteotoxicity in Heart Disease and Proteasome Function

Beside direct oxidative modification, proteasome activity seems negatively affected by accumulation of aberrant myocardial proteins, which may form aggregates (Fig. 7.1) (Bahrudin et al. 2008; Liu et al. 2006; Schlossarek et al. 2014). Proteins aberrant due to mutation, incorrect translation, folding, or stress-induced damage may be cytotoxic/proteotoxic for cardiomyocytes and contribute to their functional impairment (Sandri and Robbins 2013). Such proteins involved in the development of human cardiomyopathies are desmin and cardiac myosin-binding protein Mybpc3 (Goldfarb and Dalakas 2009; McLendon and Robbins 2011; Schlossarek et al. 2011). Both negatively impact proteasome activity (Bahrudin et al. 2008; Liu et al. 2006; Sarikas et al. 2005), and their cellular levels are proteasome activity dependent (Li et al. 2011a; Vignier et al. 2009). Several investigations of human heart failure report accumulation of ubiquitinated deposits/aggregates in cardiomyocytes and link this observation with their cell death and replacement by fibrotic tissue (Hein et al. 2003; Knaapen et al. 2001; Kostin et al. 2003). This suggests that similar mechanisms may contribute to heart failure as in neurodegenerative diseases associated with ubiquitin deposits. Overall, it also substantiates the notion that enhancement of proteasome function has the potential to improve cardiac disease conditions in part by reducing accumulation of ubiquitin deposits, which are cytotoxic.

7.1.10 Regulation of Cardiac Proteasome Activity by Posttranslational Modification

Beside of seemingly unspecific posttranslational modifications (PTMs) induced by oxidative stress, a quickly growing number of specific PTMs are being published (Cui et al. 2014). Many of them remain to be characterized functionally. In particular phosphorylation of proteasome subunits derived from myocardial tissue has been studied in more detail (Fig. 7.1). Several kinases are proposed to regulate cardiac proteasome function in vitro as well as in vivo. Protein kinase A (PKA) is among the foremost candidates. As a central mediator for β-adrenoreceptor signaling in the heart, PKA activity is modulated in basically every cardiac condition requiring neurohumoral adaptation of cardiac output (Hunter and Chien 1999; Lefkowitz et al. 2000; Lohse et al. 2003). In vitro, PKA phosphorylates cardiac 20S proteasome subunits at multiple sites and increases 20S proteasome activity (Lu et al. 2008; Zong et al. 2006). Several sites are already phosphorylated endogenously, and some of them are quantitatively and qualitatively different in a comparison with 20S proteasome complexes derived from liver (Drews et al. 2007a; Lu et al. 2008). Thus, proteasome subpopulations not only differ between organs in the ratio of incorporated inducible subunits but also in PTM.

Acute intracoronary infusion of isoproterenol or forskolin in a canine model increased 26S proteasome activity (Asai et al. 2009). The authors associate increased activity with elevated PKA-dependent 26S proteasome assembly. Elevated 19S with 20S proteasome subunit assembly at stable 20S proteasome complex levels and increased 26S proteasome activity were not observed acutely, but after chronic treatment of isoproterenol for 7 days in hypertrophic hearts of mice (Drews et al. 2010). The delayed observation may be dosage or species dependent. In parallel, 20S proteasome function is partially decreased, which can be recovered by activation of endogenous PKA (Drews et al. 2010). In that model, the rescue effect is specific for 20S proteasome activity only. Continuous β-adrenoreceptor stimulation causes desensitization of downstream signaling (Rockman et al. 2002), which may be attributed to reduced 20S proteasome activity after isoproterenol administration for 7 days. In contrast, phosphorylation sites in 26S proteasome complexes may be protected from phosphatases and hence more stable, which putatively maintains increased 26S proteasome activity. In another context, phosphorylation of subunits in 26S proteasomes was suggested to stabilize the megaprotein complex, while dephosphorylation is associated with activity loss and dissociation into 19S and 20S proteasome particles (Bose et al. 2001; Mason et al. 1996; Satoh et al. 2001). In failing human hearts, a similar suggestion has been provided as a contributing factor for diminished 26S proteasome activities (Day et al. 2013). Presence of doubly capped 20s proteasomes (30S proteasome complexes) and phosphorylated 20S proteasome subunit α7 were detected at lower levels than in nonfailing hearts.

The activities of several additional kinases were reported to be associated with stimulated cardiac proteasome activity, including those of protein kinase C ε (PKCε) (Churchill et al. 2010), protein kinase G (PKG) (Ranek et al. 2013), and protein kinase C-related kinase 1 (PRK1/PKN) (Takagi et al. 2010). Assuming an enhancement of proteasome activity in human heart failure may improve cardiomyocyte function, effective strategies for modulating PTMs of proteasome complexes towards this end would be highly valuable. Within this context, it should be noted that all mentioned kinases have been effectively shown to mitigate ischaemia/reperfusion injury (Asai et al. 2009; Churchill et al. 2010; Takagi et al. 2010) or desmin-related cardiomyopathy (Ranek et al. 2013) in vivo in parallel to stimulating proteasome activity.

7.1.11 Causality Between Altered Proteasome Activity and Substrate Degradation

Canonical substrates subject to degradation by 26S proteasomes are poly-ubiquitinated proteins (Finley 2009). In general, studies reporting analyses on proteasome activity are largely based on reporter peptides. Therefore, it is important to quantify poly-ubiquitinated proteins in parallel. Under most conditions discussed so far, differences in proteasome activity correspond to altered levels in the pool of ubiquitinated proteins. For example, diminished 26S proteasome activity in human heart failure is paralleled by accumulation of ubiquitinated proteins (Predmore et al. 2010), and elevated proteasome activity in early isoproterenol-induced cardiac remodeling is concomitant with a decreased pool of ubiquitinated proteins (Drews et al. 2010). If the results do not match, additional regulatory mechanisms, such as altered ubiquitin ligation and deubiquitination, are likely.

7.1.12 Reporter Proteins for Proteasome Activity

Endogenously expressed reporter proteins, which are constitutively targeted by the UPS, enable assessment of the intracellular proteolytic capacity of 26S proteasomes (Bence et al. 2001; Dantuma et al. 2000). One reporter protein that has been introduced in transgenic mice and studied in cardiomyocytes and heart is a fusion protein of the green fluorescent protein (GFP) and the CL1 sequence (Kumarapeli et al. 2005). The CL1 sequence encodes a 16-amino-acid-long nonnatural peptide called degron, which confers ubiquitin-dependent degradation to proteins (Bence et al. 2001; Gilon et al. 1998). Sequence motifs promoting degradation via proteasomes occur also naturally (Varshavsky 1996), but represent only a part of the pathway for proteasome-dependent degradation and hence limit corresponding conclusions. Furthermore, it has been reported that GFP-CL1 fusion proteins are more prone to aggregation than GFP by itself (Link et al. 2006). Since aggregation changes the quality and accessibility of a protein substrate, it may impact its degradation by the proteasome and potentially even cellular response mechanisms (e.g. autophagy of ubiquitin deposits as discussed later). Additional supporting evidence will aide in corroborating a more general hypothesis about the proteolytic capacity of proteasome complexes. Such evidence may be gained by measuring proteasome activity via reporter peptides. For example, enhancement of proteasome activity by PA28α overexpression in cardiomyocytes was shown by decreased levels of the GFP-CL1 reporter and increased chymotrypsin-like proteasome activity, measured by reporter peptides (Li et al. 2011b).

7.1.13 Endogenous Proteasome Substrates

A frequently cited manuscript reports up to 90 % of intracellular proteins are subject to proteasome degradation (Rock et al. 1994). This statement is frequently misunderstood and, e.g., interpreted in a manner suggesting 9 out of 10 proteins are degraded by proteasomes. However, the study measured protein turnover by pulse-chase experiments. Thus, it applies only to those proteins expressed under the investigated conditions, and their identity was not part of the study. In addition, the authors distinguish between short-lived (20–30 % of all protein mass degraded within one hour; >70 % affected by proteasome inhibition) and long-lived proteins (3–4 % of all protein mass degraded per hour; >98 % affected after preincubated proteasome inhibition) (Rock et al. 1994). Furthermore, the study focused on LB27.4 B lymphoblasts, an immortalized cell line. Primary cardiac cells, such as cardiomyocytes, likely will have lower turnover rates and abundance of polyubiquitinated proteins, due to their limited potential for proliferation and growth.

Nevertheless, a quickly growing number of proteins are reported to be subject to proteasome degradation, and many of them are expressed in cardiac tissue und cultured cardiomyocytes, including structural, contractile, and signaling proteins (Drews and Taegtmeyer 2014; Pagan et al. 2013).

7.1.14 UPS Control of the Primary Cardiomyocyte Structure and Function

Myofibrillar proteins under control of the UPS include beta/slow myosin heavy chain (beta/slow MHC) and MHCIIa (Fig. 7.1) (Fielitz et al. 2007). Since increasing expression of beta/slow MHC is associated with pathological heart failure in humans (Miyata et al. 2000), diminished proteasome activity in these hearts may promote its higher abundance. Notably, impairing beta/slow MHC degradation by proteasomes via knockout of muscle RING finger 1 (MuRF1) and MuRF3 E3 ubiquitin ligases leads to hypertrophic cardiomyopathy with myofibrillar disarray and reduced systolic function (Fielitz et al. 2007). MuRFs are considered muscle-specific E3 ligases and associated with regulating cardiac muscle mass (Willis et al. 2007, 2009). However, impaired degradation of beta/slow MHC in the aforementioned study may not be the only underlying cause for the development of hypertrophic cardiomyopathy in the absence of MuRF1 and MuRF3. Several additional myocardial proteins have been reported to be substrates of the same E3 ligases and are also targeted for proteasome degradation, including PKCε and troponin I (Arya et al. 2004; Kedar et al. 2004). Since PKCε and troponin I are implicated in hypertrophic signaling and cardiomyocyte contractility as well, the aforementioned results illustrate the challenge in formulating linear cause and effect conclusions regarding the UPS. The requirement for knocking out both MuRF1 and MuRF3 for the development of a cardiac phenotype (Fielitz et al. 2007) further exemplifies that targeting of substrate proteins toward proteasome degradation relies on a system of modular redundancy. In addition, targeted substrates may be recovered before degradation by deubiquitination (Fig. 7.1). Besides MuRFs, Atrogin-1/MAFbx is a muscle-specific E3 ligase that is associated with regulating muscle mass and acts through modulation of hypertrophic signaling (Baskin et al. 2014; Li et al. 2004, 2007).

7.1.15 UPS Control of Cardiomyocyte Signaling

The E3 ligase Atrogin-1 regulates cardiac signaling via calcineurin, Akt, and nuclear factor-κB among others (Fig. 7.1) (Li et al. 2004, 2007; Usui et al. 2011). Consequently, it is a major regulator of cardiac tissue mass. Further cardiac signaling proteins underlying proteasome-dependent degradation include members of the G-protein-coupled receptor (GPCR) signaling cascades, such as the β2-adrenoreceptor, the stimulating G-protein Gαs, and the G-protein-coupled receptor kinase 2 (GRK2) (Naviglio et al. 2004; Salcedo et al. 2006; Shenoy et al. 2001). The β2-adrenoreceptor may be an exception, as intracellular proteins are primarily targeted by the UPS, whereas cell surface GPCRs seem to be primarily targeted for lysosomal degradation (Dores and

Trejo 2012). Notably, ubiquitination is a targeting signal for lysosomal degradation as well (Bonifacino and Traub 2003). At this point, it should be emphasized that there exist different types of ubiquitination, such as mono-, multi-, and poly-ubiquitination, and that poly-ubiquitin chains can be linked via lysines in distinct positions. The canonical signal for degradation by 26S proteasome complexes is a poly-ubiquitin chain in which lysines at position 48 are covalently linked to the carboxyl-terminus of another ubiquitin molecule (Thrower et al. 2000).

7.1.16 Upstream Signaling for Ubiquitination of Members in Cardiac Signal Transduction

Several regulator of G-protein signaling (RGS) proteins are regulated via the N-end rule pathway (see Sect. 7.1.12) (Hu et al. 2005). Many substrate proteins require additional processing, damage or aberrant folding, before they are targeted by the UPS (Fig. 7.1). One of the most prominent examples is oxygen-dependent hydroxylation of hypoxia-inducible factors (HIF) (Semenza 2012). For the RGS proteins, S-nitrosylation of N-terminal cysteine followed by further oxidation and arginylation initiates the N-end rule pathway, which targets these proteins for ubiquitin-dependent degradation (Hu et al. 2005). Such additional requirements before ubiquitination of substrate proteins more than likely prevent their undue degradation and the ensuing waste of cellular resources. Although the identity of UPS substrate proteins is rapidly revealed in particular by high-throughput approaches based on proteomics (Drews et al. 2007b), the identity of the corresponding E3 ligases, which covalently link ubiquitin to them, is more often than not unknown. One of the reasons contributing to this unbalanced knowledge is to be found in the previously discussed redundancy, which means that more than one E3 ligase recognizes a substrate. Thus, understanding of gain- or loss-of-function experiments of E3 ligases will be impeded by adaptive targeting of compensatory counterparts.

7.1.17 Impact of Pharmaceutical Intervention with Proteasome Activity on the Heart

Several studies in animal models of cardiac disease phenotypes suggest that moderate inhibition of proteasome activity ameliorates early left and right ventricular remodeling and improves cardiac function (Drews 2014). In part, proteasome inhibition is also associated with reduced inflammation in models of cardiac ischaemia, but adverse effects seem to prevail (Calise and Powell 2012). The main target of proteasome inhibition is the chymotrypsin-like activity of 20S proteasomes by reversible and irreversible binding to the active site (Fig. 7.1), but inhibitors also affect the other proteolytic sites or even non-proteasomal degradation (Gaczynska and Osmulski 2014; Groll and Huber 2004; Kisselev and Goldberg 2001; Kisselev et al. 2012). Since the ratio of active sites and their activities are dependent on 20S proteasome subpopulations (Drews et al. 2007a), different proteasome inhibitors distinctly affect the activity profiles of subpopulations (Kloss et al. 2010). The distinct inhibition of proteasome subpopulations in part explains inhibitor-dependent responses in models of cardiac disease (Calise and Powell 2012; Drews 2014). The composition of proteasome subpopulations is tissue dependent (Drews et al. 2007a; Gomes et al. 2009). Therefore, preferential inhibition of proteasome subpopulations may be advantageous to minimize adverse effects. Proteasome inhibitors are approved for treatment of multiple myeloma in humans (Herndon et al. 2013; Kane et al. 2003). In this context, an increasing number of case studies report cardiotoxicity, which is reversible within weeks to months upon discontinuation of pharmacotherapy based on proteasome inhibition (Drews and Taegtmeyer 2014). Disease-related (oxidative and inflammatory) stress, aging, and subclinical features of heart disease are associated with cardiotoxicity of proteasome inhibitors in these patients (Drews and Taegtmeyer 2014). Therefore, corresponding regulatory mechanisms negatively impacting proteasome activity discussed in previous sections may leave them more prone to proteasome inhibition. Thus, avoiding targeting of cardiac proteasomes in the expanding clinical applications of proteasome inhibition would be preferential.

7.1.18 Concluding Remarks on the UPS with Respect to Cardiac Physiology and Pathophysiology

While our understanding of the underlying mechanisms for proteasome regulation is improving, comprehension of the consequences is still limited. Altogether, it is apparent that human heart disease with established heart failure is accompanied by reduced proteasome activity and increased ubiquitin accumulation. Mechanical unloading of human hearts not only improves cardiac function but also ubiquitin-proteasome-dependent degradation. Furthermore, enhancement of proteasome activity in animal models with reduced cardiac function ameliorates disease outcome, whereas further proteasome inhibition causes deterioration. Thus, the UPS likely suffers impairment in cardiac disease and enhancement seems beneficial. Several studies in animals successfully demonstrate stimulation of proteasome activity in the heart by genetic or posttranslational modification. Disease stage and duration for stimulating proteasome activity will require careful characterization though, because early hypertrophic remodeling of the myocardium is accompanied by increased proteasome activity, and here proteasome inhibition causes mitigated or reversed remodeling. Finally, gained knowledge about the role of the UPS in heart disease impacts therapeutic approaches in anticancer chemotherapy based on proteasome inhibition, because further reduction of proteasome activity in patients with reduced cardiac function may be fatal.

7.2 Autophagy in Heart Disease

Like the ubiquitin-proteasome pathway, autophagy is essential for protein degradation in the heart under baseline conditions. For example, mice with a cardiac-specific inducible deficiency in the autophagy-related gene 5 (Atg5) develop cardiomyopathy marked by left ventricular dilatation and contractile dysfunction without additional challenge (Nakai et al. 2007). The observed cardiac phenotype in Atg5 knockout mice is characterized by increased cross-sectional cardiomyocyte area (hypertrophy), sarcomeric disarray, aggregation as well as accumulation of mitochondria, endoplasmatic reticulum stress, and apoptosis. Interestingly, accumulation of ubiquitinated proteins and increased proteasome activity was observed as well, indicating overlapping and potentially compensatory substrate degradation by the UPS and autophagy. Indeed in several cardiac disease phenotypes, opposite regulation of autophagy and proteasome degradation is reported (Hein et al. 2003; Kassiotis et al. 2009; Sun et al. 2013), suggesting the pathways may at least in part cover a shared pool of substrates.

7.2.1 Similarities Between Autophagy and the Ubiquitin-Proteasome Pathway

Interpretation of studies modulating autophagy in the context of cardiac disease requires profound understanding of the role of molecular components participating in the pathway. Multiple types of autophagy have been defined (Levine and Kroemer 2008) with macroautophagy being the best understood and so far ascribed the most important role in the heart (Nakai et al. 2007; Sandri and Robbins 2013). Macroautophagy is a process exceeding the traditional perception about degradation by far, as it is capable of digesting complete organelles. Therefore, macroautophagy lacks the precision of the UPS to target individual proteins and is rather likely essential for bulk degradation. However, macroautophagy is still a specific process that requires molecular signaling for initiation (Fig. 7.2). Surprisingly, initiation signaling for macroautophagy is highly similar to that in the ubiquitin-proteasome pathway, because small ubiquitin-like proteins, such as Atg12 and microtubule-associated protein 1 light chain 3 (LC3), require activation, conjugation, and ligation via E1, E2, and E3 enzymes much like ubiquitin (Levine and Kroemer 2008). Like for ubiquitination, additional upstream signaling events precede Atg12 and LC3 ligation.

7.2.2 Early Steps in Macroautophagy: Vesicle Nucleation in Cardiac Remodeling and Heart Failure

For macroautophagy, Beclin-1 and class III phosphoinositide 3-kinase (PI3K)-dependent vesicle nucleation mark initiation of the process, and hence their expression is utilized as markers to study activation or deactivation of the pathway (Fig. 7.2). For example, Beclin-1 RNA and protein expression is elevated in human heart failure among other markers for macroautophagy and seems to normalize upon mechanical unloading by left ventricular assist devices (Kassiotis et al. 2009). Results like these suggest that autophagy and heart failure are linked and that autophagy may be elevated in a compensatory manner to support cellular function in heart failure. Within this context, it may be speculated further that autophagy compensates at least in part

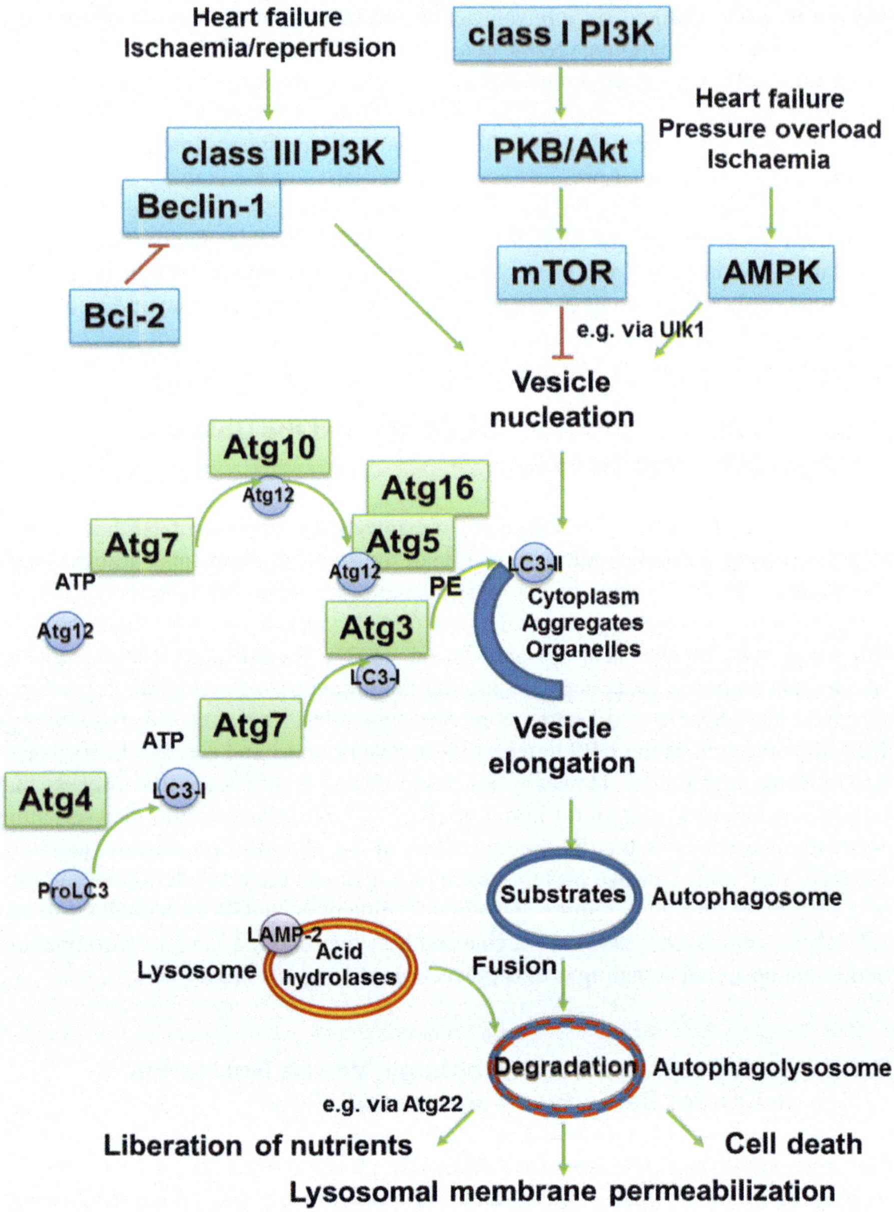

Fig. 7.2 Regulation of macroautophagy in cardiac disease. Stimulation of autophagy in heart failure is mediated by Beclin-1 and class III phosphoinositide 3-kinases (PI3K) or AMP-activated protein kinase (AMPK). In contrast, autophagy is inhibited via mTOR during compensated hypertrophic remodeling or other stimulators of class I PI3K-protein kinase B (PKB/Akt) signaling. Autophagosome formation requires vesicle nucleation and elongation, a process that is similar to ubiquitination and utilizes the ubiquitin-like modifiers Atg12 and LC3. The autophagic flux further requires fusion with lysosomes for successful degradation. Nutrients are released by permeases such as Atg22, but disturbances may lead to lysosomal membrane permeabilization (LMP) and uncontrolled release of contents, including acid hydrolases. LMP or excessive autophagy may lead to cell death. Indicated are proteins, which are utilized widely as markers for autophagy

for the described reduction in proteasome activity in heart failure (Predmore et al. 2010). In fact, opposite regulation of autophagy and the UPS is observed after mechanical unloading using left ventricular assist devices in humans, describing concomitant gains in proteasome expression and activity with improving cardiac function (Kassiotis et al. 2009; Wohlschlaeger et al. 2010). In a murine model of heart failure induced by transaortic constriction, modulating autophagy via Beclin-1 expression diminished pathological remodeling in heterozygous Beclin-1 knockout mice and accentuated pathological remodeling in Beclin-1 transgenic mice (Zhu et al. 2007). Therefore, it was concluded that induced autophagy in this model is rather maladaptive than compensatory. On the other hand, Beclin-1 heterozygous mice may be more stress tolerant due to alternative compensatory mechanisms and Beclin-1 overexpression may require more careful titration of the protein. At least, autophagy is associated with cell death (type II), which can be suppressed by inhibitors of autophagy, and is considered to promote cell death via excessive degradation (Levine and Yuan 2005; Shimizu et al. 2004).

7.2.3 Vesicle Elongation via LC3: A Useful Intracellular Reporter in Combination with GFP

Following vesicle nucleation is vesicle elongation, which gradually segregates part of the cytosol, protein aggregates, pathogens, or complete organelles (e.g., mitophagy) into autophagosomes (Fig. 7.2) (Gutierrez et al. 2004; Priault et al. 2005; Ravikumar et al. 2004). Vesicle elongation requires conjugation of the ubiquitin-like protein Atg12 to Atg5. Together with Atg16, the protein complex promotes conjugation of the ubiquitin-like protein LC3 to phosphatidylethanolamine (PE) (Walczak and Martens 2013). At this point, it should be noted that Atg5 knockout mice die briefly after birth potentially due to starvation, as nutrients can be liberated less efficiently with the defect in macroautophagy (Kuma et al. 2004). For this and other reasons, macroautophagy is considered a pro-survival pathway under conditions when the availability of nutrients is limited. Conjugation to PE converts LC3-I into LC3-II, which can be separated by SDS-PAGE and detected by Western blotting. Furthermore, LC3-II is associated with autophagic vesicles in the cells. Fusion of LC3 to GFP has been utilized successfully to estimate autophagy in a lineage of murine atrial cardiomyocytes (HL-1 cells) in vitro (Hamacher-Brady et al. 2006) as well as transgenic mice expressing the fusion protein in cardiomyocytes in vivo (Mizushima et al. 2004; Zhu et al. 2007). LC3 accumulation in autophagosomes may occur due to several reasons, which roughly fall into two categories: elevated formation or diminished removal. Under conditions with normal autophagic flux, autophagosomes are efficiently processed and disappear quickly (Mizushima et al. 2010). However, cardiac disease phenotypes may not only alter the amount of substrates subjected to autophagy and formation of autophagosomes but also their removal. Therefore, accumulation of LC3-GFP reporter and similar marker proteins may be attributed to impaired autophagosome processing as well. As a consequence, the cause for accumulated autophagosomes requires further analysis to differentiate between increased and partially halted autophagy.

7.2.4 Fusion of Autophagosomes with Lysosomes, Degradation, and Recycling: Part of the Autophagic Flux

Autophagy is a lysosomal degradation pathway. As such, the content of autophagosomes is degraded, liberated, and available for recycling after fusion of autophagosomes with lysosomes (Fig. 7.2). All steps of autophagy are addressed by the term autophagic flux. Lysosome-associated membrane protein 2 (LAMP-2) seems to be critical for fusion as LAMP-2 deficiency in humans as well as genetically-induced LAMP-2 deficiency in mice are associated with accumulation of autophagic vacuoles, cardiomyopathy, and myopathy (Danon's disease) (Nishino et al. 2000; Tanaka et al. 2000). Fused autophagosomes are called autophagolysosomes. Their content is degraded by proteases, including lysosomal cathepsins (Appelqvist et al. 2013). Investigations of cardiomyocytes and mice with cathepsin L deficiency showed increased hypertrophy induced by phenylephrine in vitro or transaortic banding in vivo (Sun et al. 2013). In parallel, accumulation of autophagosomes, deteriorated cardiac function, and increased mortality were observed. The results illustrate the importance to differentiate between increased autophagic vacuole formation and removal by additional experimental means. For example, the analytical utilization of lysosomal inhibitors in a study of HL-1 cells in vitro enabled to reveal that both formation and downstream lysosomal degradation of autophagosomes are impaired under the condition of simulated ischaemia/reperfusion (Hamacher-Brady et al. 2006). Furthermore, enhanced autophagic flux following simulated ischaemia/reperfusion and reduced activation of pro-apoptotic Bax was observed after overexpression of Beclin-1. Thus, accumulation of autophagosomes after ischaemia/reperfusion injury appears to be a compensatory mechanism with the capability of further enhancement to improve its cardioprotective potential. Additional experimental approaches to study autophagic flux have been reviewed comprehensively (Mizushima et al. 2010). In particular, early investigations of autophagy relied on the analysis of autophagic vacuoles by microscopy or electron microscopy. Thus, it cannot be deduced from the provided data, which type of autophagy was observed and whether autophagy was increased or halted. Therefore, such observations are discussed as autophagy in general.

Release of amino acids for recycling is mediated by permeases, such as Atg22 (Yang et al. 2006). However, accidental release of contents, such as cathepsins, may occur under conditions of stress or excessive autophagy and potentially induce cell death, e.g., via direct or indirect activation of caspases (Boya and Kroemer 2008). The process reminds of mitochondrial permeabilization, is referred to as lysosomal membrane permeabilization (LMP), and provides an explanation for adverse effects of overexpressing autophagy-related proteins under certain conditions.

7.2.5 Regulation of Autophagy via mTOR and Consequences in Heart Disease

Autophagy is considered a pro-survival pathway under conditions when the availability of nutrients is limited. As such, autophagy requires links to cellular signaling of low energy. One of the best characterized mechanisms is negative regulation of autophagy via the mammalian target of rapamycin (mTOR; Fig. 7.2) (Kim and Guan 2015). Since rapamycin inhibits mTOR activity, pharmaceutical therapy based on rapamycin will be also associated with increased autophagy. Rapamycin administration in mice subjected to transaortic constriction reduces cardiac hypertrophy and improves cardiac function (McMullen et al. 2004). However, autophagy was not investigated in the study. Furthermore, rapamycin is modulating a multitude of pathways in addition to autophagy (Laplante and Sabatini 2009). Therefore, establishment of a direct causality between rapamycin-induced autophagy and altered cardiac remodeling is challenging.

Negative regulation of autophagy via mTOR can be exacted by several tyrosine kinase receptors via class I PI3K-protein kinase B (PKB/Akt) signaling, such as insulin or insulin-like growth factor (IGF) receptors (Levine and Kroemer 2008; Shintani and Klionsky 2004). Besides of energy-dependent negative regulation of autophagy via mTOR, mechanisms for energy-dependent activation of autophagy exist. AMP-activated protein kinase (AMPK) is a cellular sensor for available energy, which plays an important role in the myocardial response to ischaemia, pressure overload, and heart failure (Zaha and Young 2012). AMPK promotes autophagy by activating the mammalian autophagy-initiating kinase Ulk1 under glucose starvation, and high mTOR activity prevents Ulk1 activation under nutrient sufficiency (Kim et al. 2011). Indeed, glucose deprivation of rat cardiomyocyte-derived H9c2 cells and primary rat neonatal cardiomyocytes induces accumulation of autophagic vacuoles and LC3-II, which can be pharmaceutically intervened with by inhibitors of autophagy (Aki et al. 2003; Matsui et al. 2007). However, while autophagy inhibition under glucose starvation reduces cell death in H9c2 cells (Aki et al. 2003), it increases cell death in primary cardiomyocytes (Matsui et al. 2007). In H9c2 cells, IGF-1 and class I PI3K overexpression both accelerate glucose starvation-induced autophagic cell death (Aki et al. 2003). In primary cardiomyocytes, AMPK activation and mTOR inactivation mediate glucose starvation-induced autophagy (Matsui et al. 2007). As previously mentioned, AMPK plays a role in myocardial ischaemia, and the study indicates that it potentially mediates ischaemia-induced autophagy. In contrast, ischaemia-/reperfusion-induced autophagy seems to be mediated by Beclin-1 and not AMPK (Matsui et al. 2007).

7.2.6 Dynamic Regulation of Autophagy in Hypertrophic Remodeling and Heart Failure

An early study on the nature of cardiomyocyte death associated with dilated cardiomyopathy already revealed the presence of prominent cytoplasmic vacuolization in cardiomyocytes with marked scarcity of myofibrils (Shimomura et al. 2001). The vacuoles positively reacted with immunostaining for cathepsin D and LAMP-1. Furthermore, vacuolization was observed in proximity to regions suggesting tissue replacement by fibrosis. Therefore, the authors concluded that autophagic degeneration is one of the mechanisms of myocardial cell death (Shimomura et al. 2001).

Short-term stimulation of rat hearts with isoproterenol is associated with a diminished volume fraction of autophagic vacuoles (Pfeifer et al. 1987). In contrast, acute depression of cardiac activity by propranolol or verapamil in rats increased the volume fraction of autophagic vacuoles cardiac tissue (Bahro and Pfeifer 1987). The authors concluded that the acute regulation of autophagy follows the cardiac requirements to promote or inhibit cardiomyocyte growth depending on the workload. In general agreement with this notion is that Atg5-deficient cardiomyocytes are more prone to acute isoproterenol-induced cell death than isolated adult wild-type cardiomyocytes (Nakai et al. 2007). In hearts of mice subjected for 1–4 weeks to transaortic constriction, LC3-II levels are initially decreased, but after 4 weeks increased, suggesting initially diminished and later stimulated autophagic activity (Nakai et al. 2007). The same study reported deteriorated remodeling with premature heart failure in mice with Atg5-dependent deficiency in autophagy. Therefore, a certain level of autophagy seems to be required for normal cardiac remodeling, and the upregulation was suggested to be part of an adaptive response with protective qualities. In contrast, severe transaortic constriction was reported to be associated with an acute increase in Beclin-1, LC3-II, Lamp-1, and cathepsin D (Zhu et al. 2007). The authors note that their model is associated with pressure-overload-induced heart failure. Thus, the different dynamic responses of autophagy to pressure overload in the two aforementioned studies may be attributed to a compensated versus decompensated state of the myocardium. In the decompensated state, Beclin-1 haploinsufficiency is associated with a smaller reduction in fractional shortening after 3 weeks of severe transaortic constriction (Zhu et al. 2007). As a consequence, elevated autophagy in cardiac remodeling of the decompensated myocardium was suggested to be maladaptive. Similarly, no difference in autophagy in humans with compensated cardiac hypertrophy was observed in comparison with a control group, whereas a 10-fold higher number of cardiomyocytes was reported to be affected by increased autophagy in the decompensated myocardium with heart failure (Hein et al. 2003). It should be noted that the authors quantified autophagy via cellular ubiquitin deposits, which they linked to the presence of autophagic vacuoles and referred to as ubiquitin-related autophagic cell death. The relationship of autophagy and ubiquitin accumulation is discussed in Sect. 7.3. Altogether, autophagy seems reduced at early stages of hypertrophic remodeling as seen in animal models and stable in humans with established compensated hypertrophy. In contrast, heart failure is associated with elevated signs of autophagy in animal models and humans. At least in the compensated state, autophagy appears to be beneficial (Nakai et al. 2007).

7.2.7 Acute Autophagic Response to Myocardial Ischaemia/Reperfusion Injury

Acute myocardial injury due to global ischaemia/reperfusion is associated with an increased number of autophagic vacuoles in parallel with mitochondrial degeneration in isolated rabbit hearts (Decker and Wildenthal 1980). The authors noted that lysosomal autophagy is dependent on the duration of hypoxia and that ischaemic insults of up to 40 minutes seem to be followed by reparative autophagy, whereas apparently abortive repair ensues after longer durations. As such, autophagy subsequent to ischaemia/reperfusion injury appears rather compensatory or protective.

This notion is corroborated by in vitro experiments showing that enhancement of macroautophagy protects against ischaemia/reperfusion injury in cultured cardiomyocytes (Hamacher-Brady et al. 2006). Similar observations that autophagic vacuoles accumulate after chronic ischaemia were reported for the human hibernating myocardium (Elsasser et al. 2004). As for heart failure, autophagic vacuoles were observed in parallel with ubiquitin accumulation, cellular degeneration, and nuclear disassembly. The link between ubiquitin and autophagy is discussed in Sect. 7.3.

The impact of chronic ischaemia on autophagy in the myocardium of pigs was analyzed by an unbiased survey of the cardiac proteome and Western blotting (Yan et al. 2005). The study revealed that several proteins required for autophagy, such as cathepsins, heat shock cognate protein Hsc73, Beclin-1, and LC3-II, are successively increased with the number of ischaemic episodes, whereas the incidence of apoptosis declines in parallel. Therefore, autophagy seems not only beneficial after acute ischaemia/reperfusion injury but also in the chronically ischaemic myocardium. In contrast, a study on myocardial ischaemia/reperfusion injury in mice suggests that induction of autophagy during ischaemia is beneficial, whereas it may be detrimental during reperfusion as indicated by reduced ischaemia/reperfusion injury in Beclin-1-haploinsufficient mice (Matsui et al. 2007). Again, adaptive mechanisms may contribute to cardioprotection in Beclin-1-haploinsufficient mice. Furthermore, Beclin-1-dependent activation of autophagy can be inhibited by Bcl-2 in murine hearts (Pattingre et al. 2005). Expression of a Beclin-1 Bcl-2/-x(L)-binding domain mutant in cardiomyocytes is associated with decreased autophagic flux and no protection against ischaemia/reperfusion injury (Hamacher-Brady et al. 2006). Thus, precise titration of proteins regulating autophagy may be required to differentiate between cardioprotection and maladaptation. Differences in extent of ischaemic region, duration for ischaemia/reperfusion phases, as well as species pose a challenge for an all-encompassing conclusion. Still, it appears that at least during ischaemia, autophagy is beneficial. With increasing injury during reperfusion, autophagy may contribute to further damage rather than it is beneficial, potentially in part because the damage is already excessive.

7.3 About the Role of Ubiquitin for Autophagy in Heart Disease

In the previous sections, studies are discussed, which describe ubiquitin-related autophagy in human failing hearts as well as in the human hibernating myocardium (Elsasser et al. 2004; Hein et al. 2003). Additional evidence and the potential link between autophagy and ubiquitin are discussed in the following sections.

7.3.1 Autophagy-Associated Ubiquitin Accumulation in Human Heart Disease

In cardiac tissue of patients in the terminal stage of heart failure as a consequence of ischaemic cardiomyopathy or dilated cardiomyopathy, autophagic cell death as demonstrated by granular cytoplasmic ubiquitin inclusions was reported to be the

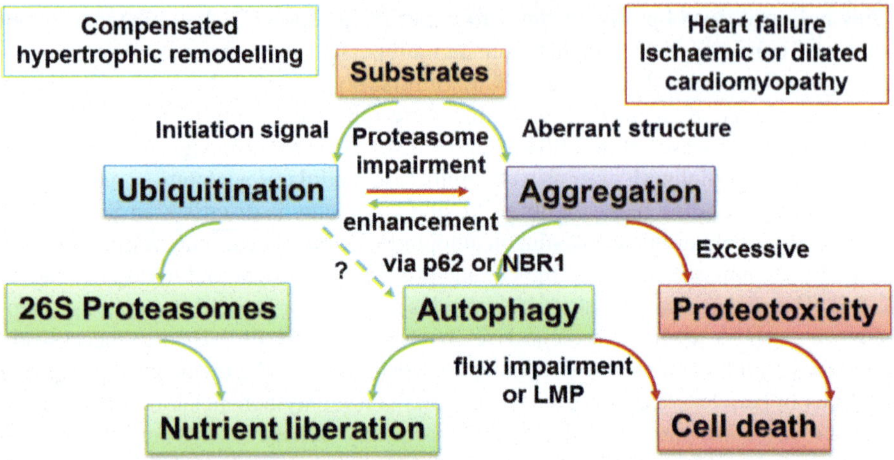

Fig. 7.3 Compensatory degradation of shared substrate pools by proteasomes and autophagy. In healthy hearts or compensated hypertrophic remodeling, substrates are ubiquitinated for canonical degradation by proteasomes. Larger amounts of aberrant proteins tend to aggregation and can be degraded via the ubiquitin-proteasome system or autophagy, but potentially require harmonized regulation of both pathways. Insufficient degradation by either one can be compensated to a certain degree by the other. Recognition of ubiquitinated proteins for autophagy is in part mediated by p62 or NBR1 and may not require protein aggregation. In failing hearts, excessive aggregation is associated with proteotoxicity and cell death. *LMP* lysosomal membrane permeabilization

primary cause of cell death (Knaapen et al. 2001). The conclusion linking granular cytoplasmic ubiquitin inclusions to autophagic cell death is based on similar observations in neurodegenerative diseases (Alves-Rodrigues et al. 1998). Herein, we potentially see compensation by macroautophagy for a prior failure of the UPS to efficiently degrade an excessive amount of ubiquitinated proteins visible as deposits (Fig. 7.3), as suggested for neurodegeneration. Accordingly, granular cytoplasmic ubiquitin inclusions may indicate failed compensation as much as induction of autophagic cell death.

In human idiopathic dilated cardiomyopathy, increased accumulation of autophagic vacuoles and monodansylcadaverine staining was observed in conjunction with increased ubiquitin and decreased cathepsin D immunodetection (Kostin et al. 2003). Although monodansylcadaverine was later demonstrated to have higher affinity for lysosomes (Mizushima et al. 2010), integrating the information suggests that the autophagic flux is diminished in these patients due to reduced cathepsin D levels and hence causes the accumulation of autophagosomes. Furthermore, it appears that autophagosomes target ubiquitinated substrate deposits. Localization of ubiquitin deposits in cells without signs of apoptosis was suggested to be an indicator for autophagic rather than apoptotic cell death (Kostin et al. 2003). Again, the alternative hypothesis for failed compensation of macroautophagy for an impaired UPS may be the case, which would be supported by the largely diminished abundance of cathepsin D in regions of ubiquitin deposits.

7.3.2 Compensatory Regulation of Autophagy and the Ubiquitin-Proteasome Pathway

In a study linking increases in ubiquitin-related autophagic cell death with human heart failure, but not compensated hypertrophic remodeling, the authors speculate "that ubiquitin binds contractile or membrane proteins destined for degradation but that because of proteasomal insufficiency, the complexes are accumulated and might cause nuclear fragmentation" (Hein et al. 2003). In this context, it should be noted that in animal models with early compensatory remodeling, autophagy is apparently decreased (Nakai et al. 2007), whereas proteasome activity is elevated (Depre et al. 2006; Drews et al. 2010). Disruption of the precisely balanced autophagy versus UPS regulation by cathepsin L deficiency results in augmented hypertrophic remodeling in response to phenylephrine and transaortic constriction as well as exacerbated heart failure (Sun et al. 2013). In parallel, higher levels of ubiquitinated proteins in cathepsin L-deficient mice indicate insufficient proteasome activity albeit its upregulation. Thus, autophagy and the UPS may be oppositely fine-tuned to gain cardiac muscle mass, but autophagy assists impaired or overwhelmed proteasome function during disease progression and liberates urgently needed nutrients due to reduced cardiac output and adverse cardiac remodeling (Fig. 7.3).

Unison stimulation of both proteasomal degradation and autophagy are described for mTOR1 inhibition (Zhao et al. 2015). In hypertrophic remodeling with preserved cardiac function, proteasome activity is elevated while ubiquitinated protein levels are decreased (Drews et al. 2010). In the study reporting uniform regulation by mTOR1 inhibition, cellular levels of lysine 48-linked ubiquitinated proteins are increased, while proteasome content or activity is unchanged (Zhao et al. 2015). Conversely, reduced ubiquitination and autophagy in compensatory hypertrophic remodeling of the heart may be in part affected by mTOR1 activation to promote cardiomyocyte growth.

In neonatal rat cardiomyocytes harboring a mutation in αB-crystallin, intracellular protein aggregation and reduced autophagy are observed (Pattison et al. 2011). Overexpression of Atg7 in this model stimulates basal autophagy and attenuates accumulation of aggregated proteins. Induced overexpression of Atg7 also stimulates autophagy in vivo and ameliorated desmin-related cardiomyopathy due to αB-crystallin mutation (Bhuiyan et al. 2013). In the same animal model of desmin-related cardiomyopathy, stimulation of the proteasome activity ameliorates the cardiac phenotype (Li et al. 2011a; Ranek et al. 2013), indicating that autophagy and the UPS target a shared pool of substrates and can compensate for each other. The question remains how autophagy recognizes UPS substrates or UPS insufficiency.

7.3.3 Recognition of Ubiquitin and Other Substrates by Autophagy

More recent models of autophagy incorporate selective recognition mechanisms of ubiquitinated substrates via proteins (Fig. 7.3), such as p62 and NBR1 (Kirkin et al. 2009; Pankiv et al. 2007). The link between ubiquitination and autophagy seems to rely on binding domains for ubiquitin and LC3, which mediate targeting of large deposits of ubiquitinated proteins, including those titled aggresomes (Kirkin et al. 2009; Pankiv et al. 2007). Similarly, endoplasmatic reticulum and mitochondria seem to be targeted for autophagy via proteins, such as Bnip3 or Nix (Hanna et al. 2012; Novak et al. 2010). Further specialization is suggested by the involvement of chaperones, which was titled chaperone-assisted selective autophagy (CASA) (Arndt et al. 2010). In CASA, BAG-3, Hsc70, and HspB8 in conjunction with the chaperone-associated ubiquitin ligase CHIP putatively target damaged Z-disc proteins for degradation, which are subsequently recognized by the autophagic ubiquitin adaptor p62 (Arndt et al. 2010). The study links impaired CASA with Z-disc disintegration and progressive muscle weakness.

7.3.4 Concluding Remarks

Altogether, harmonized regulation of autophagy and the ubiquitin-proteasome pathway may be essential to maintain balanced degradation of a partially shared substrate pool. Alternatively, the two pathways may act purely in a compensatory manner under stress and disease conditions. Current knowledge indicates that harmonized regulation exists for at least specialized parts of each pathway, such as ubiquitin-dependent degradation (Fig. 7.3). Furthermore, the capability for reciprocal compensation of proteasome or autophagy impairment points at a shared substrate pool.

References

Aki T, Yamaguchi K, Fujimiya T, Mizukami Y (2003) Phosphoinositide 3-kinase accelerates autophagic cell death during glucose deprivation in the rat cardiomyocyte-derived cell line H9c2. Oncogene 22:8529–8535

Alves-Rodrigues A, Gregori L, Figueiredo-Pereira ME (1998) Ubiquitin, cellular inclusions and their role in neurodegeneration. Trends Neurosci 21:516–520

Angeles A, Fung G, Luo H (2012) Immune and non-immune functions of the immunoproteasome. Front Biosci (Landmark Ed) 17:1904–1916

Appelqvist H, Waster P, Kagedal K, Ollinger K (2013) The lysosome: from waste bag to potential therapeutic target. J Mol Cell Biol 5:214–226

Arndt V, Dick N, Tawo R, Dreiseidler M, Wenzel D, Hesse M, Furst DO, Saftig P, Saint R, Fleischmann BK, Hoch M, Hohfeld J (2010) Chaperone-assisted selective autophagy is essential for muscle maintenance. Curr Biol CB 20:143–148

Arya R, Kedar V, Hwang JR, McDonough H, Li HH, Taylor J, Patterson C (2004) Muscle ring finger protein-1 inhibits PKC{epsilon} activation and prevents cardiomyocyte hypertrophy. J Cell Biol 167:1147–1159

Asai M, Tsukamoto O, Minamino T, Asanuma H, Fujita M, Asano Y, Takahama H, Sasaki H, Higo S, Asakura M, Takashima S, Hori M, Kitakaze M (2009) PKA rapidly enhances proteasome assembly and activity in in vivo canine hearts. J Mol Cell Cardiol 46:452–462

Bahro M, Pfeifer U (1987) Short-term stimulation by propranolol and verapamil of cardiac cellular autophagy. J Mol Cell Cardiol 19:1169–1178

Bahrudin U, Morisaki H, Morisaki T, Ninomiya H, Higaki K, Nanba E, Igawa O, Takashima S, Mizuta E, Miake J, Yamamoto Y, Shirayoshi Y, Kitakaze M, Carrier L, Hisatome I (2008)

Ubiquitin-proteasome system impairment caused by a missense cardiac myosin-binding protein C mutation and associated with cardiac dysfunction in hypertrophic cardiomyopathy. J Mol Biol 384:896–907

Baskin KK, Rodriguez MR, Kansara S, Chen W, Carranza S, Frazier OH, Glass DJ, Taegtmeyer H (2014) MAFbx/Atrogin-1 is required for atrophic remodeling of the unloaded heart. J Mol Cell Cardiol 72:168–176

Basler M, Kirk CJ, Groettrup M (2013) The immunoproteasome in antigen processing and other immunological functions. Curr Opin Immunol 25:74–80

Bence NF, Sampat RM, Kopito RR (2001) Impairment of the ubiquitin-proteasome system by protein aggregation. Science 292:1552–1555

Bhuiyan MS, Pattison JS, Osinska H, James J, Gulick J, McLendon PM, Hill JA, Sadoshima J, Robbins J (2013) Enhanced autophagy ameliorates cardiac proteinopathy. J Clin Invest 123:5284–5297

Bonifacino JS, Traub LM (2003) Signals for sorting of transmembrane proteins to endosomes and lysosomes. Annu Rev Biochem 72:395–447

Bose S, Brooks P, Mason GG, Rivett AJ (2001) Gamma-interferon decreases the level of 26 S proteasomes and changes the pattern of phosphorylation. Biochem J 353:291–297

Boya P, Kroemer G (2008) Lysosomal membrane permeabilization in cell death. Oncogene 27:6434–6451

Braunwald E (2008) Biomarkers in heart failure. N Engl J Med 358:2148–2159

Bulteau AL, Lundberg KC, Humphries KM, Sadek HA, Szweda PA, Friguet B, Szweda LI (2001) Oxidative modification and inactivation of the proteasome during coronary occlusion/reperfusion. J Biol Chem 276:30057–30063

Calise J, Powell SR (2012) The ubiquitin proteasome system and myocardial ischemia. Am J Physiol Heart Circ Physiol 304:H337–H349

Campbell B, Adams J, Shin YK, Lefer AM (1999) Cardioprotective effects of a novel proteasome inhibitor following ischemia and reperfusion in the isolated perfused rat heart. J Mol Cell Cardiol 31:467–476

Churchill EN, Ferreira JC, Brum PC, Szweda LI, Mochly-Rosen D (2010) Ischaemic preconditioning improves proteasomal activity and increases the degradation of deltaPKC during reperfusion. Cardiovasc Res 85:385–394

Cui Z, Scruggs SB, Gilda JE, Ping P, Gomes AV (2014) Regulation of cardiac proteasomes by ubiquitination, SUMOylation, and beyond. J Mol Cell Cardiol 71:32–34

Dahlmann B, Becher B, Sobek A, Ehlers C, Kopp F, Kuehn L (1993) In vitro activation of the 20S proteasome. Enzyme Protein 47:274–284

Dai DF, Chen T, Johnson SC, Szeto H, Rabinovitch PS (2012) Cardiac aging: from molecular mechanisms to significance in human health and disease. Antioxid Redox Signal 16:1492–1526

Dantuma NP, Lindsten K, Glas R, Jellne M, Masucci MG (2000) Short-lived green fluorescent proteins for quantifying ubiquitin/proteasome-dependent proteolysis in living cells. Nat Biotechnol 18:538–543

Day SM, Divald A, Wang P, Davis F, Bartolone S, Jones R, Powell SR (2013) Impaired assembly and post-translational regulation of 26S proteasome in human end-stage heart failure. Circ Heart Fail 6:544–549

Decker RS, Wildenthal K (1980) Lysosomal alterations in hypoxic and reoxygenated hearts. I. Ultrastructural and cytochemical changes. Am J Pathol 98:425–444

Depre C, Wang Q, Yan L, Hedhli N, Peter P, Chen L, Hong C, Hittinger L, Ghaleh B, Sadoshima J, Vatner DE, Vatner SF, Madura K (2006) Activation of the cardiac proteasome during pressure overload promotes ventricular hypertrophy. Circulation 114:1821–1828

Divald A, Powell SR (2006) Proteasome mediates removal of proteins oxidized during myocardial ischemia. Free Radic Biol Med 40:156–164

Divald A, Kivity S, Wang P, Hochhauser E, Roberts B, Teichberg S, Gomes AV, Powell SR (2010) Myocardial ischemic preconditioning preserves postischemic function of the 26S proteasome through diminished oxidative damage to 19S regulatory particle subunits. Circ Res 106: 1829–1838

Dores MR, Trejo J (2012) Ubiquitination of G protein-coupled receptors: functional implications and drug discovery. Mol Pharmacol 82:563–570

Drews O (2014) The left and right ventricle in the grip of protein degradation: similarities and unique patterns in regulation. J Mol Cell Cardiol 72:52–55

Drews O, Taegtmeyer H (2014) Targeting the ubiquitin-proteasome system in heart disease: the basis for new therapeutic strategies. Antioxid Redox Signal 21:2322–2343

Drews O, Wildgruber R, Zong C, Sukop U, Nissum M, Weber G, Gomes AV, Ping P (2007a) Mammalian proteasome subpopulations with distinct molecular compositions and proteolytic activities. Mol Cell Proteomics 6:2021–2031

Drews O, Zong C, Ping P (2007b) Exploring proteasome complexes by proteomic approaches. Proteomics 7:1047–1058

Drews O, Tsukamoto O, Liem D, Streicher J, Wang Y, Ping P (2010) Differential regulation of proteasome function in isoproterenol-induced cardiac hypertrophy. Circ Res 107:1094–1101

Elsasser A, Vogt AM, Nef H, Kostin S, Mollmann H, Skwara W, Bode C, Hamm C, Schaper J (2004) Human hibernating myocardium is jeopardized by apoptotic and autophagic cell death. J Am Coll Cardiol 43:2191–2199

Farout L, Friguet B (2006) Proteasome function in aging and oxidative stress: implications in protein maintenance failure. Antioxid Redox Signal 8:205–216

Fielitz J, Kim MS, Shelton JM, Latif S, Spencer JA, Glass DJ, Richardson JA, Bassel-Duby R, Olson EN (2007) Myosin accumulation and striated muscle myopathy result from the loss of muscle RING finger 1 and 3. J Clin Invest 117:2486–2495

Finley D (2009) Recognition and processing of ubiquitin-protein conjugates by the proteasome. Annu Rev Biochem 78:477–513

Gaczynska M, Osmulski PA (2014) Harnessing proteasome dynamics and allostery in drug design. Antioxid Redox Signal 21:2286–2301

Gaczynska M, Rock KL, Spies T, Goldberg AL (1994) Peptidase activities of proteasomes are differentially regulated by the major histocompatibility complex-encoded genes for LMP2 and LMP7. Proc Natl Acad Sci U S A 91:9213–9217

Gaczynska M, Goldberg AL, Tanaka K, Hendil KB, Rock KL (1996) Proteasome subunits X and Y alter peptidase activities in opposite ways to the interferon-gamma-induced subunits LMP2 and LMP7. J Biol Chem 271:17275–17280

Gilon T, Chomsky O, Kulka RG (1998) Degradation signals for ubiquitin system proteolysis in saccharomyces cerevisiae. EMBO J 17:2759–2766

Glickman MH, Ciechanover A (2002) The ubiquitin-proteasome proteolytic pathway: destruction for the sake of construction. Physiol Rev 82:373–428

Goldberg AL, Rock KL (1992) Proteolysis, proteasomes and antigen presentation. Nature 357:375–379

Goldfarb LG, Dalakas MC (2009) Tragedy in a heartbeat: malfunctioning desmin causes skeletal and cardiac muscle disease. J Clin Invest 119:1806–1813

Gomes AV, Zong C, Edmondson RD, Li X, Stefani E, Zhang J, Jones RC, Thyparambil S, Wang GW, Qiao X, Bardag-Gorce F, Ping P (2006) Mapping the murine cardiac 26S proteasome complexes. Circ Res 99:362–371

Gomes AV, Young GW, Wang Y, Zong C, Eghbali M, Drews O, Lu H, Stefani E, Ping P (2009) Contrasting proteome biology and functional heterogeneity of the 20 S proteasome complexes in mammalian tissues. Mol Cell Proteomics 8:302–315

Groll M, Huber R (2004) Inhibitors of the eukaryotic 20S proteasome core particle: a structural approach. Biochim Biophys Acta 1695:33–44

Groll M, Heinemeyer W, Jager S, Ullrich T, Bochtler M, Wolf DH, Huber R (1999) The catalytic sites of 20S proteasomes and their role in subunit maturation: a mutational and crystallographic study. Proc Natl Acad Sci U S A 96:10976–10983

Groll M, Bajorek M, Kohler A, Moroder L, Rubin DM, Huber R, Glickman MH, Finley D (2000) A gated channel into the proteasome core particle. Nat Struct Biol 7:1062–1067

Grune T, Reinheckel T, Davies KJ (1997) Degradation of oxidized proteins in mammalian cells. FASEB J 11:526–534

Gutierrez MG, Master SS, Singh SB, Taylor GA, Colombo MI, Deretic V (2004) Autophagy is a defense mechanism inhibiting BCG and Mycobacterium tuberculosis survival in infected macrophages. Cell 119:753–766

Hamacher-Brady A, Brady NR, Gottlieb RA (2006) Enhancing macroautophagy protects against ischemia/reperfusion injury in cardiac myocytes. J Biol Chem 281:29776–29787

Hanna RA, Quinsay MN, Orogo AM, Giang K, Rikka S, Gustafsson AB (2012) Microtubule-associated protein 1 light chain 3 (LC3) interacts with Bnip3 protein to selectively remove endoplasmic reticulum and mitochondria via autophagy. J Biol Chem 287:19094–19104

Hedhli N, Lizano P, Hong C, Fritzky LF, Dhar SK, Liu H, Tian Y, Gao S, Madura K, Vatner SF, Depre C (2008) Proteasome inhibition decreases cardiac remodeling after initiation of pressure overload. Am J Physiol Heart Circ Physiol 295:H1385–H1393

Hein S, Arnon E, Kostin S, Schonburg M, Elsasser A, Polyakova V, Bauer EP, Klovekorn WP, Schaper J (2003) Progression from compensated hypertrophy to failure in the pressure-overloaded human heart: structural deterioration and compensatory mechanisms. Circulation 107:984–991

Herndon T, Deisseroth AB, Kaminskas E, Kane RC, Koti KM, Rothmann MD, Habtemariam BA, Bullock J, Bray JD, Hawes JH, Palmby TR, Jee J, Adams WM, Mahayni H, Brown J, Dorantes A, Sridhara R, Farrell AT, Pazdur R (2013) U.S. food and drug administration approval: carfilzomib for the treatment of multiple myeloma. Clin Cancer Res 19:4559–4563

Herrmann J, Wohlert C, Saguner AM, Flores A, Nesbitt LL, Chade A, Lerman LO, Lerman A (2013) Primary proteasome inhibition results in cardiac dysfunction. Eur J Heart Fail 15:614–623

Hu RG, Sheng J, Qi X, Xu Z, Takahashi TT, Varshavsky A (2005) The N-end rule pathway as a nitric oxide sensor controlling the levels of multiple regulators. Nature 437:981–986

Huang S, Patterson E, Yu X, Garrett MW, De Aos I, Kem DC (2008) Proteasome inhibition 1 h following ischemia protects GRK2 and prevents malignant ventricular tachyarrhythmias and SCD in a model of myocardial infarction. Am J Physiol Heart Circ Physiol 294:H1298–H1303

Huber EM, Basler M, Schwab R, Heinemeyer W, Kirk CJ, Groettrup M, Groll M (2012) Immuno- and constitutive proteasome crystal structures reveal differences in substrate and inhibitor specificity. Cell 148:727–738

Hunter JJ, Chien KR (1999) Signaling pathways for cardiac hypertrophy and failure. N Engl J Med 341:1276–1283

Kane RC, Bross PF, Farrell AT, Pazdur R (2003) Velcade: U.S. FDA approval for the treatment of multiple myeloma progressing on prior therapy. Oncologist 8:508–513

Kassiotis C, Ballal K, Wellnitz K, Vela D, Gong M, Salazar R, Frazier OH, Taegtmeyer H (2009) Markers of autophagy are downregulated in failing human heart after mechanical unloading. Circulation 120:S191–S197

Kedar V, McDonough H, Arya R, Li HH, Rockman HA, Patterson C (2004) Muscle-specific RING finger 1 is a bona fide ubiquitin ligase that degrades cardiac troponin I. Proc Natl Acad Sci U S A 101:18135–18140

Kim YC, Guan KL (2015) mTOR: a pharmacologic target for autophagy regulation. J Clin Invest 125:25–32

Kim J, Kundu M, Viollet B, Guan KL (2011) AMPK and mTOR regulate autophagy through direct phosphorylation of Ulk1. Nat Cell Biol 13:132–141

Kirkin V, Lamark T, Sou YS, Bjorkoy G, Nunn JL, Bruun JA, Shvets E, McEwan DG, Clausen TH, Wild P, Bilusic I, Theurillat JP, Overvatn A, Ishii T, Elazar Z, Komatsu M, Dikic I, Johansen T (2009) A role for NBR1 in autophagosomal degradation of ubiquitinated substrates. Mol Cell 33:505–516

Kisselev AF, Goldberg AL (2001) Proteasome inhibitors: from research tools to drug candidates. Chem Biol 8:739–758

Kisselev AF, Akopian TN, Woo KM, Goldberg AL (1999) The sizes of peptides generated from protein by mammalian 26 and 20 S proteasomes. Implications for understanding the degradative mechanism and antigen presentation. J Biol Chem 274:3363–3371

Kisselev AF, van der Linden WA, Overkleeft HS (2012) Proteasome inhibitors: an expanding army attacking a unique target. Chem Biol 19:99–115
Kloss A, Meiners S, Ludwig A, Dahlmann B (2010) Multiple cardiac proteasome subtypes differ in their susceptibility to proteasome inhibitors. Cardiovasc Res 85:367–375
Knaapen MW, Davies MJ, De Bie M, Haven AJ, Martinet W, Kockx MM (2001) Apoptotic versus autophagic cell death in heart failure. Cardiovasc Res 51:304–312
Koitabashi N, Kass DA (2011) Reverse remodeling in heart failure – mechanisms and therapeutic opportunities. Nat Rev Cardiol 9:147–157
Kostin S, Pool L, Elsasser A, Hein S, Drexler HC, Arnon E, Hayakawa Y, Zimmermann R, Bauer E, Klovekorn WP, Schaper J (2003) Myocytes die by multiple mechanisms in failing human hearts. Circ Res 92:715–724
Kruger E, Kloetzel PM (2012) Immunoproteasomes at the interface of innate and adaptive immune responses: two faces of one enzyme. Curr Opin Immunol 24:77–83
Kuma A, Hatano M, Matsui M, Yamamoto A, Nakaya H, Yoshimori T, Ohsumi Y, Tokuhisa T, Mizushima N (2004) The role of autophagy during the early neonatal starvation period. Nature 432:1032–1036
Kumarapeli AR, Horak KM, Glasford JW, Li J, Chen Q, Liu J, Zheng H, Wang X (2005) A novel transgenic mouse model reveals deregulation of the ubiquitin-proteasome system in the heart by doxorubicin. FASEB J 19:2051–2053
Lam YA, Lawson TG, Velayutham M, Zweier JL, Pickart CM (2002) A proteasomal ATPase subunit recognizes the polyubiquitin degradation signal. Nature 416:763–767
Laplante M, Sabatini DM (2009) mTOR signaling at a glance. J Cell Sci 122:3589–3594
Lefkowitz RJ, Rockman HA, Koch WJ (2000) Catecholamines, cardiac beta-adrenergic receptors, and heart failure. Circulation 101:1634–1637
Levine B, Kroemer G (2008) Autophagy in the pathogenesis of disease. Cell 132:27–42
Levine B, Yuan J (2005) Autophagy in cell death: an innocent convict? J Clin Invest 115:2679–2688
Li HH, Kedar V, Zhang C, McDonough H, Arya R, Wang DZ, Patterson C (2004) Atrogin-1/muscle atrophy F-box inhibits calcineurin-dependent cardiac hypertrophy by participating in an SCF ubiquitin ligase complex. J Clin Invest 114:1058–1071
Li HH, Willis MS, Lockyer P, Miller N, McDonough H, Glass DJ, Patterson C (2007) Atrogin-1 inhibits Akt-dependent cardiac hypertrophy in mice via ubiquitin-dependent coactivation of Forkhead proteins. J Clin Invest 117:3211–3223
Li J, Horak KM, Su H, Sanbe A, Robbins J, Wang X (2011a) Enhancement of proteasomal function protects against cardiac proteinopathy and ischemia/reperfusion injury in mice. J Clin Invest 121:3689–3700
Li J, Powell SR, Wang X (2011b) Enhancement of proteasome function by PA28α overexpression protects against oxidative stress. FASEB J 25:883–893
Link CD, Fonte V, Hiester B, Yerg J, Ferguson J, Csontos S, Silverman MA, Stein GH (2006) Conversion of green fluorescent protein into a toxic, aggregation-prone protein by C-terminal addition of a short peptide. J Biol Chem 281:1808–1816
Liu J, Tang M, Mestril R, Wang X (2006) Aberrant protein aggregation is essential for a mutant desmin to impair the proteolytic function of the ubiquitin-proteasome system in cardiomyocytes. J Mol Cell Cardiol 40:451–454
Lohse MJ, Engelhardt S, Eschenhagen T (2003) What is the role of beta-adrenergic signaling in heart failure? Circ Res 93:896–906
Lu H, Zong C, Wang Y, Young GW, Deng N, Souda P, Li X, Whitelegge J, Drews O, Yang PY, Ping P (2008) Revealing the dynamics of the 20 S proteasome phosphoproteome: a combined CID and electron transfer dissociation approach. Mol Cell Proteomics 7:2073–2089
Mason GG, Hendil KB, Rivett AJ (1996) Phosphorylation of proteasomes in mammalian cells. Identification of two phosphorylated subunits and the effect of phosphorylation on activity. Eur J Biochem 238:453–462
Matsui Y, Takagi H, Qu X, Abdellatif M, Sakoda H, Asano T, Levine B, Sadoshima J (2007) Distinct roles of autophagy in the heart during ischemia and reperfusion: roles of AMP-activated protein kinase and Beclin 1 in mediating autophagy. Circ Res 100:914–922

McLendon PM, Robbins J (2011) Desmin-related cardiomyopathy: an unfolding story. Am J Physiol Heart Circ Physiol 301:H1220–H1228

McMullen JR, Sherwood MC, Tarnavski O, Zhang L, Dorfman AL, Shioi T, Izumo S (2004) Inhibition of mTOR signaling with rapamycin regresses established cardiac hypertrophy induced by pressure overload. Circulation 109:3050–3055

Miyata S, Minobe W, Bristow MR, Leinwand LA (2000) Myosin heavy chain isoform expression in the failing and nonfailing human heart. Circ Res 86:386–390

Mizushima N, Yamamoto A, Matsui M, Yoshimori T, Ohsumi Y (2004) In vivo analysis of autophagy in response to nutrient starvation using transgenic mice expressing a fluorescent autophagosome marker. Mol Biol Cell 15:1101–1111

Mizushima N, Yoshimori T, Levine B (2010) Methods in mammalian autophagy research. Cell 140:313–326

Nakai A, Yamaguchi O, Takeda T, Higuchi Y, Hikoso S, Taniike M, Omiya S, Mizote I, Matsumura Y, Asahi M, Nishida K, Hori M, Mizushima N, Otsu K (2007) The role of autophagy in cardiomyocytes in the basal state and in response to hemodynamic stress. Nat Med 13:619–624

Nathan JA, Spinnenhirn V, Schmidtke G, Basler M, Groettrup M, Goldberg AL (2013) Immuno- and constitutive proteasomes do not differ in their abilities to degrade ubiquitinated proteins. Cell 152:1184–1194

Naviglio S, Pagano M, Romano M, Sorrentino A, Fusco A, Illiano F, Chiosi E, Spina A, Illiano G (2004) Adenylate cyclase regulation via proteasome-mediated modulation of Galphas levels. Cell Signal 16:1229–1237

Nishino I, Fu J, Tanji K, Yamada T, Shimojo S, Koori T, Mora M, Riggs JE, Oh SJ, Koga Y, Sue CM, Yamamoto A, Murakami N, Shanske S, Byrne E, Bonilla E, Nonaka I, DiMauro S, Hirano M (2000) Primary LAMP-2 deficiency causes X-linked vacuolar cardiomyopathy and myopathy (Danon disease). Nature 406:906–910

Novak I, Kirkin V, McEwan DG, Zhang J, Wild P, Rozenknop A, Rogov V, Lohr F, Popovic D, Occhipinti A, Reichert AS, Terzic J, Dotsch V, Ney PA, Dikic I (2010) Nix is a selective autophagy receptor for mitochondrial clearance. EMBO Rep 11:45–51

Opitz E, Koch A, Klingel K, Schmidt F, Prokop S, Rahnefeld A, Sauter M, Heppner FL, Volker U, Kandolf R, Kuckelkorn U, Stangl K, Kruger E, Kloetzel PM, Voigt A (2011) Impairment of immunoproteasome function by beta5i/LMP7 subunit deficiency results in severe enterovirus myocarditis. PLoS Pathog 7:e1002233

Orino E, Tanaka K, Tamura T, Sone S, Ogura T, Ichihara A (1991) ATP-dependent reversible association of proteasomes with multiple protein components to form 26S complexes that degrade ubiquitinated proteins in human HL-60 cells. FEBS Lett 284:206–210

Pagan J, Seto T, Pagano M, Cittadini A (2013) Role of the ubiquitin proteasome system in the heart. Circ Res 112:1046–1058

Pankiv S, Clausen TH, Lamark T, Brech A, Bruun JA, Outzen H, Overvatn A, Bjorkoy G, Johansen T (2007) p62/SQSTM1 binds directly to Atg8/LC3 to facilitate degradation of ubiquitinated protein aggregates by autophagy. J Biol Chem 282:24131–24145

Pattingre S, Tassa A, Qu X, Garuti R, Liang XH, Mizushima N, Packer M, Schneider MD, Levine B (2005) Bcl-2 antiapoptotic proteins inhibit Beclin 1-dependent autophagy. Cell 122:927–939

Pattison JS, Osinska H, Robbins J (2011) Atg7 induces basal autophagy and rescues autophagic deficiency in CryABR120G cardiomyocytes. Circ Res 109:151–160

Peters JM, Cejka Z, Harris JR, Kleinschmidt JA, Baumeister W (1993) Structural features of the 26 S proteasome complex. J Mol Biol 234:932–937

Peters JM, Franke WW, Kleinschmidt JA (1994) Distinct 19 S and 20 S subcomplexes of the 26 S proteasome and their distribution in the nucleus and the cytoplasm. J Biol Chem 269:7709–7718

Pfeifer U, Fohr J, Wilhelm W, Dammrich J (1987) Short-term inhibition of cardiac cellular autophagy by isoproterenol. J Mol Cell Cardiol 19:1179–1184

Powell SR, Divald A (2010) The ubiquitin-proteasome system in myocardial ischaemia and preconditioning. Cardiovasc Res 85:303–311

Powell SR, Wang P, Katzeff H, Shringarpure R, Teoh C, Khaliulin I, Das DK, Davies KJ, Schwalb H (2005) Oxidized and ubiquitinated proteins may predict recovery of postischemic cardiac function: essential role of the proteasome. Antioxid Redox Signal 7:538–546

Powell SR, Davies KJ, Divald A (2007) Optimal determination of heart tissue 26S-proteasome activity requires maximal stimulating ATP concentrations. J Mol Cell Cardiol 42:265–269

Predmore JM, Wang P, Davis F, Bartolone S, Westfall MV, Dyke DB, Pagani F, Powell SR, Day SM (2010) Ubiquitin proteasome dysfunction in human hypertrophic and dilated cardiomyopathies. Circulation 121:997–1004

Priault M, Salin B, Schaeffer J, Vallette FM, di Rago JP, Martinou JC (2005) Impairing the bioenergetic status and the biogenesis of mitochondria triggers mitophagy in yeast. Cell Death Differ 12:1613–1621

Ranek MJ, Terpstra EJ, Li J, Kass DA, Wang X (2013) Protein kinase G positively regulates proteasome-mediated degradation of misfolded proteins. Circulation 128:365–376

Ranek MJ, Zheng H, Huang W, Kumarapeli AR, Li J, Liu J, Wang X (2015) Genetically induced moderate inhibition of 20S proteasomes in cardiomyocytes facilitates heart failure in mice during systolic overload. J Mol Cell Cardiol 85:273–281

Ravikumar B, Vacher C, Berger Z, Davies JE, Luo S, Oroz LG, Scaravilli F, Easton DF, Duden R, O'Kane CJ, Rubinsztein DC (2004) Inhibition of mTOR induces autophagy and reduces toxicity of polyglutamine expansions in fly and mouse models of Huntington disease. Nat Genet 36:585–595

Rivett AJ (1989) The multicatalytic proteinase. Multiple proteolytic activities. J Biol Chem 264:12215–12219

Rock KL, Gramm C, Rothstein L, Clark K, Stein R, Dick L, Hwang D, Goldberg AL (1994) Inhibitors of the proteasome block the degradation of most cell proteins and the generation of peptides presented on MHC class I molecules. Cell 78:761–771

Rockman HA, Koch WJ, Lefkowitz RJ (2002) Seven-transmembrane-spanning receptors and heart function. Nature 415:206–212

Salcedo A, Mayor F Jr, Penela P (2006) Mdm2 is involved in the ubiquitination and degradation of G-protein-coupled receptor kinase 2. EMBO J 25:4752–4762

Sandri M, Robbins J (2013) Proteotoxicity: an underappreciated pathology in cardiac disease. J Mol Cell Cardiol 71C:3–10

Sarikas A, Carrier L, Schenke C, Doll D, Flavigny J, Lindenberg KS, Eschenhagen T, Zolk O (2005) Impairment of the ubiquitin-proteasome system by truncated cardiac myosin binding protein C mutants. Cardiovasc Res 66:33–44

Satoh K, Sasajima H, Nyoumura KI, Yokosawa H, Sawada H (2001) Assembly of the 26S proteasome is regulated by phosphorylation of the p45/Rpt6 ATPase subunit. Biochemistry 40:314–319

Schlossarek S, Mearini G, Carrier L (2011) Cardiac myosin-binding protein C in hypertrophic cardiomyopathy: mechanisms and therapeutic opportunities. J Mol Cell Cardiol 50:613–620

Schlossarek S, Frey N, Carrier L (2014) Ubiquitin-proteasome system and hereditary cardiomyopathies. J Mol Cell Cardiol 71:25–31

Seifert U, Bialy LP, Ebstein F, Bech-Otschir D, Voigt A, Schroter F, Prozorovski T, Lange N, Steffen J, Rieger M, Kuckelkorn U, Aktas O, Kloetzel PM, Kruger E (2010) Immunoproteasomes preserve protein homeostasis upon interferon-induced oxidative stress. Cell 142:613–624

Semenza GL (2012) Hypoxia-inducible factors in physiology and medicine. Cell 148:399–408

Shenoy SK, McDonald PH, Kohout TA, Lefkowitz RJ (2001) Regulation of receptor fate by ubiquitination of activated beta 2-adrenergic receptor and beta-arrestin. Science 294:1307–1313

Shimizu S, Kanaseki T, Mizushima N, Mizuta T, Arakawa-Kobayashi S, Thompson CB, Tsujimoto Y (2004) Role of Bcl-2 family proteins in a non-apoptotic programmed cell death dependent on autophagy genes. Nat Cell Biol 6:1221–1228

Shimomura H, Terasaki F, Hayashi T, Kitaura Y, Isomura T, Suma H (2001) Autophagic degeneration as a possible mechanism of myocardial cell death in dilated cardiomyopathy. Jpn Circ J 65:965–968

Shintani T, Klionsky DJ (2004) Autophagy in health and disease: a double-edged sword. Science 306:990–995

Shringarpure R, Grune T, Mehlhase J, Davies KJ (2003) Ubiquitin conjugation is not required for the degradation of oxidized proteins by proteasome. J Biol Chem 278:311–318

Stansfield WE, Moss NC, Willis MS, Tang R, Selzman CH (2007) Proteasome inhibition attenuates infarct size and preserves cardiac function in a murine model of myocardial ischemia-reperfusion injury. Ann Thorac Surg 84:120–125

Stansfield WE, Tang RH, Moss NC, Baldwin AS, Willis MS, Selzman CH (2008) Proteasome inhibition promotes regression of left ventricular hypertrophy. Am J Physiol Heart Circ Physiol 294:H645–H650

Steinberg SF (2013) Oxidative stress and sarcomeric proteins. Circ Res 112:393–405

Strehl B, Seifert U, Kruger E, Heink S, Kuckelkorn U, Kloetzel PM (2005) Interferon-gamma, the functional plasticity of the ubiquitin-proteasome system, and MHC class I antigen processing. Immunol Rev 207:19–30

Sun M, Ouzounian M, de Couto G, Chen M, Yan R, Fukuoka M, Li G, Moon M, Liu Y, Gramolini A, Wells GJ, Liu PP (2013) Cathepsin-L ameliorates cardiac hypertrophy through activation of the autophagy-lysosomal dependent protein processing pathways. J Am Heart Assoc 2:e000191

Szalay G, Meiners S, Voigt A, Lauber J, Spieth C, Speer N, Sauter M, Kuckelkorn U, Zell A, Klingel K, Stangl K, Kandolf R (2006) Ongoing coxsackievirus myocarditis is associated with increased formation and activity of myocardial immunoproteasomes. Am J Pathol 168:1542–1552

Takagi H, Hsu CP, Kajimoto K, Shao D, Yang Y, Maejima Y, Zhai P, Yehia G, Yamada C, Zablocki D, Sadoshima J (2010) Activation of PKN mediates survival of cardiac myocytes in the heart during ischemia/reperfusion. Circ Res 107:642–649

Tanaka K, Ii K, Ichihara A, Waxman L, Goldberg AL (1986) A high molecular weight protease in the cytosol of rat liver. I. Purification, enzymological properties, and tissue distribution. J Biol Chem 261:15197–15203

Tanaka Y, Guhde G, Suter A, Eskelinen EL, Hartmann D, Lullmann-Rauch R, Janssen PM, Blanz J, von Figura K, Saftig P (2000) Accumulation of autophagic vacuoles and cardiomyopathy in LAMP-2-deficient mice. Nature 406:902–906

Thrower JS, Hoffman L, Rechsteiner M, Pickart CM (2000) Recognition of the polyubiquitin proteolytic signal. EMBO J 19:94–102

Tian Z, Zheng H, Li J, Li Y, Su H, Wang X (2012) Genetically induced moderate inhibition of the proteasome in cardiomyocytes exacerbates myocardial ischemia-reperfusion injury in mice. Circ Res 111:532–542

Tsukamoto O, Minamino T, Okada K, Shintani Y, Takashima S, Kato H, Liao Y, Okazaki H, Asai M, Hirata A, Fujita M, Asano Y, Yamazaki S, Asanuma H, Hori M, Kitakaze M (2006) Depression of proteasome activities during the progression of cardiac dysfunction in pressure-overloaded heart of mice. Biochem Biophys Res Commun 340:1125–1133

Unno M, Mizushima T, Morimoto Y, Tomisugi Y, Tanaka K, Yasuoka N, Tsukihara T (2002) The structure of the mammalian 20S proteasome at 2.75 A resolution. Structure (Camb) 10:609–618

Usui S, Maejima Y, Pain J, Hong C, Cho J, Park JY, Zablocki D, Tian B, Glass DJ, Sadoshima J (2011) Endogenous muscle atrophy F-box mediates pressure overload-induced cardiac hypertrophy through regulation of nuclear factor-kappaB. Circ Res 109:161–171

Van Kaer L, Ashton-Rickardt PG, Eichelberger M, Gaczynska M, Nagashima K, Rock KL, Goldberg AL, Doherty PC, Tonegawa S (1994) Altered peptidase and viral-specific T cell response in LMP2 mutant mice. Immunity 1:533–541

Varshavsky A (1996) The N-end rule: functions, mysteries, uses. Proc Natl Acad Sci U S A 93:12142–12149

Verma R, Aravind L, Oania R, McDonald WH, Yates JR 3rd, Koonin EV, Deshaies RJ (2002) Role of Rpn11 metalloprotease in deubiquitination and degradation by the 26S proteasome. Science 298:611–615

Vignier N, Schlossarek S, Fraysse B, Mearini G, Kramer E, Pointu H, Mougenot N, Guiard J, Reimer R, Hohenberg H, Schwartz K, Vernet M, Eschenhagen T, Carrier L (2009) Nonsense-mediated mRNA decay and ubiquitin-proteasome system regulate cardiac myosin-binding protein C mutant levels in cardiomyopathic mice. Circ Res 105:239–248

Walczak M, Martens S (2013) Dissecting the role of the Atg12-Atg5-Atg16 complex during autophagosome formation. Autophagy 9:424–425

Weekes J, Morrison K, Mullen A, Wait R, Barton P, Dunn MJ (2003) Hyperubiquitination of proteins in dilated cardiomyopathy. Proteomics 3:208–216

Willis MS, Ike C, Li L, Wang DZ, Glass DJ, Patterson C (2007) Muscle ring finger 1, but not muscle ring finger 2, regulates cardiac hypertrophy in vivo. Circ Res 100:456–459

Willis MS, Rojas M, Li L, Selzman CH, Tang RH, Stansfield WE, Rodriguez JE, Glass DJ, Patterson C (2009) Muscle ring finger 1 mediates cardiac atrophy in vivo. Am J Physiol Heart Circ Physiol 296:H997–H1006

Wohlschlaeger J, Sixt SU, Stoeppler T, Schmitz KJ, Levkau B, Tsagakis K, Vahlhaus C, Schmid C, Peters J, Schmid KW, Milting H, Baba HA (2010) Ventricular unloading is associated with increased 20s proteasome protein expression in the myocardium. J Heart Lung Transplant 29:125–132

Yan L, Vatner DE, Kim SJ, Ge H, Masurekar M, Massover WH, Yang G, Matsui Y, Sadoshima J, Vatner SF (2005) Autophagy in chronically ischemic myocardium. Proc Natl Acad Sci U S A 102:13807–13812

Yang Z, Huang J, Geng J, Nair U, Klionsky DJ (2006) Atg22 recycles amino acids to link the degradative and recycling functions of autophagy. Mol Biol Cell 17:5094–5104

Zaha VG, Young LH (2012) AMP-activated protein kinase regulation and biological actions in the heart. Circ Res 111:800–814

Zhao J, Zhai B, Gygi SP, Goldberg AL (2015) mTOR inhibition activates overall protein degradation by the ubiquitin proteasome system as well as by autophagy. Proc Natl Acad Sci U S A 112:15790–15797

Zhu H, Tannous P, Johnstone JL, Kong Y, Shelton JM, Richardson JA, Le V, Levine B, Rothermel BA, Hill JA (2007) Cardiac autophagy is a maladaptive response to hemodynamic stress. J Clin Invest 117:1782–1793

Zong C, Gomes AV, Drews O, Li X, Young GW, Berhane B, Qiao X, French SW, Bardag-Gorce F, Ping P (2006) Regulation of murine cardiac 20S proteasomes: role of associating partners. Circ Res 99:372–380

Ischemia and Reperfusion

8

Jacqueline Heger

Abstract

Ischemia, the interruption of blood perfusion of the heart, and reperfusion, the revascularization after an ischemic event, are both events that trigger certain signaling mechanisms and cellular adaptation to cardiomyocytes. Ischemia itself induces metabolic and structural changes to cardiomyocytes that are prone to reperfusion-induced stress, i.e., by inducing plasmalemmal fragility. Reperfusion restores immediately energy resources, but due to ischemia-dependent calcium overload, it generates the risk of hypercontracture. At the same time, oxidative stress further stresses postischemic cardiomyocytes. The well-described cellular changes during ischemia and subsequent reperfusion allowed researchers in the past to develop distinct protocols that reduce infarct sizes and protect at least part of non-irreversible damaged cardiomyocytes. Signaling events specifically targeted by such protocols, however, also affect subsequent wound healing processes. An optimal reperfusion strategy must therefore be directed by three different aims: (i) the reduction of infarct size to protect as much as myocardial tissue as possible; (ii) the immediate recovery of full pump function to perfuse not only the heart properly but also all other organs, i.e., the kidney; and (iii) signaling pathways must be activated that direct post-infarct remodeling into a direction that stabilizes pump function for longer time periods. In this chapter, a detailed description of ischemia-dependent, reperfusion-dependent, and post-ischemic adaptations of cardiomyocytes is given.

J. Heger
Institute of Physiology, Justus-Liebig-Universität Gießen; Physiologisches Institut, Aulweg 129, Giessen 35392, Germany
e-mail: Jacqueline.Heger@physiologie.med.uni-giessen.de

K.-D. Schlüter (ed.), *Cardiomyocytes – Active Players in Cardiac Disease*,
DOI 10.1007/978-3-319-31251-4_8

8.1 Introduction

Cardiomyocytes are dependent on continuous supply of oxygen. The heart consumes 8–15 ml O_2/min/100 g tissue under resting conditions with further increase during exercise (Giordano 2005). Electromechanical coupling of cardiomyocytes that is the prerequisite for the working heart needs an efficient energy metabolism. This high-energy requirement cannot be facilitated by anaerobic glycolysis. ATP is therefore generated by oxidative phosphorylation (OxPhos) in mitochondria which represent 30 % of cardiomyocytes volume and produce 90 % of ATP (Piquereau et al. 2013). The driving force for the synthesis of ATP is the electrochemical gradient that is effectuated by citric acid cycle or tricarboxylic acid cycle (TCA) whose enzymes are activated by mitochondrial calcium (Choi et al. 2014). Energy-delivering substrates are free fatty acids, glucose, and lactate that are converted to acetyl CoA. The TCA cycle removes electrons from acetyl CoA to form $FADH_2$ and $NADH+H^+$ that are released in OxPhos. The electrons run through the complexes I–IV of the electron transport chain to generate a proton transport gradient using oxygen as electron acceptor (Berg et al. 2002). F_1F_0 ATP synthase also known as mitochondrial complex V or ATP synthase catalyzes the ultimate coupling step of OxPhos (Long et al. 2015) to generate ATP from ADP and inorganic phosphate (Berg et al. 2002). During every heartbeat 2 % of ATP is hydrolyzed. With a heartbeat of 60/min, the whole ATP content of the cardiomyocytes will be consumed within 1 min (Kühn 2011).

Therefore the supply with ATP is vital for cardiomyocytes. The maintenance of cardiomyocytes with oxygen, glucose, and fatty acids to generate ATP is primarily facilitated via the bloodstream through coronary vessels. If a coronary artery is occluded due to pathophysiological events, parts of the heart are no longer supplied with oxygen, mitochondria cannot produce ATP, and the cardiomyocytes become ischemic, which leads to pathophysiological and metabolic changes. The absence of oxygen rests oxidative phosphorylation, leading to mitochondrial membrane depolarization, ATP depletion, and inhibition of myocardial contractile function (Hausenloy and Yellon 2013).

The absence or entire block of oxygen supply is characteristic for ischemic heart diseases (IHD) like myocardial infarction (MI) that are still the leading cause of death. MI was identified in autopsy throughout the nineteenth and twentieth century, but the clinicians of this time were hesitant to ascribe the myocardial necrosis to interruption of the blood supply (Jennings 2013). It is recently as 35 years ago that clinicians considered the occlusion of a coronary to be causally involved in the evolution of MI (DeWood et al. 1980). Therefore, timely reperfusion of the occluded vessel by percutaneous coronary intervention (PCI) or thrombolytic intervention is needed to reduce the infarct size depicted by necrosis of the cells (Chiong et al. 2011). But the reperfusion itself contributes to myocardial injury called ischemia–reperfusion injury (IRI) resulting in hypercontracture of cardiomyocytes and opening of mitochondrial permeability transition pore (MPTP) leading to cardiomyocyte death, cardiac failure, and arrhythmias (Sluijter et al. 2014).

Interventions that can reduce IRI have been investigated during the last decades. Short episodes of ischemia before or after ischemia results in ischemic pre- or post-conditioning. This leads to metabolic changes in the cardiomyocytes that reduce infarct sizes. Cardiomyocyte's fate is therefore governed not only by the length of ischemia but also by the time before and after ischemia. This chapter will summarize the cellular and metabolic changes during ischemia and after reperfusion. Furthermore, mechanisms leading to hypercontracture and cardiomyocyte death in ischemia–reperfusion (I–R) will be described. In addition, cardioprotective signaling pathways that have been identified in preconditioning, postconditioning, and/or remote conditioning will be presented displaying mitochondria as a target of protection.

8.2 Changes During Ischemia

The lack of oxygen and metabolic substrates supply results in a series of abrupt biochemical and metabolic changes within the heart. In the cardiomyocytes, several alterations can be observed: deficiency in energy, decrease in pH, sodium and calcium overload, and ROS production (Fig. 8.1). Within 8–10 s of occlusion, a change from aerobic to anaerobic glycolysis takes place (Jennings et al. 1990). Creatine phosphate stores, the easily mobilized high-energy phosphate source of cardiomyocytes, will be consumed within 1–3 min of ischemia to maintain a steady production of ATP (Jennings et al. 1981). It comes to ATP depletion in the cell: ATP production decreases to less than 10 % of the rate of OxPhos (Kobayashi and Neely 1979). The mitochondrial F_1F_0 synthase during normoxic conditions produces ATP, works during ischemia in reverse mode, and consumes much of the ATP generated by glycolysis, resulting in ATP hydrolysis and an increase in mitochondrial inorganic phosphate (Hausenloy and Yellon 2013). The energy is used to generate mitochondrial membrane potential ($\Delta\psi$), which facilitates the uptake of Ca^{2+} into the mitochondria via mitochondrial calcium uniporter (MCU) (Pan et al. 2013).

Long-lasting energy depletion results in damage of the cytoskeleton (Steenbergen et al. 1987b) and sarcolemmal proteins like dystrophin or spectrin (Armstrong et al. 2001), leading to cytoskeletal and sarcolemmal fragility.

The switch to anaerobic glycolysis resulted in the accumulation of lactate that is produced in the ischemic period and quadrupled within a minute (Jennings et al. 1986). H^+ concentration increases due to the augmented production of protons that are mainly produced from glycolytic ATP turnover and CO_2 retention and insufficient removal (Dennis et al. 1991). Increase in ion H^+ concentration during the initial minutes of ischemia decreases intracellular pH. The resulting acidosis can reach pH_i values of about 6.4 within the first 10–20 min of ischemia demonstrated in isolated pig as well as rat hearts (Fleet et al. 1985; Inserte et al. 2009). The combination of ATP depletion and acidosis inhibits Na^+–K^+-ATPase, causing a decrease of Na^+ extrusion and increase of intracellular Na^+ (Garcia-Dorado et al. 2012). To maintain pH homeostasis, Na^+–H^+ exchanger (NHE), H^+–lactate symporter, and Na^+–HCO_3^- symporter are activated, leading to extrusion of H^+ and increase of

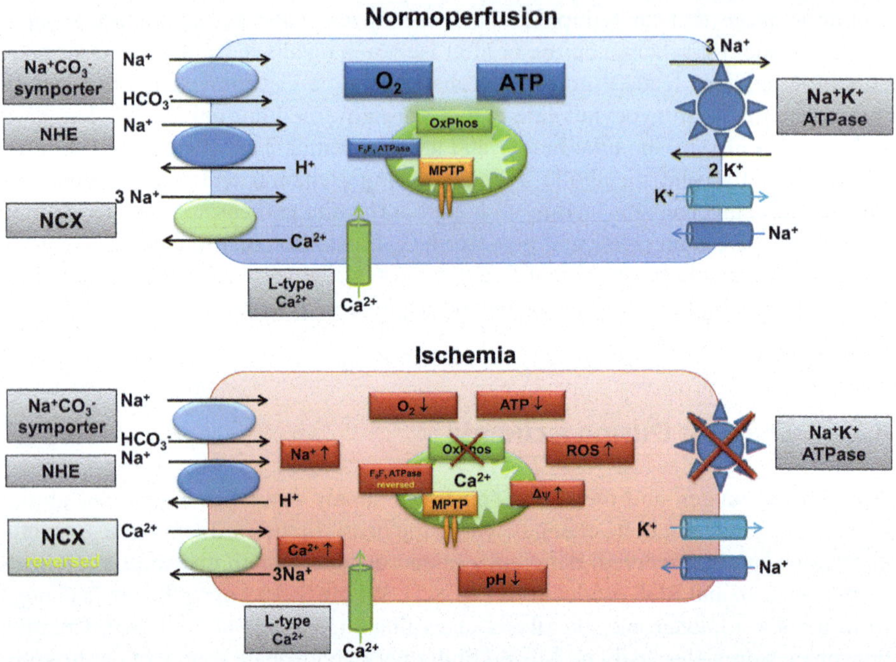

Fig. 8.1 Under normoperfusion ATP is generated by oxidative phosphorylation (OxPhos) by utilization of oxygen. F_1F_0 ATP synthase catalyzes the ultimate coupling step of OxPhos to generate ATP from ADP that is used by Na^+–K^+-ATPase to exchange Na^+ and K^+. Ca^{2+} that during excitation-coupling streams into the cell is removed by NCX. During ischemia the lack of oxygen ($O_2\downarrow$) results in ATP (ATP↓) deficiency. This results in acidosis (pH↓). ATP depletion and acidosis inhibits Na^+ –K^+-ATPase and increase of intracellular Na^+. To maintain pH homeostasis NHE and $Na^+CO_3^-$ symporter pumps Na^+ into the cell. This leads to sodium overload (Na^+ ↑). To extrude the high amount of Na^+ the NCX works in reverse mode and pumps Ca^{2+} into the cell. The mitochondrial F_1F_0 synthase works during ischemia in reverse mode and consumes much of the ATP generated by glycolysis using the energy to generate mitochondrial membrane potential (Δψ), which facilitates the uptake of Ca^{2+} into the mitochondria resulting in ROS production

intracellular HCO_3^- to buffer intracellular H^+ (Frelin et al. 1985; Vandenberg et al. 1993; Piper et al. 1996). Ultimately, ischemia causes intra- and extracellular acidification (Piper et al. 1998). As a consequence the transporters pump Na^+ into the cell further contributing to an overload of Na^+ in the cardiomyocytes. Na^+ as well as products of glycolysis accumulate and increase osmolality. Delayed in time, calcium concentration increases in the cells (Steenbergen et al. 1987a). This is attributed to a reverse mode of Na^+–Ca^{2+} exchanger (NCX) pumping three Na^+ out of the cardiomyocytes and one Ca^{2+} ion into the cell (Ladilov et al. 1999). But also Na^+-independent Ca^{2+} entries are described like the L-type Ca^{2+} channel and contribute to subsequent Ca^{2+} overload (Sun et al. 2006).

Cell shortening of cardiomyocytes is impaired during ischemia since calcium sensitivity of the contractile machinery is reduced at low pH (Ladilov et al. 1995; Marban and Kusuoka 1987). Ischemia induces dephosphorylation of phospholamban (PLB) and impairs activity of sarcoplasmic reticulum Ca^{2+}-ATPase (SERCA) due to low ATP concentrations that likewise contribute to Ca^{2+} overload (Shintani-Ishida and Yoshida 2011; Garcia-Dorado et al. 2012). The reduced ATP levels also initiates rigor contracture and hallmarks the beginning of Ca^{2+} overload.

This rise in mitochondrial Ca^{2+} is a trigger for the opening of the mitochondrial permeability transition pore (MPTP), but the low pH during ischemia prevents MPTP opening (Murphy and Steenbergen 2008). At the same time, an increase in Ca^{2+} induces Ca^{2+}-dependent dehydrogenase activation, a decline in NADH, and electron flux through the electron transport chain, resulting in increased ROS production (Chiong et al. 2011). There is a positive relation between $\Delta\psi$ and ROS production by electron transport. Although the levels of ROS that are produced during ischemia are low, they can lead to progressive damage to the electron transport chain (complex I and complex III) due to inefficient transfer of electrons, which is more rapid in subsarcolemmal mitochondria (SSM) located beneath the plasma membrane than in interfibrillar mitochondria (IFM) present between the myofibrils (Murphy and Steenbergen 2008; Lesnefsky et al. 2004).

Ischemia-induced changes in cardiomyocytes are not restricted to the energy-depleted cells. Cardiomyocytes are tightly interconnected through highly specialized regions of the plasma membrane called gap junctions (Garcia-Dorado et al. 2002) that are necessary to coordinate cell function. Gap junctions – normally located in intercalated disks of neighboring cardiomyocytes – are composed of two hemichannels or connexones. One hemichannel is built by six connexins. The highest expressed connexin in cardiomyocytes is connexin 43 (Cx43). Single hemichannels are usually closed and located in the plasma membrane. They can open by cell depolarization or high intracellular calcium concentration so that they can open during ischemia (Wang et al. 2013). Opening of hemichannels leads to influx of sodium and calcium (Li et al. 2001)and the release of ATP, leading to energy depletion, calcium overload, and cell death. In addition, the communication of gap junctions is reduced mediated by acidosis and calcium increase (Schulz et al. 2015).

In a nutshell, ischemia as a consequence of oxygen and substrate deficiency results in ATP depletion, acidification, hyperosmolality, Na^{+} and Ca^{2+} overload, as well as cytoskeletal and sarcolemmal fragility.

8.3 Reperfusion and Its Outcome

Early reopening of the occluded vessel is of outmost importance in the treatment of MI. Unexpectedly reperfusion leads to acute IRI whose cause arises from the time of reperfusion that in principal can still be inhibited in this phase (Piper et al. 1999). With the onset of reperfusion, cell death occurs that is known as reperfusion injury (Piper et al. 1998). After temporary energy depletion as well as Na^{+} and Ca^{2+} overload during the ischemic phases, reperfusion activates several processes in cardiomyocytes that are caused by burst of ROS, re-energization of the cell, rapid normalization of tissue pH, and tissue osmolality (Piper et al. 1998) (Fig. 8.2). The combination of reduced communication of gap junctions and opening of

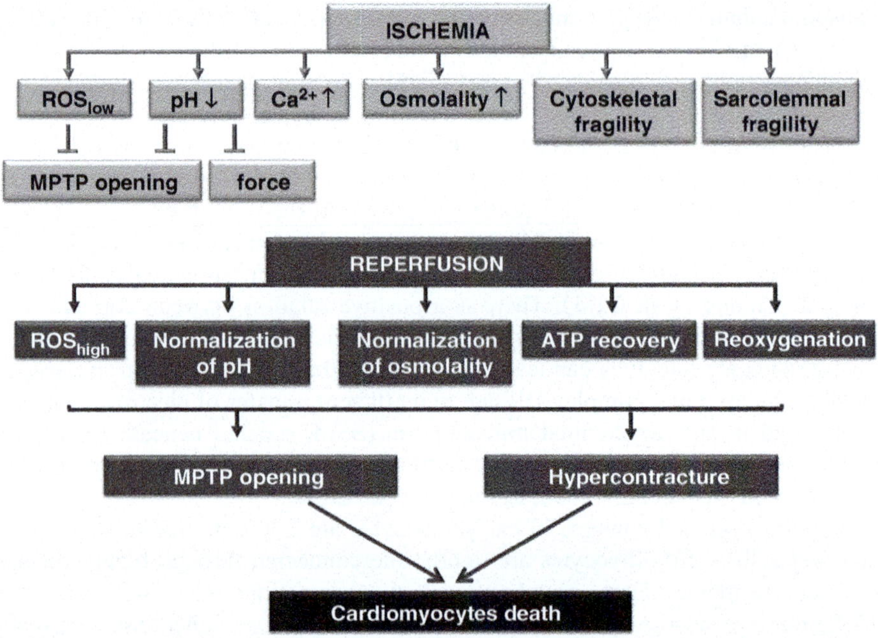

Fig. 8.2 Ischemia as a consequence of oxygen and substrate deficiency results in acidification (pH ↓), hyperosmolality, Ca^{2+} overload, and low ROS production, as well as cytoskeletal and sarcolemmal fragility. Although calcium is high, low pH and low ROS inhibit the opening of MPTP as well as force. Reperfusion amounts to reoxygenation and ATP recovery and rapid normalization of osmolality and pH leading to high ROS production. This causes MPTP opening and hypercontracture and in the end in cardiomyocytes death

hemichannels contributes to irreversible IRI mediated by the transfer of cell death (Schulz et al. 2015), which is also known as spreading of injury (García-Dorado et al. 2004).

8.3.1 Metabolic Changes

Reperfusion results in a fast recovery of pH (Garcia-Dorado et al. 2012). Upon reperfusion the extracellular acidification is first normalized generating a gradient between cytosol and interstitium that activates NHE and Na^+–HCO_3^- symporter to extrude H^+ from the cytosol (Piper et al. 1998). During the first minutes of reperfusion, cytosolic Na^+ overload offers a big driving force for Ca^{2+} influx through the reverse mode of NCX, resulting in further Ca^{2+} influx through the reverse mode operation of NCX (Garcia-Dorado et al. 2012; Wei et al. 2007). Recovery of energy production upon the resupply of oxygen and metabolic substrates rapidly reactivates the Na^+–K^+-ATPase and the Ca^{2+}-ATPase of the sarcoplasmic reticulum (SERCA) (Piper et al. 1998). The extrusion of excess calcium is thus mediated by Na^+–Ca^{2+} exchanger (NCX) in forward mode whose activity depends on sodium gradient established by Na^+–K^+-ATPase (Piper et al. 1999). The normalization of Ca^{2+} homeostasis is split. At the outset Ca^{2+} drops because it is temporarily sequestered into the SR by SERCA. When the capacity of SR is exceeded, Ca^{2+} is released into the cytosol and starts Ca^{2+} cycles with continuous intake and delivery of calcium between cytosol and SR (Siegmund et al. 1997). These oscillations only stop when NCX extrudes surplus calcium from the cardiomyocytes (Siegmund et al. 1992).

8.3.2 Hypercontracture of Cardiomyocytes

During reperfusion, cardiomyocyte death is an uncontrolled activation of the contractile machinery leading to cardiomyocyte hypercontracture that is initiated by normalization of pH and recovery of energy (Rodríguez-Sinovas et al. 2007). The provision of oxygen initializes OxPhos, providing ATP for the activation of the myofibrils and provokes a sustained maximal force development and consecutive mechanical cell injury since cytosolic Ca^{2+} concentrations are still increased (Siegmund et al. 1993). Mechanisms elucidating these changes in morphology can be attributed to Ca^{2+} oscillations resulting in high peak concentration of cytosolic Ca^{2+} that leads to sustained and uncontrolled activation of the contractile apparatus but also to a Ca^{2+}-independent rigor-associated activation of contraction (Piper et al. 2004). Since cardiomyocytes are linked to each other, hypercontracture leads to mechanical stress, resulting in sarcolemmal rupture and necrosis (Rodríguez-Sinovas et al. 2007). Hypercontracture spreads to neighboring cardiomyocytes via gap junctions, thereby passaging Na^+ with subsequent reverse mode of NCX leading to cell death (García-Dorado et al. 2004).

Na^+ overload during ischemia contributed to hyperosmolality. Reperfusion induces a gradient between the cytosol and the already washed out extracellular space (Garcia-Dorado and Oliveras 1993). The subsequent influx of water increases cell volume provoking mechanical stress of the sarcolemma resulting in rupture of the membrane (Schlüter et al. 1996). Sarcolemmal rupture mediated by hypercontracture is enhanced by cytoskeletal fragility (Rodríguez-Sinovas et al. 2007). The sarcolemmal structural protein dystrophin links the transmembrane dystroglycan complex and extracellular laminin receptors to intracellular F-actin (Armstrong et al. 2001). During reperfusion and after normalization of pH, calcium overload of cardiomyocytes activates diverse Ca^{2+}-dependent enzymes like calpains – non-lysosomal neutral cysteine proteases – that hydrolyzes substrate proteins at the membrane (Inserte et al. 2012). The key protein and backbone of the membrane cytoskeleton is α-fodrin that is cleaved by calpain leading to membrane fragility (Rodríguez-Sinovas et al. 2007). Another sarcolemmal protein that is cleaved by calpain is the protein linker ankyrin (Inserte et al. 2005), which connects integral membrane proteins to the spectrin-based cytoskeleton (Rubtsov and Lopina 2000).

It has been shown that a cluster of four amino acids is essential for binding of ankyrin to the α subunit of Na^+–K^+-ATPase (Jordan et al. 1995). During reperfusion proteolysis of ankyrin and fodrin by calpain detaches the Na^+–K^+-ATPase from ankyrin. This results in loss of pump activity preventing the normalization of cytosolic Na^+ and furthermore leading to Ca^{2+} increase through reverse mode of NCX (Inserte et al. 2005; 2012). Calcium handling is also compromised since Ca^{2+} handling proteins like NCX, SERCA, and RYR are cleaved by calpain (Kar et al. 2009; Singh et al. 2004). Moreover, calpain can cleave Bid in the absence of detectable caspase activation thereby targeting the mitochondria, causing dysfunction and release of proapoptotic factors (Chen et al. 2002). In conclusion, hypercontracture can support cell death by multiple steps.

8.3.3 Cardiomyocyte and Cell Death

The adult heart is composed of binucleated, terminally differentiated cardiomyocytes (Paradis et al. 2014), whose number determines heart function. Loss of cardiomyocytes impedes efficient contraction and is the prime reason of mortality in patients with ischemic heart disease (Inserte and Garcia-Dorado 2015). After transient ischemia cardiomyocytes die during the first minutes of reperfusion due to contraction band necrosis that results in sarcolemmal rupture (Rodríguez-Sinovas et al. 2007). For the whole heart, it was shown that cardiomyocyte death during I–R can develop in form of necrosis, apoptosis, autophagy, and necroptosis (Ibáñez et al. 2015) (for details, see Chap. 9). This is not surprising since cardiomyocytes respond to stimuli of neighboring cells like endothelial cells and fibroblasts and receive external stimuli via microcirculation.

8.3.3.1 Necrosis and I–R

Studies with isolated cardiomyocytes of adult rat showed that long-term hypoxia up to 18 h and subsequent reoxygenation up to 12 h resulted in necrosis (Taimor et al. 1999). The necrotic cardiomyocytes are characterized by myofibrillar contraction bands, damage of cardiomyocyte sarcolemma, as well as enlarged and ruptured mitochondria. Most of these marks are aggravated and manifested by reperfusion (Ibáñez et al. 2015). Impaired Ca^{2+} regulation leads to Ca^{2+} oscillation and hypercontracture. Recovery of pH activates calpain that digests sarcolemma and cytoskeleton. Normalization of extracellular osmolality by reperfusion results in intracellular edema. This osmotic cell swelling arises from intracellular hyperosmolarity due to high cytosolic concentrations of Na^+ and Ca^{2+} (Schlüter et al. 1996). ROS generation also contributes to sarcolemmal disruption leading to necrosis. Upon reperfusion, restoration of oxygen and ATP production rapidly improves mitochondrial membrane potential that is positively correlated with ROS production (Murphy and Steenbergen 2008). These changes stimulate the necessary ion gradient for more Ca^{2+} influx into the mitochondria that causes the opening of MPTP and mitochondrial swelling, leading finally to cellular necrosis (Chiong et al. 2011).

8.3.3.2 Apoptosis and I–R

Apoptosis also plays a role in myocardial infarction. Ischemia alone by permanent coronary occlusion for 2.25 h resulted in apoptosis of cardiomyocytes, and it could be demonstrated that 45 min of ischemia followed by reperfusion enhanced apoptosis, revealing apoptosis as early as 1 h after reperfusion (Fliss and Gattinger 1996). There is a causal connection between cardiomyocyte apoptosis and myocardial infarction, which involves both death receptor and mitochondrial pathway (Konstantinidis et al. 2012) since Fas deficiency as well as Bax deficiency reduces infarct size and diminishes loss of cardiomyocytes (Nakamura et al. 2000; Hochhauser et al. 2007).

Impaired calcium regulation can lead to cardiomyocyte death. Calcium is stored in the sarcoplasmic reticulum (SR), but is also absorbed by mitochondria, which are located in vicinity to the SR to avoid cytotoxicity (Orogo and Gustafsson 2013). Calcium homeostasis is regulated by the antiapoptotic Bcl-2 and the apoptotic Bax and BAK proteins that are also located in the SR and can control apoptosis not only at the mitochondria (Oakes et al. 2003). Cardiomyocytes of Bax–Bak double knockout mice as well as Bcl-2-overexpressing cells have an increased calcium release from SR into the cytosol and decreased mitochondrial Ca^{2+} uptake. Both transgenic mice show smaller infarct size after I–R (Hochhauser et al. 2003; Chen et al. 2001). Depending on their localization – SR or mitochondria – BH3-only proteins, a subdivision of proapoptotic Bcl-2 proteins, can either induce apoptosis or necrosis (Orogo and Gustafsson 2013).

8.3.3.3 Necroptosis and I–R

Regulated necrosis also plays a role in I–R since necrostatin (Nec-1) reduces infarct size (Oerlemans et al. 2013). In isolated cardiomyocytes necrostatin can delay opening of the mitochondrial permeability transition pore (MPTP) (Smith et al. 2007), and cyclophylin D, a regulatory component of the MPTP, is required for Nec-1 to exhibit a cardioprotective effect in mice in vivo. The mechanism by which Nec-1 inhibits MPTP opening is not clear but is speculated that it depends on the activation of the reperfusion injury salvage kinase (RISK) (Lim et al. 2007) (see below).

8.3.3.4 Autophagy and I–R

Autophagy has distinct roles during ischemia and reperfusion but seems to be more protective than harmful (Ibáñez et al. 2015). Whereas ischemia stimulates autophagy through an AMPK-dependent mechanism, I–R stimulates autophagy through a Beclin 1-dependent but AMPK-independent mechanism. Autophagy may therefore be protective during ischemia, whereas it may be detrimental during reperfusion. About the cellular mechanism of autophagy, see Chap. 9.

It is suggested that the impact of regulated forms of cardiomyocyte cell death in I–R injury is possibly linked to their specific signaling pathways that facilitates a targeted use of therapies, albeit from the patient's point of view, cardiomyocyte death is similarly relevant whatever the mechanism is (Ibáñez et al. 2015).

8.3.4 ROS Production

OxPhos is the endogenous source of mitochondrial ROS production that is a key mediator of both detrimental and protective responses in I–R (Kalogeris et al. 2014). A burst of ROS that occurs during reperfusion can start a severe progression of oxidative stress in the mitochondria, which induces MPTP opening with ROS-induced ROS release, leading to detrimental effects on cardiomyocytes through decay in bioenergetics and mitochondrial dysfunction (Chen and Zweier 2014; Zorov et al. 2000). Chronic exposure of cardiomyocytes to ROS leads to impairment of excitation–contraction coupling, which contributes to cardiac remodeling by inducing cardiac hypertrophy, apoptosis, necrosis, and fibrosis (Chen and Zweier 2014; Giordano 2005). ROS can damage cardiomyocytes through a number of mechanisms including peroxidation of cell membrane and organelle lipids, oxidizing of DNA, and activation of matrix metalloproteinases and calpains. In addition ROS may also lead to cell dysfunction and death by indirect mechanisms via interacting with NO, fatty acids, or free iron to form peroxynitrite, peroxyl radicals, and hydroxyl radicals (Kalogeris et al. 2014). A more detailed view on ROS and cardiomyocytes is summarized in Chap. 10.

8.3.5 The Mitochondrial Permeability Transition Pore (MPTP)

Mitochondria are the power source of the cardiomyocytes providing them with ATP by OxPhos which is required for the maintenance of permeability barrier of the inner mitochondrial membrane and the perpetuation of ionic gradients that is decisive for the survival of the cell (Halestrap and Richardson 2015). The mitochondria are an important mediator and regulator of all forms of cell death in ischemia–reperfusion. The lack of oxygen during ischemia leads to mitochondrial membrane depolarization and ATP depletion, reperfusion initiates opening of MPTP.

Opening of the MPTP is the crucial mediator of lethal reperfusion injury leading to cardiomyocyte death (Ong et al. 2015). Calcium and ROS are triggers for MPTP opening, which are both raised during ischemia and reperfusion. MPTP is a high-conductance channel that is formed from the F_1F_0 ATP synthase (Bernardi and di Lisa 2015). Opening of MPTP makes the inner mitochondrial membrane permeable for any molecules up to 1500 dalton. This leads to inhibition of OxPhos and ATP depletion (Ong et al. 2015). MPTP opening occurs during the first minutes of reperfusion, is triggered by a secondary mitochondrial Ca^{2+} overload caused by SR-driven Ca^{2+} oscillations, and promotes itself Ca^{2+} oscillation. Both, MPTP opening and Ca^{2+} oscillation, contribute to ROS formation, mitochondrial dysfunction, cardiomyocytes hypercontracture, and necrosis (Abdallah et al. 2011). Opening of MPTP also leads to swelling of mitochondria resulting from colloidal osmotic pressure that causes expansion of the matrix, unfolding of cristae, and rupture of the outer membrane (Halestrap and Richardson 2015).

During ischemia MPTP opening is inhibited by acidosis. However, ROS production, pH normalization, and Ca^{2+} overload during reperfusion lead to the opening of MPTP (Rodríguez-Sinovas et al. 2007). Sustained opening of this channel results in rupture of outer membrane and release of cytochrome C into the cytosol, thereby activating proteolytic processes (Heusch 2015). MPTP opening is typical for necrosis and is triggered by Ca^{2+} and can be further potentiated by ROS, alkalosis, and depletion of ATP (Konstantinidis et al. 2012). Thereby, necrosis can happen as a primary event or deduced from apoptosis. If necrosis takes place as a primary event, MPTP opens before cytochrome C release. During apoptosis due to caspase activation, MPTP opens simultaneous or after cytochrome C release (Ricci et al. 2004; Konstantinidis et al. 2012). Thus, MPTP is a major regulator of both apoptotic and necrotic cell death. Recent studies have shown that inhibition of MPTP by knockdown of cyclophilin D (a regulator of the MPTP) results in a significant reduction in infarct size in ischemia–reperfusion, but loss of cyclophilin D did not block BAX-mediated apoptosis (Murphy and Steenbergen 2008). This supports the novel role for BAK proteins that can control apoptosis not only at the mitochondria but also at the SR in response to Ca^{2+}-dependent stimuli (Oakes et al. 2003) (see above).

8.4 Protection of Cardiomyocytes from I–R Injury

Since reperfusion contributes to cell damage, protection is necessary to minimize the number of lost cardiomyocytes contributing to infarct size, which is a central determining factor for the prognosis of patients after myocardial infarction (Heusch 2015). In the last 30 years, a lot of studies have been done to elucidate mechanisms contributing to cardioprotection that manipulate I–R by decelerating injury (Murphy and Steenbergen 2008). The main focus is laid on conditioning – the recruitment of endogenous pathways that protect cells against I–R damage. Conditioning can be performed before (preconditioning), during (perconditioning), or after ischemia (postconditioning), either directly at the heart or in a distal tissue (remote preconditioning) (Bell and Yellon 2012).

In 1986 Murray et al. showed for the first time the protective effect of ischemic preconditioning. Four 5min occlusions of a coronary artery, each separated by 5 min of reperfusion, followed by a sustained 40 min occlusion before 4 days of reperfusion limited infarct size to 25 %. Ischemic postconditioning was first reported in 2003 (Zhao et al. 2003). Zhang and colleagues applied three cycles of 30 s reperfusion and 30 s occlusion of coronary artery after the initial ischemic insult and before 3 h of reperfusion. Remote ischemic preconditioning was discovered in 1993. Brief episodes of ischemia–reperfusion applied in remote myocardium (Przyklenk et al. 1993) or later in distant organs like the kidney (Gho et al. 1996) or skeletal muscle (Birnbaum et al. 1997) lead to cardioprotection in I–R. A lot of triggers, mediators, and signaling pathways, as well as potential effectors of conditioning, have been identified (reviewed in Heusch 2015). Many of the signaling kinases seems to be common to all methods of conditioning (Bell and Yellon 2012; Ferdinandy et al. 2014), but some can also differ between pre- and postconditioning (Heusch et al. 2006). Some of these protective pathways will be highlighted in the following chapters.

8.4.1 NO–cGMP–PKG Pathway

Nitric oxide (NO) release can be generated by one of three nitric oxide synthases, which are all expressed in cardiomyocytes (Massion et al. 2003). Agonists of

G_i-coupled receptors can activate PI3K–Akt, thereby phosphorylating and activating eNOS (Tong et al. 2000) and resulting in NO production. NO activates soluble guanylate cyclase that synthesizes cGMP from GTP leading to activation of PKG. The cGMP–PKG pathway affects the recovery of intracellular pH, Ca^{2+} oscillation, and MPTP opening (Inserte and Garcia-Dorado 2015). A supposed mechanism of cardioprotection includes PKG regulation of Ca^{2+} homeostasis through the modification of sarcoplasmic reticulum Ca^{2+} uptake mechanisms and PKG-induced opening of ATP-sensitive K_{ATP} channels during ischemia and/or reperfusion (Burley et al. 2007) (see Fig. 8.3).

The cGMP–PKG pathway delays the normalization of intracellular pH by inhibiting the NHE (Inserte et al. 2011a). PKG activation delays PLB phosphorylation via upregulation of phosphodiesterase (PDE)2 that decreases PKA activation, thereby reducing also SERCA activity (Inserte and Garcia-Dorado 2015). This leads to a mitigation of hypercontracture and cell death by hampering Ca^{2+} oscillation and regulates calcium homeostasis by permitting more effective cytosolic Ca^{2+} extrusion through NCX (Inserte et al. 2014).

During preconditioning, I–R cycles adenosine and bradykinin are released from cardiomyocytes (Schulz et al. 1998). Bradykinin can activate K_{ATP} channels with subsequent ROS release via NO–cGMP–PKG (Oldenburg et al. 2004). PKG transmits the cardioprotective signal from the cytosol to the inner mitochondrial membrane by a pathway that includes PKCε (Costa et al. 2005). NO can also lead to posttranslational reversible modification of proteins like *S*-nitrosation that can protect thiol groups from further oxidation (Murphy and Steenbergen 2008). It was shown that S-nitrosation of complex I reduces reperfusion injury by inhibiting MPTP. S-nitrosation of Cx43 located at the mitochondria is increased during ischemic preconditioning and rises mitochondrial permeability for potassium and increases ROS formation. Both processes are relevant for inhibition of MPTP opening during IRI (Soetkamp et al. 2014). The role of NO in cardiomyocytes will be further given attention in Chap. 10.

8.4.2 Reperfusion Injury Salvage Kinase (RISK) Pathway

Activation of the pro-survival kinase cascades phosphatidylinositol-3-OH kinase (PI3K)–Akt and/or p42–p44 MAP kinase (=extracellular signal-regulated kinases Erk1–Erk2) during reperfusion have been shown to confer protection against reperfusion injury (Hausenloy et al. 2004 b). Activation of PI3K is mediated via activation of receptors like G protein–coupled receptors (GPCR) and receptors of growth factors, leading to phosphorylation of AKT and to protection against IRI (Datta et al. 1996). ERK is one of the three major mammalian mitogen-activated protein kinases. It is important for the survival of cells by protecting them from programmed cell death (Yue et al. 2000). Both kinases are able to phosphorylate and inactivate proapoptotic proteins (Hausenloy et al. 2004b). Bax is inhibited in a way that its conformational change that normally allows translocation to the mitochondria and release of cytochrome C does not occur (Yamaguchi and Wang 2001; Tsuruta et al.

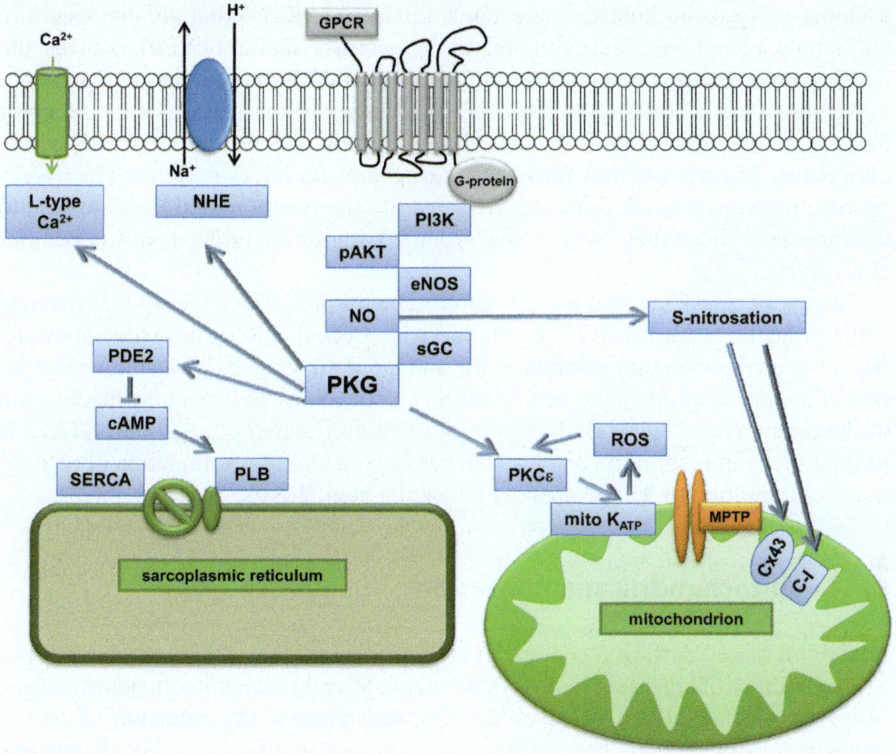

Fig. 8.3 NO–cGMP–PKG pathway. Agonists of G_i-coupled receptors can activate PI3K–Akt thereby phosphorylating and activating eNOS resulting in NO production. NO activates soluble guanylate cyclase that synthesizes cGMP from GTP leading to activation of PKG. PKG induced via PKCε opening of ATP-sensitive K_{ATP} channels with subsequent ROS release. In addition, PKG activation delays PLB phosphorylation via upregulation of PDE2 that decrease cAMP levels thereby reducing SERCA activity. Furthermore, PKG delays the normalization of intracellular pH by inhibiting the Na^+–H^+ exchanger. NO can also lead to posttranslational reversible modification of proteins like *S*-nitrosation of complex I and Cx43

2002). Phosphorylation of Bad ends in binding to the regulatory protein 14-3-3 resulting in release of antiapoptotic proteins (Zha et al. 1996). In addition, Akt can confer protection independent of Bad by inhibition of cytochrome C release and maintenance of mitochondrial membrane potential (Kennedy et al. 1999). Furthermore, Akt phosphorylates eNOS and increases cellular NO levels that inhibit MPTP opening (Dimmeler et al. 1999; Balakirev et al. 1997).

8.4.3 SAFE Pathway

The survivor activating factor enhancement (SAFE) pathway comprises the activation of STAT3, which is activated during ischemia and ischemia–reperfusion by a family of tyrosine kinases – the Janus kinases (JAKs) – that are connected to membrane receptors. The cytokine tumor necrosis factor (TNFα) induces the SAFE pathway and is essential for protection by both ischemic pre- and postconditioning (Lecour 2009). Binding of TNFα leads to dimerization of the receptor thereby activating receptor-associated JAK proteins. Subsequently, JAK phosphorylates the receptor, thus forming docking sites for STAT proteins. The STATs become phosphorylated, dimerize, detach from the receptor, and translocate into the nucleus, where they bind to the promoter region of target genes (Boengler et al. 2008).

Targets of STAT3, a member of the aforementioned STAT family that mediate cardioprotection, are involved in cell survival (Lecour 2009): in preconditioning, STAT3 is involved in upregulation of the antiapoptotic gene Bcl-2 and downregulation of the proapoptotic gene Bax (Hattori et al. 2001) as well as in the inactivation of the proapoptotic factor Bad by phosphorylation (Lecour et al. 2005). STAT3 is located in the mitochondria and leads to cardioprotection by stimulation of respiration and inhibition of MPTP opening (Boengler et al. 2010).

8.4.4 Mitochondria and Protection

Activation of cell surface receptors on the sarcolemma of cardiomyocytes transmit a large number of signaling pathways that seem to end in the mitochondrial targets. Mitochondria can release low levels of ROS that signals to key mediators of cardioprotection leading to reduced infarct size through inhibition of MPTP opening (Ferdinandy et al. 2014).

Connexin 43 (Cx43) is not only the major constituent protein of gap junctions and non-junctional hemichannels in ventricular cardiomyocytes but is also localized in cardiac inner membranes of subsarcolemmal mitochondria (SSM) (Schulz et al. 2007). Cx43 is involved in ROS formation and is a signaling molecule in preconditioning but not postconditioning (Schulz et al. 2007). Neither the RISK nor the SAFE signaling pathway is involved in protection of Cx43 to myocardial ischemia–reperfusion injury (Sánchez et al. 2013). Cx43 translocates to the mitochondria via heat shock protein 90-dependent TOM pathway (Rodríguez-Sinovas et al. 2007). Preconditioning increases the import of Cx43 into SSMs and leads to its S-nitrosation. Mitochondrial Cx43 facilitates the influx of K^+ and increases the respiration of complex I contributing to ROS formation (Boengler et al. 2012; 2013; Heinzel et al. 2005).

In response to ischemia, the mitochondria undergo fragmentation, a process that is dependent on the mitochondrial fission protein dynamin-related protein 1 (Drp1) (Ong et al. 2015). Treatment of adult cardiomyocytes with mitochondrial division inhibitor-1 diminished cell death and blocked MPTP opening after I–R and reduced MI size in mice subjected to coronary artery occlusion and reperfusion. So far it is not known whether ischemic pre- and postconditioning exerts its cardioprotective effect via modulating the morphology of the mitochondria (Ferdinandy et al. 2014).

8.4.5 MPTP and Protection

The beneficial effects of reperfusion are limited since 50 % of the final myocardial infarct size is attributed to necrotic cardiomyocytes. These cells were still viable at the end of ischemia but damaged by reperfusion (Bulluck and Hausenloy 2015). Thus protecting cardiomyocytes from cell death is a major therapeutic target. MPTP is a key in the protection against necrotic and programmed necrotic death (Piot et al. 2008; Lim et al. 2007). This conclusion is based on findings that MPTP is an effector of conditioning in I–R (Chiong et al. 2011).

Transient opening of MPTP can be protective by limiting mitochondrial matrix calcium load and facilitating ROS signaling through mitochondrial uncoupling (Hausenloy et al. 2004a). The mechanism of ROS generation is mediated via opening of mitochondrial ATP-sensitive K^+ channel (mitoK_{ATP}) that causes matrix alkalization (Andrukhiv et al. 2006). This increases levels of ROS in cardiomyocytes that are derived from complex I (NADH:ubiquinone oxidoreductase).

Several preconditioning as well as postconditioning studies have shown that MPTP opening is suppressed at the beginning of reperfusion thereby leading to cardioprotection that could be facilitated by indirect MPTP inhibition through modulating several factors like ROS, Ca^{2+}, and pH or by direct MPTP inhibition (Ong et al. 2015). Prevention of MPTP opening by pre- or postconditioning is mediated through decrease in mitochondrial ROS production at the end of ischemia (Clarke et al. 2008; Sun et al. 2005). Activation of mitoK_{ATP} leading to increased ROS production can also inhibit MPTP opening via PKCε that is phosphorylated by PKG after translocation to the outer mitochondrial membrane (Costa et al. 2005; Costa and Garlid 2008).

One mediator that is involved in cardioprotection by limiting MPTP opening is GSK3β (Ferdinandy et al. 2014). GSK3β is active in its non-phosphorylated form and can be inactivated through phosphorylation. Activation of receptor tyrosine kinases or certain G protein–coupled receptors induces phosphorylation of GSK3β (Juhaszova et al. 2004). The mechanisms by which inhibition of GSK3β delays MPTP opening are still unclear, but it is conceivable that GSK3β changes MPTP by modifying phosphorylation of target molecules in the mitochondria (Murphy and Steenbergen 2008).

Activation of the RISK pathway at the onset of reperfusion has been described to reduce MPTP opening as well (Hausenloy et al. 2005). Activation of these kinases can lead to activation of the mitoK_{ATP} channel which reduces membrane potential thereby reducing the uptake of mitochondrial Ca^{2+} and subsequently the generation of ROS (Murata et al. 2001; Andrukhiv et al. 2006).

8.5 Concluding Remarks

Long-lasting ischemia and its implications, like acidosis, energy depletion, increased osmolality, and calcium overload, result in damage of cardiomyocytes during reperfusion. This damage is referred to normalization of pH, regeneration of ionic homeostasis, and recovery of energy (Piper et al. 1998). Protection of cardiomyocytes against IRI is based on reduction of cell swelling and ROS production, inhibition of contraction, and spreading of cell death that is facilitated by activation of sarcolemmal receptors (Heusch 2015; Murphy and Steenbergen 2008; Vinten-Johansen and Shi W 2011).

So far, all conditioning methods investigated in animal models observe a decrease in IRI. Ischemic preconditioning conveys a robust protection against cardiomyocyte death (Schulz et al. 2001) and delays infarct development (Heusch 2013), but it is inapplicable in clinical settings due to the feasibility of the method (Rodríguez-Sinovas et al. 2007). Postconditioning is a practicable method that induces activation of protein kinases (Ovize et al. 2010). It can also be performed with isolated cardiomyocytes indicating the genuine protection in cardiomyocytes. Postconditioning exerts its protective effects by delaying normalization of pH (Inserte et al. 2011b), preservation of mitochondrial function, and preventing the opening of the MPTP (Argaud et al. 2005). Remote conditioning is an elegant protection method and can be easily performed in the clinics. The protective signal is transferred from the remote organ to the heart by neuronal and humoral mediators (Heusch 2015). It seems to activate similar signaling pathways like pre- and postconditioning (Heusch 2015). However, pharmacologic conditioning that activates single protective signaling pathways or inhibit targets that contribute to IRI have been in the majority of cases negative or inconsistent for the drugs tested or show side effects of inhibitors like some MPTP blockers (Rodríguez-Sinovas et al. 2007; Atar et al. 2015). Beyond that, comorbidities like diabetes, hypertension, atherosclerosis, or drugs can influence the outcome of protection, but also sex and age (Ferdinandy et al. 2014). All these factors interfere with the protective signaling pathways and alter the expression and/or activity of proteins involved in cardioprotection.

Therefore, novel therapeutic strategies are discussed that may protect cardiomyocytes and thus the heart against IRI (Sluijter et al. 2014). Interestingly, circadian genes that are important for sleep and wake cycling are associated with cardiovascular disease (Storch et al. 2002). One of these circadian genes is period (Per), since infarct size in Per2 knockout mice is larger than in wild-type controls after ischemia–reperfusion (Eckle et al. 2012). Another therapeutically interesting object is the regulation of gene expression by epigenetically mediated events or nonprotein-coding RNAs or miRNAs (Sluijter et al. 2014; Fiedler and Thum 2013). Several miRNAs are involved in MI such as miR-1 and miR-214 that control the gene expression for NCX in cardiomyocytes (Fiedler and Thum 2013). In addition, ROS upregulates miR-1 and depresses expression of PKCε and heat shock protein 60 thereby enhancing apoptosis (Pan et al. 2012). So, interference in miRNA signaling might have a therapeutic potential.

Finally, cardioprotective signaling molecules that target cardiomyocytes can be expressed and secreted by the cardiomyocyte itself, but can also be produced by other cells like endothelial cells, smooth muscle cells, or fibroblast at least in the whole heart (Ovize et al. 2010). This is in particular of interest, since cardiomyocytes are indeed the largest cell type by volume within the myocardium, but they are not the predominant cell type in number (Bell and Yellon 2012).

References

Abdallah Y, Kasseckert SA, Iraqi W, Said M, Shahzad T, Erdogan A, Neuhof C, Gündüz D, Schlüter KD, Tillmanns H, Piper HM, Reusch HP, Ladilov Y (2011) Interplay between Ca^{2+} cycling and mitochondrial permeability transition pores promotes reperfusion-induced injury of cardiac myocytes. J Cell Mol Med 15:2478–2485

Andrukhiv A, Costa AD, West IC, Garlid KD (2006) Opening mitoKATP increases superoxide generation from complex I of the electron transport chain. Am J Physiol 291: H2067–H2074

Argaud L, Gateau-Roesch O, Raisky O, Loufouat J, Robert D, Ovize M (2005) Postconditioning inhibits mitochondrial permeability transition. Circulation 111:194–197

Armstrong SC, Latham CA, Shivell CL, Ganote CE (2001) Ischemic loss of sarcolemmal dystrophin and spectrin: correlation with myocardial injury. J Mol Cell Cardiol 33:1165–1179

Atar D, Arheden H, Berdeaux A, Bonnet JL, Carlsson M, Clemmensen P, Cuvier V, Danchin N, Dubois-Randé JL, Engblom H, Erlinge D, Firat H, Halvorsen S, Hansen HS, Hauke W, Heiberg E, Koul S, Larsen AI, Le Corvoisier P, Nordrehaug JE, Paganelli F, Pruss RM, Rousseau H, Schaller S, Sonou G, Tuseth V, Veys J, Vicaut E, Jensen SE (2015) Effect of intravenous TRO40303 as an adjunct to primary percutaneous coronary intervention for acute ST-elevation myocardial infarction: MITOCARE study results. Eur Heart J 36:112–119

Balakirev MY, Khramtsov VV, Zimmer G (1997) Modulation of the mitochondrial permeability transition by nitric oxide. Eur J Biochem 246:710–718

Bell RM, Yellon DM (2012) Conditioning the whole heart-not just the cardiomyocyte. J Mol Cell Cardiol 53:24–32

Berg JM, Tymoczko JL, Stryer L (2002) Biochemistry, 5th edn. W H Freeman, New York. ISBN-10: 0-7167-3051-0.

Bernardi P, Di Lisa F (2015) The mitochondrial permeability transition pore: molecular nature and role as a target in cardioprotection. J Mol Cell Cardiol 78:100–106

Birnbaum Y, Hale SL, Kloner RA (1997) Ischemic preconditioning at a distance: reduction of myocardial infarct size by partial reduction of blood supply combined with rapid stimulation of the gastrocnemius muscle in the rabbit. Circulation 96:1641–1646

Boengler K, Hilfiker-Kleiner D, Drexler H, Heusch G, Schulz R (2008) The myocardial JAK/STAT pathway: from protection to failure. Pharmacol Ther 120:172–185

Boengler K, Hilfiker-Kleiner D, Heusch G, Schulz R (2010) Inhibition of permeability transition pore opening by mitochondrial STAT3 and its role in myocardial ischemia/reperfusion. Basic Res Cardiol 105:771–785

Boengler K, Ruiz-Meana M, Gent S, Ungefug E, Soetkamp D, Miro-Casas E, Cabestrero A, Fernandez-Sanz C, Semenzato M, Di Lisa F, Rohrbach S, Garcia-Dorado D, Heusch G, Schulz R (2012) Mitochondrial connexin 43 impacts on respiratory complex I activity and mitochondrial oxygen consumption. J Cell Mol Med 16:1649–1655

Boengler K, Ungefug E, Heusch G, Leybaert L, Schulz R (2013) Connexin 43 impacts on mitochondrial potassium uptake. Front Pharmacol 4:73. doi:10.3389/fphar.2013.00073

Bulluck H, Hausenloy DJ (2015) Ischaemic conditioning: are we there yet? Heart 101:1067–1077

Burley DS, Ferdinandy P, Baxter GF (2007) Cyclic GMP and protein kinase-G in myocardial ischaemia-reperfusion: opportunities and obstacles for survival signaling. Br J Pharmacol 152:855–869

Chen YR, Zweier JL (2014) Cardiac mitochondria and reactive oxygen species generation. Circ Res 114:524–537

Chen Z, Chua CC, Ho YS, Hamdy RC, Chua BH (2001) Overexpression of Bcl-2 attenuates apoptosis and protects against myocardial I/R injury in transgenic mice. Am J Physiol 280:H2313–H2320

Chen M, Won DJ, Krajewski S, Gottlieb RA (2002) Calpain and mitochondria in ischemia/reperfusion injury. J Biol Chem 277:29181–29186

Chiong M, Wang ZV, Pedrozo Z, Cao DJ, Troncoso R, Ibacache M, Criollo A, Nemchenko A, Hill JA, Lavandero S (2011) Cardiomyocyte death: mechanisms and translational implications. Cell Death Dis 2:e244. doi:10.1038/cddis.2011.130

Choi E, Cha MJ, Hwang KC (2014) Roles of Calcium Regulating MicroRNAs in Cardiac Ischemia-Reperfusion Injury. Cells 3:899–913

Clarke SJ, Khaliulin I, Das M, Parker JE, Heesom KJ, Halestrap AP (2008) Inhibition of mitochondrial permeability transition pore opening by ischemic preconditioning is probably mediated by reduction of oxidative stress rather than mitochondrial protein phosphorylation. Circ Res 102:1082–1090

Costa AD, Garlid KD (2008) Intramitochondrial signaling: interactions among mitoKATP, PKCepsilon, ROS, and MPT. Am J Physiol 295:H874–H882

Costa AD, Garlid KD, West IC, Lincoln TM, Downey JM, Cohen MV, Critz SD (2005) Protein kinase G transmits the cardioprotective signal from cytosol to mitochondria. Circ Res 97:329–336

Datta K, Bellacosa A, Chan TO, Tsichlis PN (1996) Akt is a direct target of the phosphatidylinositol 3-kinase. Activation by growth factors, v-src and v-Ha-ras, in Sf9 and mammalian cells. J Biol Chem 271:30835–30839

Dennis SC, Gevers W, Opie LH (1991) Protons in ischemia: where do they come from; where do they go to? J Mol Cell Cardiol 23(9):1077–1086

DeWood MA, Spores J, Notske R, Mouser LT, Burroughs R, Golden MS, Lang HT (1980) Prevalence of total coronary occlusion during the early hours of transmural myocardial infarction. N Engl J Med 303:897–902

Dimmeler S, Fleming I, Fisslthaler B, Hermann C, Busse R, Zeiher AM (1999) Activation of nitric oxide synthase in endothelial cells by Akt-dependent phosphorylation. Nature 399:601–605

Eckle T, Hartmann K, Bonney S, Reithel S, Mittelbronn M, Walker LA, Lowes BD, Han J, Borchers CH, Buttrick PM, Kominsky DJ, Colgan SP, Eltzschig HK (2012) Adora2b-elicited Per2 stabilization promotes a HIF-dependent metabolic switch crucial for myocardial adaptation to ischemia. Nat Med 18:774–782

Ferdinandy P, Hausenloy DJ, Heusch G, Baxter GF, Schulz R (2014) Interaction of risk factors, comorbidities, and comedications with ischemia/reperfusion injury and cardioprotection by preconditioning, postconditioning, and remote conditioning. Pharmacol Rev 66:1142–1174

Fiedler J, Thum T (2013) MicroRNAs in myocardial infarction. Arterioscler Thromb Vasc Biol 33:201–205

Fleet WF, Johnson TA, Graebner CA, Gettes LS (1985) Effect of serial brief ischemic episodes on extracellular K^+, pH, and activation in the pig. Circulation 72:922–932

Fliss H, Gattinger D (1996) Apoptosis in ischemic and reperfused rat myocardium. Circ Res 79:949–956

Frelin C, Vigne P, Lazdunski M (1985) The role of the Na^+/H^+ exchange system in the regulation of the internal pH in cultured cardiac cells. Eur J Biochem 149:1–4

Garcia-Dorado D, Oliveras J (1993) Myocardial oedema: a preventable cause of reperfusion injury? Cardiovasc Res 27:1555–1563

Garcia-Dorado D, Ruiz-Meana M, Padilla F, Rodriguez-Sinovas A, Mirabet M (2002) Gap junction-mediated intercellular communication in ischemic preconditioning. Cardiovasc Res 55:456–465

García-Dorado D, Rodríguez-Sinovas A, Ruiz-Meana M (2004) Gap junction-mediated spread of cell injury and death during myocardial ischemia-reperfusion. Cardiovasc Res 61:386–401

Garcia-Dorado D, Ruiz-Meana M, Inserte J, Rodriguez-Sinovas A, Piper HM (2012) Calcium-mediated cell death during myocardial reperfusion. Cardiovasc Res 94:168–180

Gho BC, Schoemaker RG, van den Doel MA, Duncker DJ, Verdouw PD (1996) Myocardial protection by brief ischemia in noncardiac tissue. Circulation 94:2193–2200

Giordano FJ (2005) Oxygen, oxidative stress, hypoxia, and heart failure. J Clin Invest 115:500–508

Halestrap AP, Richardson AP (2015) The mitochondrial permeability transition: a current perspective on its identity and role in ischaemia/reperfusion injury. J Mol Cell Cardiol 78:129–141

Hattori R, Maulik N, Otani H, Zhu L, Cordis G, Engelman RM, Siddiqui MA, Das DK (2001) Role of STAT3 in ischemic preconditioning. J Mol Cell Cardiol 33:1929–1936
Hausenloy DJ, Yellon DM (2013) Myocardial ischemia-reperfusion injury: a neglected therapeutic target. J Clin Invest 123:92–100
Hausenloy D, Wynne A, Duchen M, Yellon D (2004a) Transient mitochondrial permeability transition pore opening mediates preconditioning-induced protection. Circulation 109:1714–1717
Hausenloy DJ, Mocanu MM, Yellon DM (2004b). Cross-talk between the survival kinases during early reperfusion: its contribution to ischemic preconditioning. Cardiovasc Res:305–12.
Hausenloy DJ, Tsang A, Mocanu MM, Yellon DM (2005) Ischemic preconditioning protects by activating prosurvival kinases at reperfusion. Am J Physiol Heart Circ Physiol 288:H971–H976
Heinzel FR, Luo Y, Li X, Boengler K, Buechert A, García-Dorado D, Di Lisa F, Schulz R, Heusch G (2005). Impairment of diazoxide-induced formation of reactive oxygen species and loss of cardioprotection in connexin 43 deficient mice. Circ Res 97:583–586
Heusch G (2013) Cardioprotection: chances and challenges of its translation to the clinic. Lancet 381:166–175
Heusch G (2015) Molecular basis of cardioprotection: signal transduction in ischemic pre-, post-, and remote conditioning. Circ Res 116:674–699
Heusch G, Büchert A, Feldhaus S, Schulz R (2006) No loss of cardioprotection by postconditioning in connexin 43-deficient mice. Basic Res Cardiol 101:354–356
Hochhauser E, Kivity S, Offen D, Maulik N, Otani H, Barhum Y, Pannet H, Shneyvays V, Shainberg A, Goldshtaub V, Tobar A, Vidne BA (2003) Bax ablation protects against myocardial ischemia-reperfusion injury in transgenic mice. Am J Physiol 284:H2351–H2359
Hochhauser E, Cheporko Y, Yasovich N, Pinchas L, Offen D, Barhum Y et al (2007) Bax deficiency reduces infarct size and improves long-term function after myocardial infarction. Cell Biochem Biophys 47:11–20
Ibáñez B, Heusch G, Ovize M, Van de Werf F (2015) Evolving therapies for myocardial ischemia/reperfusion injury. J Am Coll Cardiol 65:1454–1471
Inserte J, Garcia-Dorado D (2015) The cGMP/PKG pathway as a common mediator of cardioprotection: translatability and mechanism. Br J Pharmacol 172:1996–2009
Inserte J, Garcia-Dorado D, Hernando V, Soler-Soler J (2005) Calpain-mediated impairment of Na^+/K^+-ATPase activity during early reperfusion contributes to cell death after myocardial ischemia. Circ Res 97:465–473
Inserte J, Barba I, Hernando V, Garcia-Dorado D (2009) Delayed recovery of intracellular acidosis during reperfusion prevents calpain activation and determines protection in postconditioned myocardium. Cardiovasc Res 81:116–122
Inserte J, Barba I, Poncelas-Nozal M, Hernando V, Agulló L, Ruiz-Meana M, Garcia-Dorado D (2011a) cGMP/PKG pathway mediates myocardial postconditioning protection in rat hearts by delaying normalization of intracellular acidosis during reperfusion. J Mol Cell Cardiol 50:903–909
Inserte J, Ruiz-Meana M, Rodríguez-Sinovas A, Barba I, Garcia-Dorado D (2011b) Contribution of delayed intracellular pH recovery to ischemic postconditioning protection. Antioxid Redox Signal 14:923–939
Inserte J, Hernando V, Garcia-Dorado D (2012) Contribution of calpains to myocardial ischaemia/reperfusion injury. Cardiovasc Res 96:23–31
Inserte J, Hernando V, Ruiz-Meana M, Poncelas-Nozal M, Fernández C, Agulló L, Sartorio C, Vilardosa U, Garcia-Dorado D (2014) Delayed phospholamban phosphorylation in post-conditioned heart favours Ca^{2+} normalization and contributes to protection. Cardiovasc Res 103:542–553
Jennings RB (2013) Historical perspective on the pathology of myocardial ischemia/reperfusion injury. Circ Res 13:428–438
Jennings RB, Reimer KA, Steenbergen C (1986). Myocardial ischemia revisited. The osmolar load, membrane damage, and reperfusion. J Mol Cell Cardiol 18:769–80
Jennings RB, Reimer KA, Hill ML, Mayer SE (1981) Total ischemia in dog hearts, in vitro. 1. Comparison of high energy phosphate production, utilization, and depletion, and of adenine

nucleotide catabolism in total ischemia in vitro vs. severe ischemia in vivo. Circ Res 49:892–900

Jennings RB, Murry CE, Steenbergen C Jr, Reimer KA (1990) Development of cell injury in sustained acute ischemia. Circulation 82(3 Suppl):II2–II12

Jordan C, Püschel B, Koob R, Drenckhahn D (1995) Identification of a binding motif for ankyrin on the alpha-subunit of Na+, K+–ATPase. J Biol Chem 270:29971–22975

Juhaszova M, Zorov DB, Kim SH, Pepe S, Fu Q, Fishbein KW, Ziman BD, Wang S, Ytrehus K, Antos CL, Olson EN, Sollott SJ (2004) Glycogen synthase kinase-3beta mediates convergence of protection signaling to inhibit the mitochondrial permeability transition pore. J Clin Invest 113:1535–1549

Kalogeris T, Bao Y, Korthuis RJ (2014) Mitochondrial reactive oxygen species: a double edged sword in ischemia/reperfusion vs. preconditioning. Redox Biol 2:702–714

Kar P, Chakraborti T, Samanta K, Chakraborti S (2009) mu-Calpain mediated cleavage of the Na+/Ca2+ exchanger in isolated mitochondria under A23187 induced Ca^{2+} stimulation. Arch Biochem Biophys 482:66–76

Kennedy SG, Kandel ES, Cross TK, Hay N (1999) Akt/Protein kinase B inhibits cell death by preventing the release of cytochrome c from mitochondria. Mol Cell Biol 19:5800–5810

Kobayashi K, Neely JR (1979) Control of maximum rates of glycolysis in rat cardiac muscle. Circ Res 44:166–175

Konstantinidis K, Whelan RS, Kitsis RN (2012) Mechanisms of cell death in heart disease. Arterioscler Thromb Vasc Biol 32:1552–1562

Kühn H (2011) Herz-Kreislauf-System. In: Püschel GP, Kietzmann T, Doenecke D, Kühn H, Höhne W, Christ B, Koolman J (ed) Taschenlehrbuch Biochemie. Georg Thieme, Verlag

Ladilov YV, Siegmund B, Piper HM (1995) Protection of reoxygenated cardiomyocytes against hypercontracture by inhibition of Na^+/H^+ exchange. Am J Physiol 268:H1531–H1539

Ladilov Y, Haffner S, Balser-Schäfer C, Maxeiner H, Piper HM (1999) Cardioprotective effects of KB-R7943: a novel inhibitor of the reverse mode of Na^+/Ca^{2+} exchanger. Am J Physiol 276:H1868–H1876

Lecour S (2009) Multiple protective pathways against reperfusion injury: a SAFE path without action? J Mol Cell Cardiol 46:607–609

Lecour S, Suleman N, Deuchar GA, Somers S, Lacerda L, Huisamen B, Opie LH (2005) Pharmacological preconditioning with tumor necrosis factor-alpha activates signal transducer and activator of transcription-3 at reperfusion without involving classic prosurvival kinases (Akt and extracellular signal-regulated kinase). Circulation 112:3911–3918

Lesnefsky EJ, Chen Q, Moghaddas S, Hassan MO, Tandler B, Hoppel CL (2004) Blockade of electron transport during ischemia protects cardiac mitochondria. J Biol Chem 279:47961–47967

Li F, Sugishita K, Su Z, Ueda I, Barry WH (2001). Activation of connexin-43 hemichannels can elevate [Ca(2+)]i and [Na(+)]i in rabbit ventricular myocytes during metabolic inhibition. J Mol Cell Cardiol 33, 2145–2155

Lim SY, Davidson SM, Mocanu MM, Yellon DM, Smith CC (2007) The cardioprotective effect of necrostatin requires the cyclophilin-D component of the mitochondrial permeability transition pore. Cardiovasc Drugs Ther 21:467–469

Long Q, Yang K, Yang Q (2015) Regulation of mitochondrial ATP synthase in cardiac pathophysiology. Am J Cardiovasc Dis 5:19–32

Marban E, Kusuoka H (1987) Maximal Ca^{2+} activated force and myofilament Ca^{2+} sensitivity in intact mammalian hearts. J Gen Physiol 90:609–623

Massion PB, Feron O, Dessy C, Balligand JL (2003) Nitric oxide and cardiac function: ten years after, and continuing. Circ Res 93:388–398

Murata M, Akao M, O'Rourke B, Marban E (2001) Mitochondrial ATP-sensitive potassium channels attenuate matrix Ca(2+) overload during simulated ischemia and reperfusion: possible mechanism of cardioprotection. Circ Res 89:891–898

Murphy E, Steenbergen C (2008) Mechanisms underlying acute protection from cardiac ischemia-reperfusion injury. Physiol Rev 88:581–609

Nakamura T, Ueda Y, Juan Y, Katsuda S, Takahashi H, Koh E (2000) Fas-mediated apoptosis in adriamycin-induced cardiomyopathy in rats: in vivo study. Circulation 102:572–578

Oakes SA, Opferman JT, Pozzan T, Korsmeyer SJ, Scorrano L (2003) Regulation of endoplasmic reticulum Ca^{2+} dynamics by proapoptotic BCL-2 family members. Biochem Pharmacol 66:1335–1340

Oerlemans MI, Koudstaal S, Chamuleau SA, de Kleijn DP, Doevendans PA, Sluijter JP (2013) Targeting cell death in the reperfused heart: pharmacological approaches for cardioprotection. Int J Cardiol 165:410–422

Oldenburg O, Qin Q, Krieg T, Yang XM, Philipp S, Critz SD, Cohen MV, Downey JM (2004) Bradykinin induces mitochondrial ROS generation via NO, cGMP, PKG, and mitoKATP channel opening and leads to cardioprotection. Am J Physiol Heart Circ Physiol 286:H468–H476

Ong SB, Samangouei P, Kalkhoran SB, Hausenloy DJ (2015) The mitochondrial permeability transition pore and its role in myocardial ischemia reperfusion injury. J Mol Cell Cardiol 78:23–34

Orogo AM, Gustafsson ÅB (2013) Cell death in the myocardium: my heart won't go on. IUBMB Life 65:651–656

Ovize M, Baxter GF, Di Lisa F, Ferdinandy P, Garcia-Dorado D, Hausenloy DJ, Heusch G, Vinten-Johansen J, Yellon DM, Schulz R, Working Group of Cellular Biology of Heart of European Society of Cardiology (2010) Postconditioning and protection from reperfusion injury: where do we stand? Position paper from the Working Group of Cellular Biology of the Heart of the European Society of Cardiology. Cardiovasc Res 87:406–423

Pan Z, Sun X, Ren J, Li X, Gao X, Lu C, Zhang Y, Sun H, Wang Y, Wang H, Wang J, Xie L, Lu Y, Yang B (2012) miR-1 exacerbates cardiac ischemia-reperfusion injury in mouse models. PLoS One 7:e50515

Pan X, Liu J, Nguyen T, Liu C, Sun J, Teng Y, Fergusson MM, Rovira II, Allen M, Springer DA, Aponte AM, Gucek M, Balaban RS, Murphy E, Finkel T (2013) The physiological role of mitochondrial calcium revealed by mice lacking the mitochondrial calcium uniporter. Nat Cell Biol 15:1464–1472

Paradis AN, Gay MS, Zhang L (2014). Binucleation of cardiomyocytes: the transition from a proliferative to a terminally differentiated state. Drug Discov Today 19:602–609

Piot C, Croisille P, Staat P, Thibault H, Rioufol G, Mewton N et al (2008) Effect of cyclosporine on reperfusion injury in acute myocardial infarction. N Engl J Med 359:473–481

Piper HM, Balser C, Ladilov YV, Schäfer M, Siegmund B, Ruiz-Meana M, Garcia-Dorado D (1996) The role of Na^+/H^+ exchange in ischemia-reperfusion. Basic Res Cardiol 91:191–202

Piper HM, García-Dorado D, Ovize M (1998) A fresh look at reperfusion injury. Cardiovasc Res 38:291–300

Piper HM, Ladilov Y V, Siegmund B, Taimor G (1999) Zelluläre Ursachen des myokardialen Ischämie-Reperfusionsschadens. In: Griebenow G, Gülker H, Dominiak P, Piper HM (ed) Autonomes Nervensystem und Koronare Herzkrankheit. Novartis Pharma GmbH, Nürnberg, SBN:9783933185211, pp 21–35

Piper HM, Abdallah Y, Schafer C (2004) The first minutes of reperfusion: a window of opportunity for cardioprotection. Cardiovasc Res 61:365–371

Piquereau J, Caffin F, Novotova M, Lemaire C, Veksler V, Garnier A, Ventura-Clapier R, Joubert F (2013) Mitochondrial dynamics in the adult cardiomyocytes: which roles for a highly specialized cell? Front Physiol 4:102. doi:10.3389/fphys.2013.00102

Przyklenk K, Bauer B, Ovize M, Kloner RA, Whittaker P (1993) Regional ischemic 'preconditioning' protects remote virgin myocardium from subsequent sustained coronary occlusion. Circulation 87:893–899

Ricci JE, Muñoz-Pinedo C, Fitzgerald P, Bailly-Maitre B, Perkins GA, Yadava N, Scheffler IE, Ellisman MH, Green DR (2004). Disruption of mitochondrial function during apoptosis is mediated by caspase cleavage of the p75 subunit of complex I of the electron transport chain. Cell 117:773–86

Rodríguez-Sinovas A, Abdallah Y, Piper HM, Garcia-Dorado D (2007) Reperfusion injury as a therapeutic challenge in patients with acute myocardial infarction. Heart Fail Rev 12:207–216

Rubtsov AM, Lopina OD (2000) Ankyrins. FEBS Lett 482:1–5

Sánchez JA, Rodríguez-Sinovas A, Barba I, Miró-Casas E, Fernández-Sanz C, Ruiz-Meana M, Alburquerque-Béjar JJ, García-Dorado D (2013) Activation of RISK and SAFE pathways is not involved in the effects of Cx43 deficiency on tolerance to ischemia-reperfusion injury and preconditioning protection. Basic Res Cardiol 108:351

Schlüter KD, Jakob G, Ruiz-Meana M, Garcia-Dorado D, Piper HM (1996) Protection of reoxygenated cardiomyocytes against osmotic fragility by nitric oxide donors. Am J Physiol 271:H428–H434

Schulz R, Post H, Vahlhaus C, Heusch G (1998) Ischemic preconditioning in pigs: a graded phenomenon: its relation to adenosine and bradykinin. Circulation 98:1022–1029

Schulz R, Cohen MV, Behrends M, Downey JM, Heusch G (2001) Signal transduction of ischemic preconditioning. Cardiovasc Res 52:181–198

Schulz R, Boengler K, Totzeck A, Luo Y, Garcia-Dorado D, Heusch G (2007) Connexin 43 in ischemic pre- and postconditioning. Heart Fail Rev 12:261–266

Schulz R, Görge PM, Görbe A, Ferdinandy P, Lampe PD, Leybaert L (2015). Connexin 43 is an emerging therapeutic target in ischemia/reperfusion injury, cardioprotection and neuroprotection. Pharmacol Ther 153:90–106.

Shintani-Ishida K, Yoshida K (2011) Ischemia induces phospholamban dephosphorylation via activation of calcineurin, PKC-α, and protein phosphatase 1, thereby inducing calcium overload in reperfusion. Biochim Biophys Acta 1812:743–751

Siegmund B, Zude R, Piper HM (1992) Recovery of anoxic-reoxygenated cardiomyocytes from severe Ca^{2+} overload. Am J Physiol 263:H1262–H1269

Siegmund B, Schlüter KD, Piper HM (1993) Calcium and the oxygen paradox. Cardiovasc Res 27:1778–1783

Siegmund B, Schlack W, Ladilov YV, Balser C, Piper HM (1997) Halothane protects cardiomyocytes against reoxygenation-induced hypercontracture. Circulation 96:4372–4379

Singh RB, Chohan PK, Dhalla NS, Netticadan T (2004) The sarcoplasmic reticulum proteins are targets for calpain action in the ischemic-reperfused heart. J Mol Cell Cardiol 37:101–110

Sluijter JP, Condorelli G, Davidson SM, Engel FB, Ferdinandy P, Hausenloy DJ, Lecour S, Madonna R, Ovize M, Ruiz-Meana M, Schulz R, Van Laake LW, Nucleus of the European Society of Cardiology Working Group Cellular Biology of the Heart (2014) Novel therapeutic strategies for cardioprotection. Pharmacol Ther 144:60–70

Smith CC, Davidson SM, Lim SY, Simpkin JC, Hothersall JS, Yellon DM (2007) Necrostatin: a potentially novel cardioprotective agent? Cardiovasc Drugs Ther 21:227–233

Soetkamp D, Nguyen TT, Menazza S, Hirschhäuser C, Hendgen-Cotta UB, Rassaf T, Schlüter KD, Boengler K, Murphy E, Schulz R (2014) S-nitrosation of mitochondrial connexin 43 regulates mitochondrial function. Basic Res Cardiol 109:433

Steenbergen C, Murphy E, Levy L, London RE (1987a) Elevation in cytosolic free calcium concentration early in myocardial ischemia in perfused rat heart. Circ Res 60:700–707

Steenbergen CJ, Hill ML, Jennings RB (1987b) Cytoskeletal damage during myocardial ischemia: changes in vinculin immunofluorescence staining during total in vitro ischemia and in canine heart. Circ Res 60:478–486

Storch KF, Lipan O, Leykin I, Viswanathan N, Davis FC, Wong WH, Weitz CJ (2002) Extensive and divergent circadian gene expression in liver and heart. Nature 417:78–83

Sun HY, Wang NP, Kerendi F, Halkos M, Kin H, Guyton RA, Vinten-Johansen J, Zhao ZQ (2005) Hypoxic postconditioning reduces cardiomyocyte loss by inhibiting ROS generation and intracellular Ca^{2+} overload. Am J Physiol 288:H1900–H1908

Sun J, Picht E, Ginsburg KS, Bers DM, Steenbergen C, Murphy E (2006) Hypercontractile female hearts exhibit increased S-nitrosylation of the L-type Ca^{2+} channel alpha1 subunit and reduced ischemia/reperfusion injury. Circ Res 98:403–411Taimor G, Lorenz H, Hofstaetter B, Schlüter KD, Piper HM (1999). Induction of necrosis but not apoptosis after anoxia and reoxygenation in isolated adult cardiomyocytes of rat. Cardiovasc Res 41:147-56.

Taimor G, Lorenz H, Hofstaetter B, Schlüter KD, Piper HM (1999). Induction of necrosis but not apoptosis after anoxia and reoxygenation in isolated adult cardiomyocytes of rat. Cardiovasc Res 41:147–56

Tong H, Chen W, Steenbergen C, Murphy E (2000) Ischemic preconditioning activates phosphatidylinositol-3-kinase upstream of protein kinase C. Circ Res 87:309–315

Tsuruta F, Masuyama N, Gotoh Y (2002) The phosphatidylinositol 3-kinase (PI3K)-Akt pathway suppresses Bax translocation to mitochondria. J Biol Chem 277:14040–14047

Vandenberg JI, Metcalfe JC, Grace AA (1993) Mechanisms of pHi recovery after global ischemia in the perfused heart. Circ Res 72:993–1003

Vinten-Johansen J, Shi W (2011). Perconditioning and postconditioning: current knowledge, knowledge gaps, barriers to adoption, and future directions. J Cardiovasc Pharmacol Ther 16:260–266

Wang N, De Bock M, Decrock E, Bol M, Gadicherla A, Bultynck G, Leybaert L (2013) Connexin targeting peptides as inhibitors of voltage- and intracellular Ca2+–triggered Cx43 hemichannel opening. Neuropharmacology 75:506–516

Wei GZ, Zhou JJ, Wang B, Wu F, Bi H, Wang YM et al (2007) Diastolic Ca^{2+} overload caused by Na^{+}/Ca^{2+} exchanger during the first minutes of reperfusion results in continued myocardial stunning. Eur J Pharmacol 572:1–11

Yamaguchi H, Wang HG (2001) The protein kinase PKB/Akt regulates cell survival and apoptosis by inhibiting Bax conformational change. Oncogene 20:7779–7786

Yue TL, Wang C, Gu JL, Ma XL, Kumar S, Lee JC, Feuerstein GZ, Thomas H, Maleeff B, Ohlstein EH (2000) Inhibition of extracellular signal-regulated kinase enhances Ischemia/Reoxygenation-induced apoptosis in cultured cardiac myocytes and exaggerates reperfusion injury in isolated perfused heart. Circ Res 86:692–699

Zha J, Harada H, Yang E, Jockel J, Korsmeyer SJ (1996) Serine phosphorylation of death agonist BAD in response to survival factor results in binding to 14-3-3 not BCL-X(L). Cell 87:619–628

Zhao ZQ, Corvera JS, Halkos ME, Kerendi F, Wang NP, Guyton RA, Vinten-Johansen J (2003) Inhibition of myocardial injury by ischemic postconditioning during reperfusion: comparison with ischemic preconditioning. Am J Physiol 285:H579–H588

Zorov DB, Filburn CR, Klotz LO, Zweier JL, Sollott SJ (2000) Reactive oxygen species (ROS)-induced ROS release: a new phenomenon accompanying induction of the mitochondrial permeability transition in cardiac myocytes. J Exp Med 192:1001–1014

Mechanisms of Cardiac Cell Death 9

Gerhild Euler

Abstract

As cardiomyocytes build up the contractile machinery of the heart, every loss of cardiomyocytes in diverse cardiac diseases inevitably weakens the contractile power of the heart. Therefore, prevention of cardiomyocyte death should be one of the major aims when thinking about cardioprotection. As cell death can take different forms that are induced by various stressors and that proceed via different pathways, it is necessary to strictly differentiate between these kinds of death in order to find optimised ways of therapy. In this chapter, we will introduce necrosis, apoptosis, necroptosis and the death-related process of autophagy as the different kinds of cell death occurring in cardiomyocytes. The triggers of cardiomyocyte death and the contribution of different kinds of cell death to cardiac diseases, with a special focus on ischemia/reperfusion injury and heart failure progression, will be presented.

9.1 Introduction

Cardiomyocytes with their contractile apparatus are the main machinery driving the pump function of the heart. Therefore, loss of cardiomyocytes inevitably must go along with reductions in cardiac function. Loss of cardiomyocytes outweighs even more serious due to the fact that regeneration of cardiomyocytes is limited, and only a small amount of regeneration could arise from residential stem cells. This is the reason why prevention of cardiomyocyte death in diverse cardiac diseases is one the major therapeutic aims.

G. Euler
Institute of Physiology, Justus Liebig University, Aulweg 129, 35392 Giessen, Germany
e-mail: Gerhild.Euler@physiologie.med.uni-giessen.de

K.-D. Schlüter (ed.), *Cardiomyocytes – Active Players in Cardiac Disease*,
DOI 10.1007/978-3-319-31251-4_9

About 40 years ago, cell death was viewed as a passive and unregulated process called necrosis. Interference in this process was, therefore, considered as impossible, and prevention of cell death appeared to be feasible only by inhibition of the initiating insult. Then, Kerr and co-workers (1972) recognised a mechanism of controlled cell deletion called apoptosis. Due to its morphological features, apoptosis appeared to be an active, inherently programmed phenomenon. Twenty years later, studies in *Caenorhabditis elegans* (*C. elegans*) defined the signalling networks and identified proapoptotic genes that actively convey programmed cell death. Orthologs of defined signalling molecules in *C. elegans* were found in mammals and also humans. Thus, it became evident that cells not only die passively by necrosis but also via a defined signalling pathway, the so-called apoptotic pathway. These findings opened the possibility to intervene in an already started death signalling process and thereby to enlarge the therapeutic possibilities in prevention of cell death. But research did not stop at this point of death signalling in apoptosis. Just recently, a new kind of cell death, which uses signalling elements of apoptosis but displays features of necrosis, was identified, the so-called necroptosis (Golstein and Kroemer 2007). Another kind of cell death was characterised with the identification of autophagy. Autophagy is an intracellular catabolic recycling process of degraded proteins and was, thus, initially considered as cytoprotective. However, deregulated autophagy can also result in cell death. So, nowadays, we are confronted with four mechanisms of cell death, necrosis, apoptosis, necroptosis and deregulated autophagy, that all can emerge in mammalian cells and also in cardiomyocytes. As these kinds of cell death differentially contribute to cardiac diseases and are mediated via different pathways, the interventional possibilities in cardiomyocyte cell death have increased tremendously.

In this chapter, we want to depict the contribution of the different kinds of cardiomyocyte death to cardiac diseases, illustrate the identified pathways and finally reveal the therapeutic possibilities that go out from these findings.

9.2 Necrosis

Necrosis can be caused by mechanical forces that disrupt tissues or cells, but also by external stimuli, like toxins, ischaemic conditions or other extreme conditions. This impairs the cellular homeostasis, leading to an influx of water and extracellular ions that causes cell swelling. Consequently, spontaneous rupture of the cell membrane occurs that is accompanied by uncontrolled release of cytoplasmic contents into the extracellular space. This release then triggers massive inflammation of the surrounding tissue. Thus, necrosis is not restricted to a single cell but involves whole tissue areas.

Detection of necrosis on the cellular level utilises the disintegration of the plasma membrane. Due to this effect, dyes like trypan blue or propidium iodide, which cannot penetrate the cell membrane and, therefore, do not enter healthy cells, only stain necrotic cells (Strober 2001). Other strategies for necrosis detection use cell membrane permeable dyes like MTT (3-(4,5-dimethylthiazol-2-yl)-2,5-diphenyltetrazolium bromide). Viable cells with active metabolism convert MTT into a purple-coloured formazan product with an absorbance maximum near 570 nm, while necrotic cells have lost their ability to convert MTT into formazan. Thus, colour formation serves as a marker of viable cells (Riss et al. 2013). Furthermore, a sign of the unspecific breakdown of DNA during necrosis is blunt DNA fragments that can be detected by ligation of labelled double-stranded primers to these ends with pfu T4 ligase (Didenko and Hornsby 1996). Another possibility is the detection of enzymes or proteins that are released by the necrotic cells. Here often the release of lactate dehydrogenase (LDH) or troponin-T (Tn-T) is used (Krysko et al. 2008).

The first thing that gets into our minds when we consider cardiac necrosis is myocardial infarction. Due to a partial or complete block of coronary arteries, blood flow to the heart is reduced, preventing it from receiving enough oxygen. Going along with oxygen deprivation, an infarct scar arises that is mainly caused by necrotic cell death of cardiomyocytes. As already described in Chap. 8, oxygen deprivation leads to acidosis, Ca^{2+} overload, ROS production and MPTP opening. All these parameters are known inducers of necrosis.

To simulate ischaemic conditions in isolated ventricular cardiomyocytes, they can be cultured in an anoxic, glucose-free and acidic medium. These conditions lead to energy depletion of the cells within 2 h and necrotic cell death that increases in dependence on the time of simulated ischaemia, while no apoptotic cell death was observed (Taimor et al. 1999). This indicates that infarct size under ischaemic conditions is mainly determined by necrosis of cardiomyocytes. Primary therapeutic aim in patients with myocardial infarction is, therefore, an early reperfusion to prevent necrotic death of cardiomyocytes. However, reperfusion itself is an additional trigger of cardiomyocytes necrosis. As discussed in Chap. 8 in more detail, post-reperfusion infarcts develop within the first minutes of reperfusion and are almost exclusively formed by areas of contraction band necrosis composed of hypercontracted myocytes (Barrabés et al. 1996). This Ca^{2+}-dependent hyper-contracture as cause of sarcolemmal rupture and necrotic cell death can also be resembled in isolated cardiomyocytes under simulated ischaemia (Inserte et al. 2002). Induction of calpain in ischaemic/reperfused cardiomyocytes additionally contributes to enhanced cytoskeletal fragility (Inserte et al. 2005) and necrotic cell death. Interestingly, hypercontraction-mediated cell death can be propagated from cell to cell via gap junctions which explains a continuous disruption of cardiac tissue in the infarct area (Rodriguez-Sinovas et al. 2004). Further mechanisms of cardiomyocyte necrosis in ischaemia/reperfusion involve rigour contracture due to heavily injured mitochondria that display a massive opening of MPTP, resulting in mitochondrial depolarization, swelling and rupture. Inhibition of MPTP opening via cyclosporine A protected cardiomyocytes against reoxygenation injury (Nazareth et al. 1991). Deletion of cyclophilin D, a modulator of MPTP, decreased necrotic but not apoptotic cell death during ischaemia/reperfusion in vivo (Nakagawa et al. 2005), thereby demonstrating mediation of necrotic cell death via MPTP in ischaemic hearts.

In contrast to the massive insults of ischaemia/reperfusion injury that immediately leads to necrotic cell death, the development of cardiomyopathy or heart failure is a slow-going process that is accompanied by activation of the neuroendocrine and renin-angiotensin systems and the activation of growth factors or inflammatory processes that can be induced by preceding necrotic cell death. Although at first sight, these stimuli seem not to be as aggressive as under myocardial infarction; also in heart failure, progressive loss of cardiomyocytes by necrosis has been found. Signs of necrosis are multiple foci of replacement fibrosis across the ventricular wall, associated with a significant amount of myocyte loss, as it was found in end-stage failure in human ischaemic cardiomyopathy (Beltrami et al. 1994). In the myocardium of patients with end-stage heart failure due to idiopathic dilated or ischaemic cardiomyopathy, levels of myocyte necrosis were sevenfold greater than apoptosis and amounted to around 0.5–1.2 % of the cells (Guerra et al. 1999). Ca^{2+}-and mitochondrial-dependent cardiomyocyte necrosis has been identified as a primary mediator of heart failure. Enhanced overexpression of sarcolemmal L-type Ca^{2+} channel (LTCC) provokes Ca^{2+} influx-induced cardiomyocyte necrosis and cardiomyopathy in transgenic mice that could be inhibited by loss of cyclophilin D, a regulator of MPTP (Nakayama et al. 2007). β-adrenoceptor stimulation of LTCC overexpressing mice even enhanced the necrosis-associated heart failure progression. Also in isolated cardiomyocytes of LTCC overexpressing mice, the rate of necrotic cell death was increased and accelerated by β-adrenoceptor stimulation (Wang et al. 2010). However, in isolated feline cardiomyocytes, apoptosis was the primary kind of death under LTCC overexpression and not necrosis (Chen et al. 2005). Reasons for this may be species-specific differences of the death reaction or the use of different overexpression systems (transgenic vs. adenoviral) that may result in various degrees and kinetics of LTCC overexpression. In spite of this species controversy, LTCC can be considered as a major contributor to heart failure via induction of cardiomyocyte death, either by necrosis or apoptosis. Therefore, LTCC or the appropriate control of intracellular Ca^{2+} levels is a worthwhile therapeutic target against heart failure progression.

9.3 Apoptosis

As already mentioned before, apoptosis is an active death process that is mediated via distinct signalling pathways. Mainly, apoptosis is categorised into receptor-mediated extrinsic-induced and mitochondria-driven intrinsic-induced processes. Both pathways finally result in the activation of common apoptosis-executing molecules, namely, the activation of caspases. They are central in the apoptotic process. Caspases are cysteine proteases that need a specific peptide recognition sequence. This enables selective protein cleavage necessary for destruction of cellular proteins and induction of endonucleases for DNA fragmentation, finally accumulating in an ordered cell disassembly. In 2009, the Nomenclature Committee on Cell Death (NCCD) classified apoptosis in regard to morphological criteria, which includes reduction in cellular and nuclear volume (pyknosis), nuclear fragmentation (karyorrhexis) and plasma membrane blebbing that forms the so-called apoptotic bodies (Kroemer et al. 2009). Due to the substantial progress in the biochemical and genetic exploration of cell death, those morphological characteristics were extended by

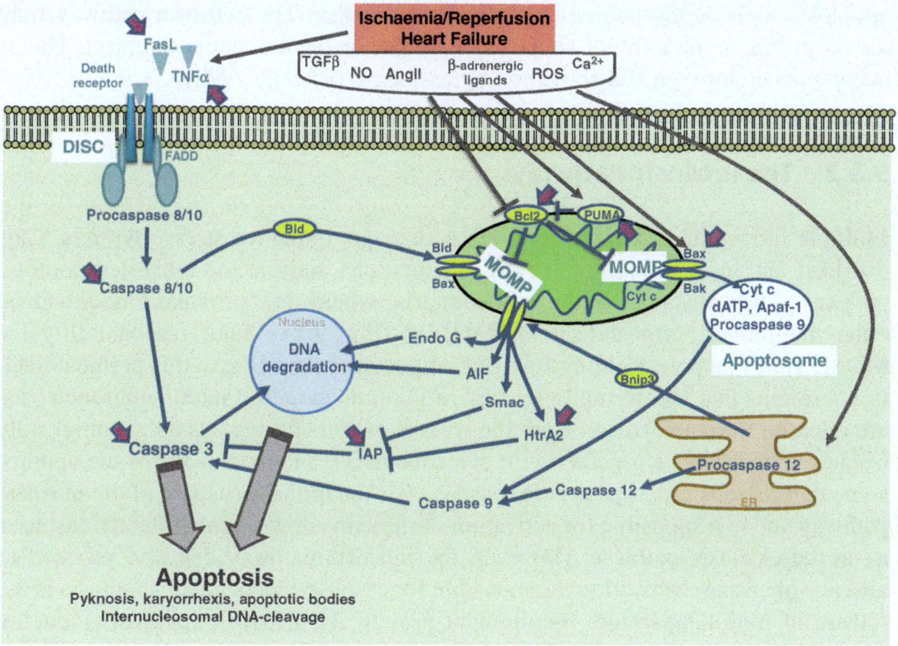

Fig. 9.1 Apoptotic signalling pathways in cardiac disease. Apoptosis is induced in ischaemia/reperfusion and heart failure progression either by extrinsic pathways via death receptors or by the intrinsic mitochondria-driven pathway. Both signalling cascades finally result in induction of effector caspases, i.e. activation of caspase 3 and the typical cellular breakdown of apoptosis. Depicted are the signalling molecules involved in this process. Pink arrows indicate the factors that were directly identified to be part of the apoptotic machinery in ischaemia/reperfusion injury or heart failure progression, because either their pharmacologic inhibition or their transgenic modulation changed the apoptotic response and cardiac disease progression. These effects are detailed discussed in the text

molecular classification parameters of cell death modalities. Among others, evidence of caspase activation or detection of specific DNA fragmentation were included (Galluzzi et al. 2012).

9.3.1 The Extrinsic Pathway

At the beginning of extrinsic pathway, death receptors, like Fas- or TNF-α receptors, are activated by binding to their specific ligands (Ashkenazi and Dixit 1998). This process initiates recruitment of adaptor proteins, like FADD, and procaspase 8 or 10. These molecules form the death-inducing signalling complex (DISC) (Kischkel et al. 1995). Autoproteolytic cleavage of the recruited procaspase 8 or 10 in the DISC results in the release and activation of the so-called initiator caspase 8 or 10, which is responsible for cleavage of other downstream effector caspases, like caspase 3 (Fig. 9.1). Effector caspases finally exert the execution phase of apoptosis with its morphological cell characteristics. The extrinsic pathway may also converge in the intrinsic signalling pathway on the mitochondrial level. This is accomplished through Bid activation by caspase 8 (Li et al. 1998).

9.3.2 The Intrinsic Pathway

Multiple intracellular stresses, like DNA damage, oxidative stress, hypoxia, Ca^{2+} overload and also the loss of survival factors, can activate the intrinsic apoptotic pathway. They all target at the mitochondria, where they provoke mitochondrial outer membrane permeabilisation (MOMP) (Fig. 9.1). Major responsibility for MOMP relies on pore-forming Bax/Bak oligomerization. Due to this permeabilisation, proteins that are normally located in the mitochondrial intermembrane space are released into the cytosol. Here the release of cytochrome c is an essential step, because cytochrome c together with cytosolic dATP and Apaf-1 forms the apoptosome that recruits procaspase 9. Procaspase 9 is the initiator caspase of the intrinsic pathway and is responsible for activation of the same downstream effector caspases as in the extrinsic pathway. However, the intrinsic pathway can also proceed in absence of caspase activation. Responsible for caspase-independent apoptosis is the release of high-temperature requirement protein A2 (HtrA2), apoptosis-inducing factor (AIF) or endonuclease G (EndoG) from the mitochondria. HtrA2 has itself serine protease activities that induce apoptotic pathways via degradation of the caspase inhibitor XIAP (Bhuiyan and Fukunaga 2008). AIF and EndoG mediate caspase-independent DNA fragmentation (Li et al. 2001).

One major control point in the apoptotic process is MOMP. As said before, MOMP relies on pore-forming Bax/Bak oligomerization. Both proteins are pro-apoptotic members of the Bcl-2 family. If expression of antiapoptotic family members, like Bcl-2 or Bcl-xl, prevails, pore formation is prevented and MOMP does not occur, and the intrinsic apoptotic pathway is blocked. In addition to these control points, the cell has tremendous other possibilities to control the apoptotic pathways, like the induction of intracellular caspase inhibitors (IAPs) or release of Smac/Diablo via MOMP that in turn inhibit IAPs. Thus, the cell has multiple control points to ensure secure and appropriate execution of the apoptotic death process.

9.3.3 Cardiomyocyte Apoptosis

One of the first reports describing the appearance of apoptotic cardiomyocytes in the heart came from the group of Gottlieb et al. (1994). They detected signs of apoptosis in rabbits after 30 min ischaemia and 4 h of reperfusion while ischaemia alone did not provoke apoptosis. These results were based on electron microscopic evidence of chromatin condensation and detection of specific internucleosomal DNA cleavage by DNA laddering and by in situ nick-end labelling with DNA polymerase I. After these findings, a lot of studies followed that described apoptosis induction in ischaemic/reperfused hearts with great variations in the number of apoptotic cardiomyocytes. Using terminal deoxynucleotidyl transferase assay (TUNEL-assay) for detection of specific DNA fragmentation, in the border zone of human infarct samples, 0.8–12 % TUNEL-positive myocytes were found (Olivetti et al. 1996; Saraste et al. 1997).

Also in animal studies or Langendorff-perfused hearts, various numbers of apoptotic cells were detected after myocardial infarction ranging between 0.07 and 14 % (Cheng et al. 1996; Kajstura et al. 1996; Yue et al. 1998; Maulik et al. 1998; Scarabelli et al. 2001). From these studies, it became evident that ischaemia per se can initiate apoptosis, not only in cardiomyocytes but also in endothelial cells or macrophages. But reperfusion is required to either enhance or complete the process. The number and location of apoptotic cardiomyocytes may thereby vary depending on the severity and length of the ischaemic insult (Scarabelli and Gottlieb 2004). Both intrinsic and extrinsic apoptotic pathways were found to be activated in ischaemic/reperfused myocardium, because reduction in Bcl-2 and enhancement of Bax-expression (Cheng et al. 1996), as well as induction of Fas (Kajstura et al. 1996), have been described. With the finding that either pharmacologic caspase inhibitors or the overexpression of the intracellular caspase inhibitor cIAP attenuates ischaemia/reperfusion injury in vivo, final evidence for contribution of apoptosis to myocardial infarction injury was provided (Yaoita et al. 1998; Holly et al. 1999; Chua et al. 2007). Although, this conclusion is somewhat reduced by the finding that caspase 3 also cleaves α-actin and troponin-T in cardiomyocytes, thereby destroying contractile structures (Communal et al. 2002). Therefore, to some extent, caspase 3 may contribute to ischaemia/reperfusion-induced cardiac dysfunction independent of apoptosis.

In mice that lack Fas (lpr mice), cardiac injury upon ischaemia/reperfusion is reduced in vivo and in isolated perfused hearts, indicating involvement of the extrinsic pathway in ischaemia/reperfusion injury (Lee et al. 2003; Jeremias et al. 2000). Contribution of the intrinsic pathway to myocardial infarction was identified in several animal models: cardiomyocyte-specific Bcl2-overexpression reduced infarct size, apoptosis and cardiac dysfunction after ischaemia/reperfusion (Chen et al. 2001), as well as Bax deficiency protected against ischaemia/reperfusion injury in mice (Hochhauser et al. 2007). Deletion of other proapoptotic Bcl family members, like PUMA (p53-upregulated modulator of apoptosis), also reduced myocardial injury following ischaemia/reperfusion (Toth et al. 2006). The proapoptotic, hypoxia-induced Bcl-2 family member Bnip3 is activated in ischaemic cardiomyocytes and contributes to mitochondria-driven apoptosis (Regula et al. 2002). Activation of Bnip3 is due to endoplasmic reticulum (ER) stress during ischaemia/reperfusion. ER stress not only influences Bcl family members but also activates initiator caspase 12 and thereby contributes to ischaemia/reperfusion-induced cardiomyocyte apoptosis (Mughal and Kirshenbaum 2011). In conclusion, there is clear evidence of apoptosis induction in ischaemic/reperfused myocardium, and major signalling pathways have been identified.

However, apoptosis does not only emerge in ischaemic heart disease. It is also a phenomenon in failing hearts. While after myocardial infarction a large number of cells undergo apoptosis, during heart failure development a modest, but sustained

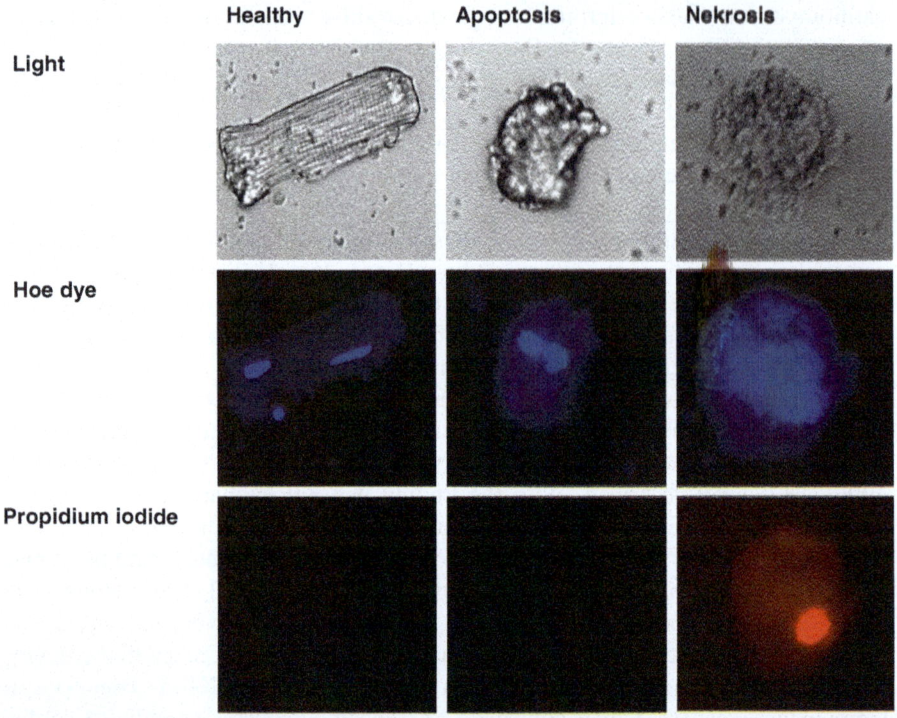

Fig. 9.2 Apoptotic and necrotic adult cardiomyocytes. Isolated ventricular cardiomyocytes of adult rat were stained with the DNA dye Hoe33528 and propidium iodide and analysed by fluorescence microscopy. Non-damaged, healthy cells are rod shaped. Their nuclei are stained by the cell membrane permeable dye Hoe33528 and present an oval morphology. Apoptotic cardiomyocytes are hypercontracted and present round nuclei with condensed chromatin, as seen by Hoe33528 staining. Propidium iodide is not cell membrane permeable and is thus excluded from apoptotic cardiomyocyte. Necrotic cells and their nuclei present a round morphology. But in contrast to apoptotic cells, necrotic cardiomyocytes are double stained, because due to their disrupted plasma membrane also propidium iodide penetrates the cell

increase in apoptotic rates is found, like in pressure overload-induced heart hypertrophy (Teiger et al. 1996; Olivetti et al. 1994) or idiopathic dilated human cardiomyopathy (Beltrami et al. 1995). That low levels of apoptosis can induce heart failure was shown by cardiac-specific overexpression of modified caspase 8 which promoted low levels of cardiomyocyte apoptosis in the heart (0.023 %). These mice developed severe cardiomyopathy and died within 6 months (Wencker et al. 2003). These examples indicate that apoptosis induction in the heart is caused by diverse initial events.

In order to determine the triggers of apoptosis in myocardial infarction, studies on isolated neonatal or adult cardiomyocytes were conducted. In Fig. 9.2 the typical morphology of apoptotic adult cardiomyocytes is depicted. Using the cell permeable DNA dye Hoe33528, the chromatin condensation in the nuclei gets visible. Interestingly, it became evident that hypoxia alone is not sufficient to cause apoptosis in neonatal or adult rat cardiomyocytes (Taimor et al. 1999; Webster et al. 1999). This indicates that additional triggers provoke apoptosis induction. Webster et al. (1999) demonstrated dependence on reductions of pH during hypoxia. But also factors that are either produced by cardiomyocytes themselves or by non-cardiomyocytes after myocardial infarction or during heart failure progression, like oxygen radicals (Inserte et al. 2000), nitric oxide (Taimor et al. 1999), transforming growth factor β-1 (TGF-β_1) (Schneiders et al. 2005), angiotensin II (Schröder et al. 2006) or β-adrenoceptor stimulation (Communal et al. 1998; Zaugg et al. 2000), have been shown to induce apoptosis in cardiomyocytes. Some of these stimuli convene on the same molecular signalling pathways in order to stimulate apoptosis: angiotensin II enhances the release of TGF-β_1 from cardiomyocytes (Schröder et al. 2006). Then, in an autocrine loop, TGF-β_1 enhances eNOS activity and NO release from cardiomyocytes that results in activation of the transcription factors SMAD and AP-1 and finally induces apoptosis (Heger et al. 2011). Both of these transcription factors, SMAD and AP-1, are essentially involved in apoptosis induction in cardiomyocytes under AngII or TGF-β_1 stimulation (Schröder et al. 2006; Schneiders et al. 2005). Interestingly, in absence of SMAD activation, AP-1 is a mediator of hypertrophic growth in cardiomyocytes under α-adrenergic stimulation (Taimor et al. 2004). Adenoviral overexpression of SMAD4 in α-adrenergic-stimulated cardiomyocytes changes the response from hypertrophic growth to apoptotic cell death (Heger et al. 2009). As cardiac hypertrophy is a predictor of heart failure development (see also Chap. 6 for details), it can be assumed that additional activation of angiotensin II or TGF-β_1-receptors in the stressed heart may turn compensated hypertrophy into decompensation and heart failure progression via enhancement of SMAD levels and apoptosis induction in cardiomyocytes. Proapoptotic target genes of TGF-β_1 still have to be defined in cardiomyocytes. However, enhanced transcription of proapoptotic members of the Bcl family, like Bim, by SMADs has been shown already in other cell types (Bakhshayesh et al. 2012; Ha Thi et al. 2013) and may also hold true for cardiomyocytes. An argument for involvement of mitochondria-mediated apoptosis in TGF-β_1-stimulated cardiomyocytes comes from the findings that cardiomyocytes from adenine nucleotide translocase 1 (ANT1)-overexpressing rats are resistant to opening of the mitochondrial permeability transition pore (MPTP) and apoptosis induction by TGF-β_1 (Heger et al. 2012). Thus, inhibition of SMAD activation in the heart may have a pivotal protective role in heart failure progression.

Besides the angiotensin II/TGFβ_1-system, stimulation of β-adrenoceptors by catecholamines plays a pivotal role in cardiomyocyte (patho)physiology. Primarily, β-adrenoceptor activation couples the cardiac function with the physiological demand by enhancing heart rate and contractility under, i.e. exercise. However, chronically high catecholamine levels promote adverse remodelling. One aspect of β-adrenergic-stimulated adverse cardiac outcomes is the induction of apoptosis that has been demonstrated in cardiomyocytes isolated from mice, rat or feline. Protein kinase A (PKA) is the main mediator of β-adrenoceptor signalling concerning effects like positive inotropy or chronotropy, but it also provokes apoptosis. β_1-adrenoceptors were identified as the main receptor type, signalling via the adenylyl cyclase/cAMP/PKA pathway to mediate apoptosis, at least in cultured adult feline ventricular myocytes. But in parallel, β-adrenoceptors activate cardioprotective signalling via the cAMP/exchange protein EPAC (Zhang et al. 2013). β_1-adrenoceptor antagonist therapies, as they are used in the treatment of heart failure patients, may, therefore, be supplemented with EPAC in order to support the cardioprotective features of β_1-adrenoceptors while having blocked the PKA-mediated adverse effects.

Another approach for protection of cardiomyocytes against apoptosis but also against other destructive stimuli on contractile function is the enhancement of the protein kinase Akt. One factor that is central for Akt-induced protection against myocyte apoptosis seems to be the activation of Pim1-kinase. Pim1 overexpression in cardiomyocytes antagonises several facets of apoptotic signalling, i.e. enhances anti-apoptotic Bcl family members (Muraski et al. 2007). Other Akt targets are the inhibition of proapoptotic Bad or caspase 9. Collectively, the actions of Akt serve to maintain mitochondrial integrity and antagonise the apoptotic cascade (Sussman 2009).

Having all this said about evidences for the involvement of apoptosis in cardiac diseases that have accumulated during the last 20 years, due to new recent findings we may have to take a second look on the old data. Reasons for this are on the one hand newly described interactions between different kinds of cell death, i.e. the mutual use of signalling molecules in apoptosis and programmed necrosis as will be discussed in the next chapter, and on the other hand novel cardioprotective mechanisms of caspase inhibition that are not related to apoptosis. One of the first descriptions in this area came from studies in an isolated working heart rat model of myocardial stunning (Ruetten et al. 2001): the caspase 3 inhibitor Ac-DEVD-CHO significantly improved postischemic contractile recovery independent of reductions in apoptotic indexes. These findings indicated that the integrity of the contractile apparatus may be adversely affected by caspase 3 activation. Indeed, proteins related to the contractile apparatus of cardiomyocytes have been defined as caspase substrate. In neonatal cardiomyocytes hypoxia-induced troponin I (TnI) degradation was blocked by caspase inhibitors (Castro et al. 2013). However, not TnI but MMP2 is the substrate of caspases, and caspase inhibition led to reduction of MMP2-mediated TnI degradation. In adult cardiomyocytes, cleavages of cTnT and actin are observed after β-adrenergic stimulation (Communal et al. 2002) that was prevented by pretreatment with the caspase inhibitor VAD-fmk. Thus, caspases, in addition to its effects in the apoptotic process, can actively contribute to degradation of proteins from the contractile machinery and thereby decrease the functional reserve in cardiomyocytes.

The positive action of caspase inhibitors in the diseased whole heart is not only based on the described effects on cardiomyocytes. Apoptosis in failing hearts is not restricted to cardiomyocytes. In a pressure overload model, it was shown that caspase inhibition preserved left ventricular function, but rather prevented apoptosis in macrophages than in cardiomyocytes (Park et al. 2013). In spite of this, isolated cardiomyocytes from these animals showed preserved contractile function and a reduced TnI degradation. Indeed, contractile performance of adult ventricular rat cardiomyocytes is not directly jeopardised by induction of proapoptotic pathways (Tastan et al. 2007). This indicates that not only apoptosis but also the distortion of

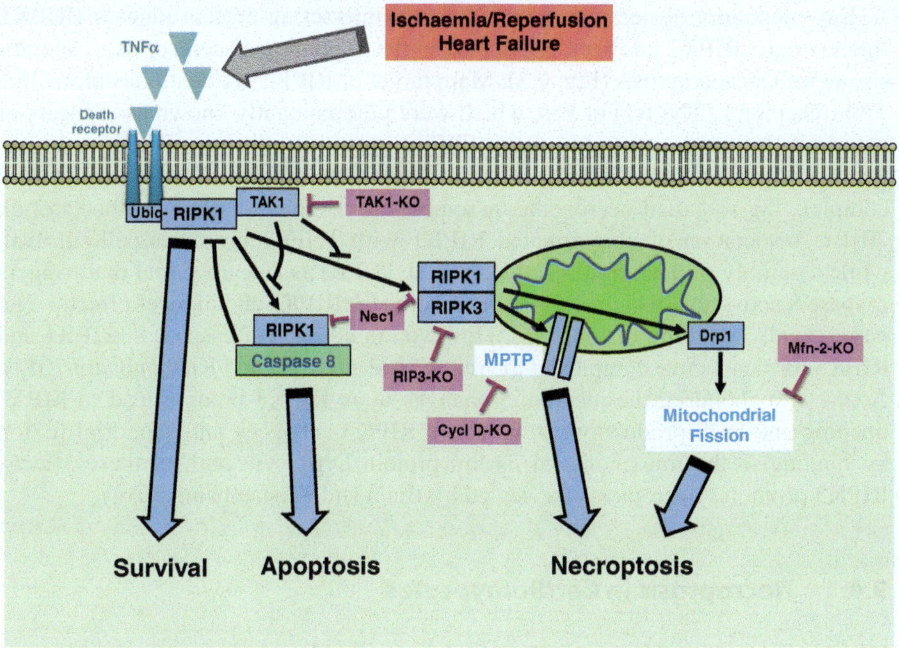

Fig. 9.3 Necroptotic signalling pathways in cardiac disease. In this signalling cascade, the predominant role of RIPK in the control of cell survival, apoptosis or necroptosis is depicted. If ubiquitinated RIPK1 is associated to the death receptor, cell survival is promoted under TNFα stimulation. Upon deubiquitination RIPK1 either associates to caspase 8 and promotes apoptosis or binds to RIPK3. The RIPK1/RIPK3 complex binds to the mitochondria and facilitates MPTP opening and mitochondrial fission. Both contribute to necroptosis. Interventions in the necroptotic pathway that reduced infarct size and cardiac remodelling are depicted in pink. These effects are outlined in the text in detail

the contractile machinery in cardiomyocytes plays a pivotal role in caspase-mediated effects. Surely both, apoptosis and destruction of myofibrillar proteins, act in concert to result in the fulminate cardiac restrains under caspase activation. Therefore, caspase inhibition can be considered as a hopeful therapeutic approach to limit ischaemia/reperfusion injury and heart failure progression.

9.4 Necroptosis

Recent findings have shed light not only on new aspects in cardiomyocyte apoptosis but elucidated also other kinds of cell death. As mentioned above, for a long time, necrosis was considered to be an exclusively passive process. However, nowadays, we recognised active signalling cascades that result in typical morphological features of necrotic cell death. As this kind of cell death is a hybrid of apoptosis and necrosis that share common overlapping signalling cascades, it is called "necroptosis".

Key mediators of necroptosis are receptor-interacting protein kinases (RIPK). Interestingly, RIPK1 has been implicated in the regulation of cell survival, apoptosis, as well as necroptosis (Fig. 9.3). Main target of RIPK1 are death receptors, like TNF-receptor 1 (TNFR1) or Fas, which were predominantly known as inducers of apoptosis. Now it became clear that RIPK1 regulates the decision if TNF-α acts as a pro-survival signal or induces cell death. Upon stimulation with TNF-α, a protein complex, the so-called necrosome, assembles at the TNFR1 (Vandenabeele et al. 2010). Association of ubiquitinated RIPK1 with TNFR1 promotes cell survival, while assembly of deubiquitinated RIPK1 at TNFR1 is a death signal that triggers caspase 8 activation, which cleaves and inactivates RIPK1 and induces effector caspases, finally resulting in apoptosis. However, in absence of caspase 8, RIPK1 and RIPK3 form an active complex at the mitochondria (Biala and Kirshenbaum 2014). At the mitochondria, the metabolic changes upon RIPK3 binding lead to MPTP opening and ROS production. In addition, RIPK3 enhances mitochondrial fission by binding to the mitochondrial fission protein Drp1. Via both of these effects, RIPK3 promotes necroptosis (reviewed by Biala and Kirshenbaum 2014).

9.4.1 Necroptosis in Cardiomyocytes

Using necrostatin-1 (Nec-1) as RIPK1 inhibitor, Smith and co-workers (2007) were the first to indicate that RIPK-inhibition is protective against ischaemia/reperfusion (I/R) injury in the myocardium, as it attenuated H_2O_2-induced necrosis in a myoblastic cell line (H9c2 cells), as well as it reduced infarct size in in vivo and in vitro models of I/R injury (Fig. 9.3). Application of Nec-1 at the beginning of reperfusion could still reduce infarct size, inflammation and oxidative stress and provoked long-term improvements in cardiac performance in rodents (Oerlemans et al. 2012). Just recently also in the pig model of I/R injury, intravenous administration of Nec-1 prior to reperfusion was an effective therapeutic strategy to significantly reduce infarct size and preserve left ventricular function (Koudstaal et al. 2015). As the cardioprotective effects of Nec-1 against ischaemia/reperfusion injury were lost in cyclophilin D knockout mice, it was concluded that necroptosis is mediated through MPTP opening (Lim et al. 2007). In neonatal rat cardiomyocytes, close co-localisation of RIPK3 and the mitochondria was described (Luedde et al. 2014). Overexpression of RIPK3 in these cells induced formation of RIPK1/RIPK3 complexes that executed necroptosis of cardiomyocytes. In contrast to other cell types, no additional stimulation with TNF-α was needed for necroptosis induction in RIPK3 overexpressing cardiomyocytes. In vivo, RIPK3 upregulation was detected in ischaemic mouse hearts, especially in the peri-infarct zone of the left ventricle. Constitutive knockout of RIPK3 prevented ROS formation and inflammation in the early phase after myocardial infarction, but did not reduce infarct size. However, adverse remodelling in the late phase after myocardial infarction was prevented in RIPK3 knockout mice, resulting in a preserved cardiac function (Luedde et al. 2014). Thus, RIPK1 and RIPK3 are mediators of adverse remodelling after myocardial infarction, thereby emphasising an important role of necroptosis in I/R injury. If these effects can solely be attributed to cardiomyocytes, or if inflammatory cells are also involved cannot be said yet. But as necroptosis via RIPK1/RIPK3 signalling was observed in isolated cardiomyocytes, they surely play a pivotal role.

As stated above, mitochondrial fission can also promote necroptosis. Interestingly, proapoptotic members of the Bcl family seem to be involved in this process in the heart. In Bax/Bak double knockout mice, fragmented cardiac mitochondria are found that are protected against Ca^{2+}-induced mPTP opening. In these mice reduced apoptotic but also primary necrotic injury after myocardial infarction is found (Whelan et al. 2012). This study introduces the concept that the Bax-driven fusion of the mitochondria renders cardiomyocytes poised to undergo necrosis, if presented with an appropriate stimulus, like myocardial infarction. In alignment with these findings that mitochondrial fusion and fission determine cardiomyocyte death are findings from Papanicolaou et al. (2011) showing that the loss of mitofusin-2 (Mfn-2), a GTPase that is essential for mitochondrial fusion, in adult cardiomyocytes protects them from a number of necroptosis-inducing stimuli and that Mfn-2 knockout hearts display better recovery following reperfusion injury.

Besides the role of necroptosis in ischaemia/reperfusion injury, it may also contribute to the loss of cardiomyocytes in heart failure. Evidence for this comes from studies on the cardiac function of TAK1 in respect to cardiac survival or death. Under normal conditions, stimulation of neonatal cardiomyocytes with TNF-α promotes association of TAK1 with RIPK1, thereby preventing formation of the RIPK1-caspase 8 associations and formation of the necrosome. Ablation of TAK1 in neonatal cardiomyocytes resulted in RIPK1-dependent necroptotic, but also apoptotic cell death under TNF-α stimulation. Furthermore, TAK1-deficient mice develop severe cardiac dysfunction and necroptosis that could be largely rescued by Nec-1 administration. These studies of Li et al. (2014) suggest that TAK1 functions as a key survival factor in the heart by directly antagonising necroptosis, which is critical for the maintenance of myocardial homeostasis and the prevention of adverse myocardial remodelling.

9.5 Autophagy

In contrast to the kinds of cell death discussed so far, autophagy is primarily a survival mechanism of cells. In times of nutrient deprivation, autophagy provides the cell with amino acids, fatty acids and energy by degradation of organelles and proteins in an intracellular recycling process. However, excessively up regulated autophagy results in an over-degradation of the cell and may result in cell death. Both, protective and detrimental roles of autophagy have been described in cardiomyocytes and myocardium. Although this field of research is quite new and harbours the most controversial points about its contributions to cell death, it should not be unspoken in this chapter of cardiomyocytes death.

Under baseline conditions, autophagy is essential for the heart, as deletion of autophagy-related gene 5 (Atg5) in the adult heart causes adverse cardiac remodelling and dysfunction (Nakai et al. 2007). Similarly, deletion of lysosomal membrane protein 2 (lamp2) results in reduced cardiac function (Tanaka et al. 2000). This demonstrates that autophagy is a homeostatic mechanism for maintenance of cardiac structure and function (Nakai et al. 2007).

However, autophagy can be enhanced due to diverse cellular stresses and may then no longer be cardioprotective. Enhancement of autophagy was observed under ischaemic conditions and even more upon reperfusion (Matsui et al. 2007). In myocardial ischaemia, energy depletion and oxygen deprivation strongly trigger autophagy. Due to the low ATP/AMP levels in such situations, adenine monophosphate-activated protein kinase (AMPK) is activated. AMPK reduces activity of the autophagy repressor mTOR (regulator of mammalian target of rapamycin) and, simultaneously, promotes activation of autophagy activating kinases (ULKs). This initiates autophagosome formation. These are double-membrane vesicles that fuse with lysosomes to form autophagolysosomes for degradation of their contents (reviewed by Gatica et al. 2015). AMPK-deficient mice develop larger infarcts under prolonged ischaemia indicating protective effects of autophagy in ischaemia-induced cardiomyocyte death, although other effects of AMPK on cellular metabolism may add to the protection (Takagi et al. 2007). Furthermore, in a preclinical model of ischaemia/reperfusion in swine, acute induction of autophagy by administration of CAPS (chloramphenicol succinate) reduced infarct size (Przyklenk et al. 2011). However, as CAPS also increases Akt activity, this may, in addition to autophagy, act cardioprotective. Upregulation of autophagy by preconditioning is essential for cardiac protection because mice with a dominant-negative mutation of the autophagy protein Atg5 lost the protective effect of preconditioning (Huang et al. 2010). The protective effects of autophagy in ischaemic hearts can be explained by the regeneration of energy and the removal of damaged proteins and organelles.

Further enhancement of autophagy during reperfusion can be brought about by Beclin-1, a protein required for early autophagosome formation and thus another positive regulator of autophagy. Beclin-1 is activated in ischaemic/reperfused myocardium and also in isolated cardiomyocytes after a single cycle of simulated ischaemia/reperfusion (Valentim et al. 2006). As inducers of Beclin-1 in reperfused myocardium, reactive oxygen species (Hariharan et al. 2011) or the interaction of Beclin-1 with Bcl family members are discussed (Matsui et al. 2007). The induction of Beclin-1 and autophagy in reperfusion seems to be detrimental, as reduction of Beclin-1 expression in cardiac myocytes by RNAi reduces I/R-induced autophagy and is associated with enhanced cell survival (Valentim et al. 2006). Inhibition of autophagy in Beclin-1+/– mice reduced reperfusion injury (Matsui et al. 2007). In conclusion, hyper-activated autophagy may result in over-degradation of cellular contents and necrotic distortion of the cell. Another detrimental role may come from the interaction of Beclin-1 with pro- and antiapoptotic Bcl family members, i.e. binding of Beclin-1 to Bcl-2 enhances apoptosis (Huang et al. 2014).

In addition to the destructive role of autophagy in reperfusion injury, there is also evidence for the involvement of autophagy in heart failure progression. Pressure overload induced by aortic banding induces heart failure and greatly increases cardiac autophagy in mice. Heterozygous disruption of the gene coding for Beclin-1 decreased cardiomyocyte autophagy and diminished pathological remodelling induced by severe pressure stress. Conversely, Beclin-1 overexpression heightened autophagic activity and accentuated pathological remodelling (Zhu et al. 2007). Catecholamines that are elevated under pressure overload in vivo enhance autophagy and hypertrophic growth of isolated cardiomyocytes. RNAi-mediated knockdown of either Atg5 or Beclin-1, two essential autophagy effectors, was capable of suppressing this ligand-induced process of autophagy and myocyte growth (Cao et al. 2011). Cardiac-specific deficiency of Atg5 early in cardiogenesis showed no cardiac phenotypes under baseline conditions, but developed cardiac dysfunction and left ventricular dilatation 1 week after treatment with pressure overload. Interestingly, cardio-specific deletion of Atg5 in adulthood resulted in heart failure progression independent of additional stimuli, indicating that timing of autophagy induction and additional cellular stresses may determine positive or detrimental roles of cardiac autophagy in heart failure (Nakai et al. 2007).

References

Ashkenazi A, Dixit VM (1998) Death receptors: signaling and modulation. Science 281(5381):1305–1308

Bakhshayesh M, Zaker F, Hashemi M, Katebi M, Solaimani M (2012) TGF-β1-mediated apoptosis associated with SMAD-dependent mitochondrial Bcl-2 expression. Clin Lymphoma Myeloma Leuk 12:138–143

Barrabés JA, Garcia-Dorado D, Ruiz-Meana M, Piper HM, Solares J, González MA, Oliveras J, Herrejón MP, Soler Soler J (1996) Myocardial segment shrinkage during coronary reperfusion in situ. Relation to hypercontracture and myocardial necrosis. Pflugers Arch 431:519–526

Beltrami CA, Finato N, Rocco M, Feruglio GA, Puricelli C, Cigola E, Quaini F, Sonnenblick EH, Olivetti G, Anversa P (1994) Structural basis of end-stage failure in ischemic cardiomyopathy in humans. Circulation 89:151–163

Beltrami CA, Finato N, Rocco M, Feruglio GA, Puricelli C, Cigola E, Sonnenblick EH, Olivetti G, Anversa P (1995) The cellular basis of dilated cardiomyopathy in humans. J Mol Cell Cardiol 27:291–305

Biala AK, Kirshenbaum LA (2014) The interplay between cell death signaling pathways in the heart. Trends Cardiovasc Med 24(8):325–331

Bhuiyan MS, Fukunaga K (2008) Activation of HtrA2, a mitochondrial serine protease mediates apoptosis: current knowledge on HtrA2 mediated myocardial ischemia/reperfusion injury. Cardiovasc Ther 26(3):224–232

Cao DJ, Wang ZV, Battiprolu PK, Jiang N, Morales CR, Kong Y, Rothermel BA, Gillette TG, Hill JA (2011) Histone deacetylase (HDAC) inhibitors attenuate cardiac hypertrophy by suppressing autophagy. Proc Natl Acad Sci U S A 108:4123–4128

Castro MM, Fuah J, Ali M, Sung M, Schulz J, Kondo MY, Fan X, Holt A, Schulz R (2013) Inhibitory effects of caspase inhibitors on the activity of matrix metalloproteinase-2. Biochem Pharmacol 86:469–475

Chen Z, Chua CC, Ho YS, Hamdy RC, Chua BH (2001) Overexpression of Bcl-2 attenuates apoptosis and protects against myocardial I/R injury in transgenic mice. Am J Physiol Heart Circ Physiol 280:H2313–H2320

Chen X, Zhang X, Kubo H, Harris DM, Mills GD, Moyer J, Berretta R, Potts ST, Marsh JD, Houser SR (2005) Ca^{2+} influx-induced sarcoplasmic reticulum Ca^{2+} overload causes mitochondrial-dependent apoptosis in ventricular myocytes. Circ Res 97:1009–1017

Cheng W, Kajstura J, Nitahara JA, Li B, Reiss K, Liu Y, Clark WA, Krajewski S, Reed JC, Olivetti G, Anversa P (1996) Programmed myocyte cell death affects the viable myocardium after infarction in rats. Exp Cell Res 226(2):316–327

Chua CC, Gao J, Ho YS, Xiong Y, Xu X, Chen Z, Hamdy RC, Chua BH (2007) Over expression of IAP-2 attenuates apoptosis and protects against myocardial ischemia/reperfusion injury in transgenic mice. Biochim Biophys Acta 1773:577–583

Communal C, Singh K, Pimentel DR, Colucci WS (1998) Norepinephrine stimulates apoptosis in adult rat ventricular myocytes by activation of the beta-adrenergic pathway. Circulation 98:1329–1334

Communal C, Sumandea M, deTombe P, Narula J, Solaro RJ, Hajjar RJ (2002) Functional consequences of caspase activation in cardiac myocytes. Proc Natl Acad Sci U S A 99:6252–6256

Didenko VV, Hornsby PJ (1996) Presence of double-strand breaks with single-base 3′ overhangs in cells undergoing apoptosis but not necrosis. J Cell Biol 135:1369–1376

Galluzzi L, Vitale I, Abrams JM, Alnemri ES, Baehrecke EH, Blagosklonny MV, Dawson TM, Dawson VL, El-Deiry WS, Fulda S, Gottlieb E, Green DR, Hengartner MO, Kepp O, Knight RA, Kumar S, Lipton SA, Lu X, Madeo F, Malorni W, Mehlen P, Nuñez G, Peter ME, Piacentini M, Rubinsztein DC, Shi Y, Simon HU, Vandenabeele P, White E, Yuan J, Zhivotovsky B, Melino G, Kroemer G (2012) Molecular definitions of cell death subroutines: recommendations of the Nomenclature Committee on Cell Death 2012. Cell Death Differ 19:107–120

Gatica D, Chiong M, Lavandero S, Klionsky DJ (2015) Molecular mechanisms of autophagy in the cardiovascular system. Circ Res 116:456–467

Golstein P, Kroemer G (2007) Cell death by necrosis: towards a molecular definition. Trends Biochem Sci 32(1):37–43

Gottlieb RA, Burleson KO, Kloner RA, Babior BM, Engler RL (1994) Reperfusion injury induces apoptosis in rabbit cardiomyocytes. J Clin Invest 94:1621–1628

Guerra S, Leri A, Wang X, Finato N, Di Loreto C, Beltrami CA, Kajstura J, Anversa P (1999) Myocyte death in the failing human heart is gender dependent. Circ Res 85(9):856–866

Ha Thi HT, Lim HS, Kim J, Kim YM, Kim HY, Hong S (2013) Transcriptional and post-translational regulation of Bim is essential for TGF-β and TNF-α-induced apoptosis of gastric cancer cell. Biochim Biophys Acta 1830:3584–3592

Hariharan N, Zhai P, Sadoshima J (2011) Oxidative stress stimulates autophagic flux during ischemia/reperfusion. Antioxid Redox Signal 14:2179–2190

Heger J, Peters SC, Piper HM, Euler G (2009) SMAD-proteins as a molecular switch from hypertrophy to apoptosis induction in adult ventricular cardiomyocytes. J Cell Physiol 220:515–523

Heger J, Warga B, Meyering B, Abdallah Y, Schlüter KD, Piper HM, Euler G (2011) TGFβ receptor activation enhances cardiac apoptosis via SMAD activation and concomitant NO release. J Cell Physiol 226(10):2683–2690

Heger J, Abdallah Y, Shahzad T, Klumpe I, Piper HM, Schultheiss HP, Schlüter KD, Schulz R, Euler G, Dörner A (2012) Transgenic overexpression of the adenine nucleotide translocase 1 protects cardiomyocytes against TGFβ1-induced apoptosis by stabilization of the mitochondrial permeability transition pore. J Mol Cell Cardiol 53(1):73–81

Hochhauser E, Cheporko Y, Yasovich N, Pinchas L, Offen D, Barhum Y, Pannet H, Tobar A, Vidne BA, Birk E (2007) Bax deficiency reduces infarct size and improves long-term function after myocardial infarction. Cell Biochem Biophys 47:11–20

Holly TA, Drincic A, Byun Y, Nakamura S, Harris K, Klocke FJ, Cryns VL (1999) Caspase inhibition reduces myocyte cell death induced by myocardial ischemia and reperfusion in vivo. J Mol Cell Cardiol 31(9):1709–1715

Huang C, Yitzhaki S, Perry CN, Liu W, Giricz Z, Mentzer RM Jr, Gottlieb RA (2010) Autophagy induced by ischemic preconditioning is essential for cardioprotection. J Cardiovasc Transl Res 3:365–373

Huang X, Qi Q, Hua X, Li X, Zhang W, Sun H, Li S, Wang X, Li B (2014) Beclin 1, an autophagy-related gene, augments apoptosis in U87 glioblastoma cells. Oncol Rep 31:1761–1767

Inserte J, Taimor G, Hofstaetter B, Garcia-Dorado D, Piper HM (2000) Influence of simulated ischemia on apoptosis induction by oxidative stress in adult cardiomyocytes of rats. Am J Physiol Heart Circ Physiol 278(1):H94–H99

Inserte J, Garcia-Dorado D, Ruiz-Meana M, Padilla F, Barrabés JA, Pina P, Agulló L, Piper HM, Soler-Soler J (2002) Effect of inhibition of Na(+)/Ca(2+) exchanger at the time of myocardial reperfusion on hypercontracture and cell death. Cardiovasc Res 55:739–748

Inserte J, Garcia-Dorado D, Hernando V, Soler-Soler J (2005) Calpain-mediated impairment of Na+/K+ -ATPase activity during early reperfusion contributes to cell death after myocardial ischemia. Circ Res 97:465–473

Jeremias I, Kupatt C, Martin-Villalba A, Habazettl H, Schenkel J, Boekstegers P, Debatin KM (2000) Involvement of CD95/Apo1/Fas in cell death after myocardial ischemia. Circulation 102:915–920

Kajstura J, Cheng W, Reiss K, Clark WA, Sonnenblick EH, Krajewski S, Reed JC, Olivetti G, Anversa P (1996) Apoptotic and necrotic myocyte cell deaths are independent contributing variables of infarct size in rats. Lab Invest 74(1):86–107

Kerr JF, Wyllie AH, Currie AR (1972) Apoptosis: a basic biological phenomenon with wide-ranging implications in tissue kinetics. Br J Cancer 26:239–257

Kischkel FC, Hellbardt S, Behrmann I, Germer M, Pawlita M, Krammer PH, Peter ME (1995) Cytotoxicity-dependent APO-1 (Fas/CD95)-associated proteins form a death-inducing signalling complex (DISC) with the receptor. EMBO J 14:5579–5588

Koudstaal S, Oerlemans MI, Van der Spoel TI, Janssen AW, Hoefer IE, Doevendans PA, Sluijter JP, Chamuleau SA (2015) Necrostatin-1 alleviates reperfusion injury following acute myocardial infarction in pigs. Eur J Clin Invest 45(2):150–159

Kroemer G, Galluzzi L, Vandenabeele P, Abrams J, Alnemri ES, Baehrecke EH, Blagosklonny MV, El-Deiry WS, Golstein P, Green DR, Hengartner M, Knight RA, Kumar S, Lipton SA, Malorni W, Nuñez G, Peter ME, Tschopp J, Yuan J, Piacentini M, Zhivotovsky B, Melino G (2009) Classification of cell death: recommendations of the Nomenclature Committee on Cell Death 2009. Cell Death Differ 16:3–11

Krysko DV, Vanden Berghe T, D'Herde K, Vandenabeele P (2008) Apoptosis and necrosis: detection, discrimination and phagocytosis. Methods 44:205–221

Lee P, Sata M, Lefer DJ, Factor SM, Walsh K, Kitsis RN (2003) Fas pathway is a critical mediator of cardiac myocyte death and MI during ischemia-reperfusion in vivo. Am J Physiol Heart Circ Physiol 284:H456–H463

Li H, Zhu H, Xu CJ, Yuan J (1998) Cleavage of BID by caspase 8 mediates the mitochondrial damage in the Fas pathway of apoptosis. Cell 94:491–501

Li LY, Luo X, Wang X (2001) Endonuclease G is an apoptotic DNase when released from mitochondria. Nature 412:95–99

Li L, Chen Y, Doan J, Murray J, Molkentin JD, Liu Q (2014) Transforming growth factor β-activated kinase 1 signaling pathway critically regulates myocardial survival and remodeling. Circulation 130:2162–2172

Lim SY, Davidson SM, Mocanu MM, Yellon DM, Smith CC (2007) The cardio protective effect of necrostatin requires the cyclophilin-D component of the mitochondrial permeability transition pore. Cardiovasc Drugs Ther 21:467–469

Luedde M, Lutz M, Carter N, Sosna J, Jacoby C, Vucur M, Gautheron J, Roderburg C, Borg N, Reisinger F, Hippe HJ, Linkermann A, Wolf MJ, Rose-John S, Lüllmann-Rauch R, Adam D, Flögel U, Heikenwalder M, Luedde T, Frey N (2014) RIP3, a kinase promoting necroptotic cell death, mediates adverse remodelling after myocardial infarction. Cardiovasc Res 103:206–216

Matsui Y, Takagi H, Qu X, Abdellatif M, Sakoda H, Asano T, Levine B, Sadoshima J (2007) Distinct roles of autophagy in the heart during ischemia and reperfusion: roles of AMP-activated protein kinase and Beclin 1 in mediating autophagy. Circ Res 100:914–922

Maulik N, Yoshida T, Das DK (1998) Oxidative stress developed during the reperfusion of ischemic myocardium induces apoptosis. Free Radic Biol Med 24(5):869–875

Mughal W, Kirshenbaum LA (2011) Cell death signalling mechanisms in heart failure. Exp Clin Cardiol 16:102–108

Muraski JA, Rota M, Misao Y, Fransioli J, Cottage C, Gude N, Esposito G, Delucchi F, Arcarese M, Alvarez R, Siddiqi S, Emmanuel GN, Wu W, Fischer K, Martindale JJ, Glembotski CC, Leri A, Kajstura J, Magnuson N, Berns A, Beretta RM, Houser SR, Schaefer EM, Anversa P, Sussman MA (2007) Pim-1 regulates cardiomyocyte survival downstream of Akt. Nat Med 13:1467–1475

Nakai A, Yamaguchi O, Takeda T, Higuchi Y, Hikoso S, Taniike M, Omiya S, Mizote I, Matsumura Y, Asahi M, Nishida K, Hori M, Mizushima N, Otsu K (2007) The role of autophagy in cardiomyocytes in the basal state and in response to hemodynamic stress. Nat Med 13:619–624

Nakagawa T, Shimizu S, Watanabe T, Yamaguchi O, Otsu K, Yamagata H, Inohara H, Kubo T, Tsujimoto Y (2005) Cyclophilin D-dependent mitochondrial permeability transition regulates some necrotic but not apoptotic cell death. Nature 434:652–658

Nakayama H, Chen X, Baines CP, Klevitsky R, Zhang X, Zhang H, Jaleel N, Chua BH, Hewett TE, Robbins J, Houser SR, Molkentin JD (2007) Ca^{2+}- and mitochondrial-dependent cardiomyocyte necrosis as a primary mediator of heart failure. J Clin Invest 117:2431–2444

Nazareth W, Yafei N, Crompton M (1991) Inhibition of anoxia-induced injury in heart myocytes by cyclosporin A. J Mol Cell Cardiol 23:1351–1354

Oerlemans MI, Liu J, Arslan F, den Ouden K, van Middelaar BJ, Doevendans PA, Sluijter JP (2012) Inhibition of RIP1-dependent necrosis prevents adverse cardiac remodeling after myocardial ischemia-reperfusion in vivo. Basic Res Cardiol 107:270

Olivetti G, Melissari M, Balbi T, Quaini F, Cigola E, Sonnenblick EH, Anversa P (1994) Myocyte cellular hypertrophy is responsible for ventricular remodelling in the hypertrophied heart of middle aged individuals in the absence of cardiac failure. Cardiovasc Res 28:1199–1208

Olivetti G, Quaini F, Sala R, Lagrasta C, Corradi D, Bonacina E, Gambert SR, Cigola E, Anversa P (1996) Acute myocardial infarction in humans is associated with activation of programmed myocyte cell death in the surviving portion of the heart. J Mol Cell Cardiol 28(9):2005–2016

Papanicolaou KN, Khairallah RJ, Ngoh GA, Chikando A, Luptak I, O'Shea KM, Riley DD, Lugus JJ, Colucci WS, Lederer WJ, Stanley WC, Walsh K (2011) Mitofusin-2 maintains mitochondrial structure and contributes to stress-induced permeability transition in cardiac myocytes. Mol Cell Biol 31:1309–1328

Park M, Vatner SF, Yan L, Gao S, Yoon S, Lee GJ, Xie LH, Kitsis RN, Vatner DE (2013) Novel mechanisms for caspase inhibition protecting cardiac function with chronic pressure overload. Basic Res Cardiol 108(1):324

Przyklenk K, Undyala VV, Wider J, Sala-Mercado JA, Gottlieb RA, Mentzer RM Jr (2011) Acute induction of autophagy as a novel strategy for cardioprotection: getting to the heart of the matter. Autophagy 7:432–433

Regula KM, Ens K, Kirshenbaum LA (2002) Inducible expression of BNIP3 provokes mitochondrial defects and hypoxia-mediated cell death of ventricular myocytes. Circ Res 91:226–231

Riss TL, Moravec RA, Niles AL, Benink HA, Worzella TJ, Minor L, Storts D, Reid Y (2004–2013) Cell viability assays. In: Sittampalam GS, Coussens NP, Nelson H et al (eds) Assay guidance manual [Internet]. Eli Lilly & Company and the National Center for Advancing Translational Sciences, Bethesda. Available from: http://www.ncbi.nlm.nih.gov/books/NBK144065

Rodriguez-Sinovas A, García-Dorado D, Pina P, Ruiz-Meana M, Soler-Soler J (2004) Effect of sarcolemmal rupture on myocardial electrical impedance during oxygen deprivation. Am J Physiol Heart Circ Physiol 288:H1396–H1403

Ruetten H, Badorff C, Ihling C, Zeiher AM, Dimmeler S (2001) Inhibition of caspase-3 improves contractile recovery of stunned myocardium, independent of apoptosis-inhibitory effects. J Am Coll Cardiol 38:2063–2070

Saraste A, Pulkki K, Kallajoki M, Henriksen K, Parvinen M, Voipio-Pulkki LM (1997) Apoptosis in human acute myocardial infarction. Circulation 95(2):320–323

Scarabelli TM, Gottlieb RA (2004) Functional and clinical repercussions of myocyte apoptosis in the multifaceted damage by ischemia/reperfusion injury: old and new concepts after 10 years of contributions. Cell Death Differ 11 Suppl 2:S144–S152

Scarabelli T, Stephanou A, Rayment N, Pasini E, Comini L, Curello S, Ferrari R, Knight R, Latchman D (2001) Apoptosis of endothelial cells precedes myocyte cell apoptosis in ischemia/reperfusion injury. Circulation 104(3):253–256

Schneiders D, Heger J, Best P, Michael Piper H, Taimor G (2005) SMAD proteins are involved in apoptosis induction in ventricular cardiomyocytes. Cardiovasc Res 67:87–96

Schröder D, Heger J, Piper HM, Euler G (2006) Angiotensin II stimulates apoptosis via TGF-beta1 signalling in ventricular cardiomyocytes of rat. J Mol Med 84:975–983

Smith CC, Davidson SM, Lim SY, Simpkin JC, Hothersall JS, Yellon DM (2007) Necrostatin: a potentially novel cardio protective agent? Cardiovasc Drugs Ther 21:227–233

Strober W (2001) Trypan blue exclusion test of cell viability. Curr Protoc Immunol Appendix 3B. doi:10.1002/0471142735

Sussman MA (2009) Mitochondrial integrity: preservation through Akt/Pim-1 kinase signalling in the cardiomyocyte. Expert Rev Cardiovasc Ther 7:929–938

Taimor G, Lorenz H, Hofstaetter B, Schlüter KD, Piper HM (1999) Induction of necrosis but not apoptosis after anoxia and reoxygenation in isolated adult cardiomyocytes of rat. Cardiovasc Res 41:147–156

Taimor G, Schlüter KD, Best P, Helmig S, Piper HM (2004) Transcription activator protein 1 mediates alpha- but not beta-adrenergic hypertrophic growth responses in adult cardiomyocytes. Am J Physiol Heart Circ Physiol 286:H2369–H2375

Takagi H, Matsui Y, Hirotani S, Sakoda H, Asano T, Sadoshima J (2007) AMPK mediates autophagy during myocardial ischemia in vivo. Autophagy 3:405–407

Tanaka Y, Guhde G, Suter A, Eskelinen EL, Hartmann D, Lullmann-Rauch R, Janssen PM, Blanz J, von Figura K, Saftig P (2000) Accumulation of autophagic vacuoles and cardiomyopathy in LAMP-2-deficient mice. Nature 406:902–906

Tastan H, Abdallah Y, Euler G, Piper HM, Schlüter K-D (2007) Contractile performance of adult ventricular rat cardiomyocytes is not directly jeopardized by NO/cGMP-dependent inhibition of pro-apoptotic pathways. J Mol Cell Cardiol 42:411–421

Teiger E, Than VD, Richard L, Wisnewsky C, Tea BS, Gaboury L, Tremblay J, Schwartz K, Hamet P (1996) Apoptosis in pressure overload-induced heart hypertrophy in the rat. J Clin Invest 97(12):2891–2897

Toth A, Jeffers JR, Nickson P, Min JY, Morgan JP, Zambetti GP, Erhardt P (2006) Targeted deletion of Puma attenuates cardiomyocyte death and improves cardiac function during ischemia-reperfusion. Am J Physiol Heart Circ Physiol 291:H52–H60

Valentim L, Laurence KM, Townsend PA, Carroll CJ, Soond S, Scarabelli TM, Knight RA, Latchman DS, Stephanou A (2006) Urocortin inhibits Beclin1-mediated autophagic cell death in cardiac myocytes exposed to ischaemia/reperfusion injury. J Mol Cell Cardiol 40:846–852

Vandenabeele P, Galluzzi L, Vanden Berghe T, Kroemer G (2010) Molecular mechanisms of necroptosis: an ordered cellular explosion. Nat Rev Mol Cell Biol 11(10):700–714

Wang W, Zhang H, Gao H, Kubo H, Berretta RM, Chen X, Houser SR (2010) {beta}1-Adrenergic receptor activation induces mouse cardiac myocyte death through both L-type calcium channel-dependent and -independent pathways. Am J Physiol Heart Circ Physiol 299:H322–H331

Webster KA, Discher DJ, Kaiser S, Hernandez O, Sato B, Bishopric NH (1999) Hypoxia-activated apoptosis of cardiac myocytes requires reoxygenation or a pH shift and is independent of p53. J Clin Invest 104:239–252

Wencker D, Chandra M, Nguyen K, Miao W, Garantziotis S et al (2003) A mechanistic role for cardiac myocyte apoptosis in heart failure. J Clin Invest 111:1497–1504

Whelan RS, Konstantinidis K, Wei AC, Chen Y, Reyna DE, Jha S, Yang Y, Calvert JW, Lindsten T, Thompson CB, Crow MT, Gavathiotis E, Dorn GW 2nd, O'Rourke B, Kitsis RN (2012) Bax regulates primary necrosis through mitochondrial dynamics. Proc Natl Acad Sci U S A 109:6566–6571

Yaoita H, Ogawa K, Maehara K, Maruyama Y (1998) Attenuation of ischemia/reperfusion injury in rats by a caspase inhibitor. Circulation 97(3):276–281

Yue TL, Ma XL, Wang X, Romanic AM, Liu GL, Louden C, Gu JL, Kumar S, Poste G, Ruffolo RR Jr, Feuerstein GZ (1998) Possible involvement of stress-activated protein kinase signaling pathway and Fas receptor expression in prevention of ischemia/reperfusion-induced cardiomyocyte apoptosis by carvedilol. Circ Res 82(2):166–174

Zaugg M, Xu W, Lucchinetti E, Shafiq SA, Jamali NZ, Siddiqui MA (2000) Beta-adrenergic receptor subtypes differentially affect apoptosis in adult rat ventricular myocytes. Circulation 102:344–350

Zhang X, Szeto C, Gao E, Tang M, Jin J, Fu Q, Makarewich C, Ai X, Li Y, Tang A, Wang J, Gao H, Wang F, Ge XJ, Kunapuli SP, Zhou L, Zeng C, Xiang KY, Chen X (2013) Cardiotoxic and cardioprotective features of chronic β-adrenergic signaling. Circ Res 112:498–509

Zhu H, Tannous P, Johnstone JL, Kong Y, Shelton JM, Richardson JA, Le V, Levine B, Rothermel BA, Hill JA (2007) Cardiac autophagy is a maladaptive response to hemodynamic stress. J Clin Invest 117:1782–1793

Oxidative Stress and Nitrosative Stress 10

Rainer Schulz and Fabio Di Lisa

Abstract

Cardiomyocytes generate multiple forms of short half-life molecules known as reactive oxygen and reactive nitrogen species (ROS) and acting as signaling molecules (RNS). However, under certain conditions the balance between formation of oxygen radicals and degradation of them is severely affected resulting in oxidative stress that significantly contributes to cardiac dysfunction. In this chapter key molecules that trigger ROS and RNS formation in cardiomyocytes are introduced as well as mechanisms by which cardiomyocytes normally balance oxidative stress. Target molecules and the subsequent functional consequence of ROS and RNS stress will be explained. In many cases, these processes are either located in mitochondria or influence the behavior of mitochondrial function. Therefore, ROS and RNS couples mitochondrial function to key regulatory processes of cardiomyocytes, such as force generation and growth control, despite their genuine function in energy metabolism.

10.1 Introduction

Cardiovascular diseases are a group of numerous pathological disorders in which signaling processes of reactive oxygen and reactive nitrogen species (ROS and RNS) play an important role (Andreadou et al. 2015; Ferdinandy et al. 2014). ROS and RNS

R. Schulz (✉)
Institute of Physiology, Justus-Liebig-Universität Gießen, Giessen, Germany
e-mail: Rainer.Schulz@physiologie.med.uni-giessen.de

F. Di Lisa
Department of Biomedical Sciences, University of Padova, Padova, Italy
e-mail: dilisa@bio.unipd.it

K.-D. Schlüter (ed.), *Cardiomyocytes – Active Players in Cardiac Disease*,
DOI 10.1007/978-3-319-31251-4_10

are collectively referred to as reactive inflammatory species (RIS). RNS are nitric oxide (NO, a free radical)-related species. NO reacts quickly with superoxide to form peroxynitrite (a diamagnetic molecule), which is involved in lipid peroxidation and protein modifications (for review, see Rassaf et al. 2014; Victorino et al. 2015). The interaction between NO and ROS generation is termed the "nitroso/redox balance."

Cardiovascular diseases are normally characterized by ROS overproduction, whereas the formation of major RNS can decrease or increase depending on the nature of the injury. Free radicals on the other hand are important signaling species and contribute to endogenous ("conditioning") or exogenous cardioprotective strategies (Di Lisa et al. 2011; Ferdiandy et al. 2014).

Many sources of ROS and RNS production have been identified: among them are NADPH oxidases (Nox), xanthine oxidase (XO), lipoxygenases (LO), mitochondria, and nitric oxide synthases (NOS) (Afanas'ev 2011; Chen and Zweier 2014; Di Lisa et al. 2009; Granger and Kvietys 2015; Luo et al. 2014; Matsushima et al. 2014).

10.2 Cytosolic Enzymes

Nox proteins constitute a family of membrane-associated, multiunit enzymes that catalyze the reduction of molecular oxygen using nicotinamide adenine dinucleotide phosphate (NADPH) as an electron donor. The Nox family is composed of seven members, Nox1–5 and Duox1/2, with only Nox1, Nox2, Nox4, and Nox5 present in cardiovascular tissues. Nox proteins primarily produce superoxide anions by a single-electron reduction; however, Nox4 primarily produces hydrogen peroxide (Matsushima et al. 2014). XO is the oxidized form of xanthine dehydrogenase and catalyzes the conversion of hypoxanthine to xanthine and xanthine to uric acid. XO is mainly localized outside the cell but under pathophysiological conditions might be expressed in cardiomyocyte itself (Zhang et al. 2015). In this reaction, the superoxide anion is formed as a byproduct. However, at low oxygen tension, XO can also produce hydrogen peroxide. Interestingly, increased oxidative stress increases the conversion of xanthine dehydrogenase to XO, which then produces additional ROS in a feed forward mechanism. XO is a major contributor of ROS formation after ischemia–reperfusion injury (Granger and Kvietys 2015).

Enzymes of the LO family catalyze the oxidation of polyunsaturated fatty acids, including those involved in the biosynthesis of inflammatory leukotriene molecules.

Uncoupling of endothelial NOS (eNOS) by reduced availability of substrates and/or cofactors (tetrahydrobiopterin and l-arginine) results in increased superoxide anion and reduced NO formation. Formed peroxynitrite further contributes to eNOS uncoupling and ROS-induced oxidation of tetrahydrobiopterin, both accelerate ROS formation (Luo et al. 2014).

Although ROS can be generated at several intracellular sites, in cardiomyocytes the major source is represented by mitochondria that are on the other hand a prominent target of oxidative stress (Chen and Zweier 2014; Di Lisa et al. 2009; Kaludercic et al. 2014a, b).

10.3 Mitochondrial Enzymes

Mitochondria contain enzymes that catalyze hydrogen peroxide generation as the obligatory product along with systems generating ROS in sporadic, possibly undesired, reactions. This is especially the case with the electron transport chain (ETC) (Murphy 2009; Chen and Zweier 2014; Zorov et al. 2014). Indeed, a minor fraction (about 0.1 %) of the electrons flowing through the ETC is suggested to cause the partial reduction of oxygen (O_2) into superoxide (Murphy 2009) that is then rapidly and efficiently further reduced into hydrogen peroxide.

ROS formation is favored by high mitochondrial membrane potential (i.e., low ATP synthesis) and large NADH(H^+) availability or when electron flow is hampered by alterations in respiratory complexes. Conversely, a decrease in ROS levels should follow the acceleration in electron flow caused by mitochondrial uncoupling (Korshunov et al. 1997).

Superoxide anions do not cross the inner mitochondrial membrane and are rapidly dismutated into the freely permeable hydrogen peroxide by manganese superoxide dismutase (Mn SOD or SOD2). Therefore, hydrogen peroxide is the main ROS produced. Its accumulation per se would hardly be detrimental, but reacting with iron (Fe^{2+}) in the Fenton reaction generates the highly reactive and toxic hydroxyl radical.

Besides respiratory chain complexes, several other mitochondrial enzymes potentially form ROS, such as the flavin containing glycerol-3-phosphate, proline, and dihydroorotate dehydrogenase at the outer leaflet of the inner mitochondrial membrane, the electron transfer flavoprotein-ubiquinone (ETF:Q) oxidoreductase system of fatty acid β-oxidation within the inner mitochondrial membrane, and pyruvate and 2-oxoglutarate dehydrogenase within the mitochondrial matrix. All these enzymes and respiratory complexes normally catalyze reactions other than ROS formation that are required for energy metabolism, cell function, and viability maintenance. Their characterization as ROS-forming enzymes has been carried out in isolated mitochondria, and thus these approaches can hardly be adapted to living cells or tissues.

The demonstration that mitochondria generate ROS in vivo is provided by studies on other mitochondrial enzymes, such as p66shc and monoamine oxidases (MAO) that generate hydrogen peroxide as a direct and obligatory product. A fraction of p66Shc has been observed within the mitochondrial intermembrane space (IMS) though the import mechanism of p66Shc into mitochondrial IMS has not been clearly identified, but stress kinase-induced phosphorylation of p66Shc at Ser36 contribute to translocation. In basal conditions, mitochondrial p66Shc associates to a high molecular weight complex of about 670 kDa and to the mitochondrial chaperon mtHsp70 (Orsini et al. 2006) and components of TIM/TOM import complex (Cosentino et al. 2008). Notably, treatment of cells with deleterious stimuli induces the dissociation of these complexes and the release of monomeric p66Shc, which then reacts with cytochrome c (Orsini et al. 2006) sequestering electrons from the ETC to generate hydrogen peroxide (Giorgio et al. 2005; Trinei et al. 2013).

MAOs are flavoenzymes anchored to the outer mitochondrial membrane through a transmembrane helix located within the carboxyl-terminal domain. They are responsible for the oxidative deamination of neurotransmitters and dietary amines. MAOs exist in two isoforms, MAO-A and MAO-B, distinguished by different substrate specificity and inhibitor sensitivity. The flavin moiety of MAO is the only redox-dependent factor necessary for catalysis, since mutations of flavin-linking residues ablate enzymatic activity. Aldehyde intermediates are toxic for the biological systems, and a decrease in aldehyde dehydrogenase activity, also due to increased oxidative stress, might further contribute to the exacerbation of damage. Formation of these by-products is a salient aspect of MAO biochemical and pharmacological profile that certainly warrants further attention. The MAO relevance in cardiac diseases has been covered in a recent review (Kaludercic et al. 2014a, b).

The list of dedicated enzymes for ROS formation in mitochondria might include Nox4 (Ago et al. 2010), although a recent work argued against its mitochondrial localization at least in the normoxic heart under physiological conditions (Hirschhäuser et al. 2015).

More recently, connexin 43 has been demonstrated to be present in mitochondria (especially of subsarcolemmal origin) and is involved in controlling mitochondrial respiratory function, ion flux, and ROS formation (for review, see Schulz et al. 2015). Figure 10.1 summarizes the most important oxidative radical sources in cardiomyocytes.

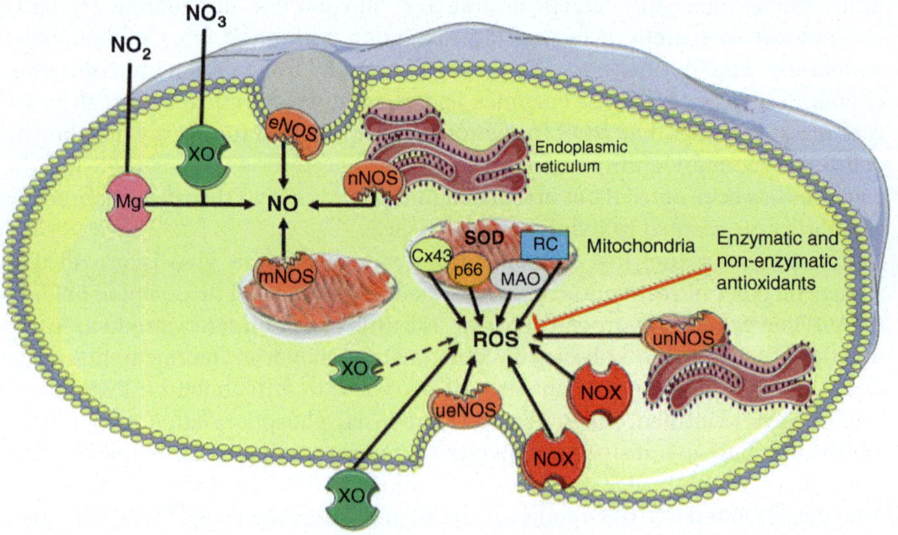

Fig. 10.1 Potential ROS and RNS sources in cardiomyocytes. ROS is formed mainly but not exclusively in mitochondria (p66, MAO, connexin 43, respiratory electron transport chain (*RC*)). Additional ROS derived from uncoupled eNOS and nNOS that display differential localization in cells (*ueNOS*, *unNOS*) and from xanthine oxidase (*XO*). NO is formed by eNOS, nNOS, and mitochondrial NOS (mtNOS). This process can be affected by XO activity and magnesium ions (*Mg*)

10.4 Antioxidant Enzymes

The damage from ROS is usually minimized by a multitude of enzymes and mechanisms that are protective in nature. Enzymatic defenses employ superoxide dismutases (SOD) (Bresciani et al. 2015), catalases (Dai and Rabinovitch 2009), peroxiredoxins (Prx), and glutathione peroxidase (Gpx) (Sies 2014). Antioxidants, such as the lipid-soluble vitamins D, E, and K, protect membranes, and the water-soluble vitamin C, neutralize ROS in the cytosol. Other defenses that allow for free radical neutralization by absorbing the reactive electron involve phenolic compounds, such as tocopherols, carotenes, ascorbate, ubiquinols and flavonoids, or glutathione.

SODs rapidly convert superoxide anions to hydrogen peroxide. Most mammalian cells contain three forms of SOD (SOD1, SOD2, and SOD3) which differ in their localization. SOD1, a Cu/Zn SOD, is expressed in the cytoplasm; genetic deletion of SOD1 leads to increased superoxide anion and peroxynitrite formation.

Catalase catalyzes the decomposition of hydrogen peroxide to water and oxygen. Catalase is a tetramer containing four heme groups. The heme iron is initially oxidized by hydrogen peroxide to form a high-valence intermediate (compound I) which through a further reaction with hydrogen peroxide forms molecular oxygen and water. Although catalase is highly abundant in many tissues, catalase null mice develop normally. Pox likely provides the necessary redundancy in hydrogen peroxide breakdown.

Poxs, which are highly expressed in the cardiovascular system, are a class of six antioxidant enzymes that are typically classified by characteristics of their hydrogen peroxide-sensitive catalytic cysteines as 2-Cys (Prx1-4), atypical 2-Cys (Prx5), and 1-Cys (Prx6).

Gpx catalyzes the decomposition of hydrogen peroxide and lipid hydroperoxides to water or corresponding alcohols. Gpx converts reduced glutathione (GSH) to its oxidized form (GSSG). The pools of GSH are replenished by the action of glutathione reductase (GSR) on GSSG. Gpx exhibits higher expression in cardiovascular tissues than catalase, and because of its ability to reduce hydrogen peroxide and lipid peroxides, Gpx is more important as protective antioxidant in the cardiovascular system than catalase and Gpx1 might become the major sink for H_2O_2 (Murphy 2009).

As in the rest of the cell also in mitochondria, ROS generation is counterbalanced by efficacious removal systems. Prx3 is responsible for more than 90% of H_2O_2 removal in mitochondria (Cox et al. 2010) but becomes rapidly oxidized under conditions of severe oxidative stress (Kumar et al. 2009). However, Prx3 overexpression confers increased resistance to ischemia–reperfusion injury (Matsushima et al. 2006). In the mitochondria, also mitochondrial manganese containing superoxide dismutase (Mn SOD or SOD2) is primarily responsible for converting the superoxide anion to hydrogen peroxide, which is then converted to water by the enzyme catalase, largely localized in peroxisomes. Hydrogen peroxide handling is also carried out by a thiol redox system centered on glutathione (GSH and GSSG in its reduced and oxidized form, respectively) and thioredoxin (Trx) (Berndt et al.

2007; Murphy 2012; Forman et al. 2014). To allow a continuous removal of hydrogen peroxide, oxidized glutathione and Trx are immediately reduced by the corresponding reductases at the expense of NADPH(H^+). Therefore, the maintenance of an optimal $NADP^+$/NADPH(H^+) ratio (by a set of mitochondrial enzymes) is crucial for both tuning ROS signaling and fueling antioxidant activities (Nickel et al. 2014). For review, see also Brown and Griendling (2015).

10.5 Nitric Oxide

Endogenous NO is produced enzymatically by a family of nitric oxide synthases (NOS), which converts oxygen and L-arginine to NO and L-citrulline. The enzymatic NO synthesis is a key determinant of the cellular nitrosative milieu, and the NOS activity of the constitutively expressed isoforms (neuronal and endothelial NOS) is transient in nature and depends on activation by calcium or phosphorylation through protein–protein interactions (for review, see Milsom et al. 2012). Also in mitochondria a functional NOS has been identified (Kirca et al. 2015).

Apart from the production of NO through NOS, it can also be produced from nitrite by myoglobin in cardiomyocytes and from nitrate by xanthine oxidase (Cantu-Medellin and Kelley 2013). Thus, an enzyme involved in the oxidative stress might also limit the lack of NO.

10.6 ROS/RNS Effects

The physiological role of ROS generation (low amounts) in a wide variety of cardiomyocyte functions is likely to depend on posttranslational modifications of proteins, especially at the level of cysteine residues, that underlie short-term responses to different stimuli. In addition, long-lasting changes are obtained by the effects of ROS on transcriptional factors, such as hypoxia-inducible factors (HIFs) and nuclear factor erythroid 2–related factor 2 (Nrf2) (Hayes and Dinkova-Kostova 2014; Semenza 2014; Yun and Finkel 2014). However, at high amounts ROS induce detrimental effects on proteins, lipids, carbohydrates, and nucleotides. ROS induce tissue dysfunction by directly damaging cells via a number of mechanisms including peroxidation of cell membrane and organelle lipids, oxidizing DNA, activation of matrix metalloproteinases and calpains, producing osmotic cell lysis, induction of no reflow, and causing opening of the mitochondrial permeability transition pore (Canton et al. 2011; Di Lisa et al. 2011). ROS may also induce cell dysfunction and death by indirect mechanisms by interacting with fatty acids or free iron to form peroxyl radicals and hydroxyl radicals (for review, see Brown and Griendling 2015). Potential ROS targets are summarized in Fig. 10.2.

ROS also act to enhance the inflammatory response to reperfusion via formation of oxidant-dependent pro-inflammatory mediators and upregulation of cytokine/chemokine and adhesion molecule expression; this is especially the case with ischemia–reperfusion injury.

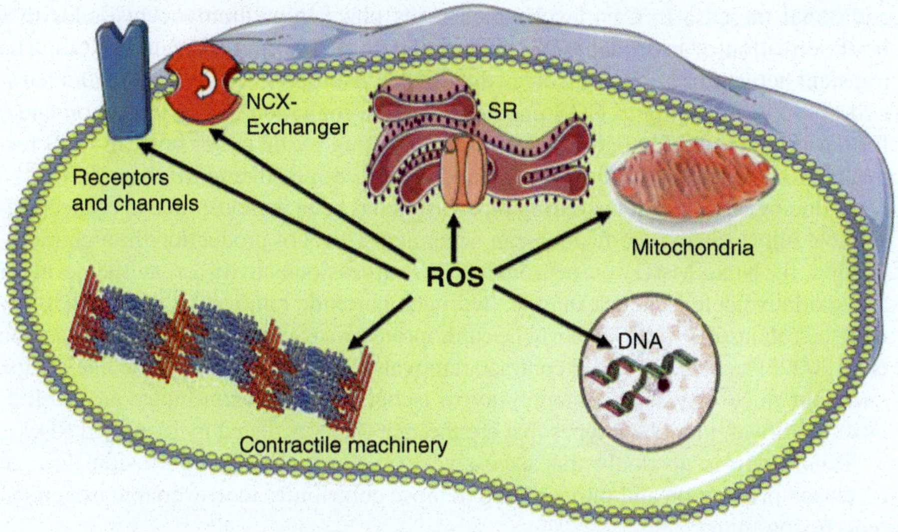

Fig. 10.2 Potential targets of ROS in cardiomyocytes. ROS affects proteins of the contractile machinery, such as tropomyosin, calcium transports in the sarcoplasmic and sarcolemmal membrane, transcription factors, and mitochondrial proteins

Nitrosative modifications, either as S-nitros(yl)ation of cysteine thiols or as nitration of tyrosine residues, are selective processes that target precise molecular sites in proteins or lipids for gain or loss of function. S-nitrosylation occurs through the covalent attachment of a diatomic nitroso group to a reactive thiol sulfhydryl in a redox-dependent fashion. Tyrosine nitration, on the other hand, results from the covalent addition of a triatomic nitro group to the phenolic ring of tyrosine residues. Both S-nitrosylation and tyrosine nitration may result from protein interactions with RNS. Protein S-nitrosylation may, however, also result from transnitrosylation involving S-nitrosoglutathione (GSNO). Peroxynitrite is one of the nitrating agents. Despite being nonenzymatic in nature, only a relatively small number of tyrosine and cysteine residues appear to be the target of nitrosative adducts (for review on the topic, see Murphy et al. 2014; Schulz and Ferdinandy 2013).

10.7 Ischemia–Reperfusion Injury

Myocardial ischemia–reperfusion (I/R) injury occurs when 30 or more minutes of ischemia is followed by reperfusion. The injury consists of cell necrosis and apoptosis plus a period of hypocontractility or myocardial stunning. Also arrhythmias, including ventricular tachycardia and fibrillation, may appear during I/R. A slight generation of ROS has been observed in ischemic or hypoxic cardiomyocytes, especially in isolated cells (van den Hook et al. 1997).

Rather than a worsening of ischemic injury, reperfusion is per se the source of specific components of cardiac damage. At the early stage of reperfusion, the additional increase in Ca^{2+} level which takes place in cardiomyocytes leads to a brief activation of neuronal NOS, endothelial NOS, and mitochondrial NOS. The transient activation of NOS involves the rapid consumption of L-arginine and tetrahydrobiopterin leading to NOS uncoupling and a reduction in NO formation and a further production of superoxide anion contributing to the major burst in ROS formation seen at the onset of reperfusion (Zweier 1988; Bolli and Marban 1999).

Reducing ROS formation during reperfusion has been demonstrated to reduce irreversible injury. Such modification can occur at the level of production through modification of Nox, MAO, or neuronal NOS expression/activity or at the level of anti-oxidative defense. For example, hydrogen peroxide removal by mitochondrially targeted catalase has been shown to be cardioprotective (Schriner et al. 2005; Anderson et al. 2009; Dai et al. 2011). In contrast, removal of anti-oxidative defense like in Gpx knockout mice increased susceptibility to ischemia–reperfusion injury and various other cardiovascular phenotypes that are thought to be mediated by increased ROS.

Thus, there is no doubt that excessive ROS formation at reperfusion through increased production and/or decreased defense contributes to irreversible ischemia–reperfusion injury.

On the contrary, ROS also play an important role in endogenous cardioprotective phenomena, namely, ischemic or pharmacological conditioning effects. Blocking ROS formation and/or scavenging ROS during the conditioning phases (being either before or immediately following the sustained, lethal ischemic episode) as well as early on during the sustained ischemia abolishes the otherwise obtained cardioprotection, i.e., reduction in the incidence of arrhythmias, the extent of myocardial infarction, or endothelial dysfunction. Some of the proteins involved in ROS formation are involved in both the triggering of cardioprotection and mediation of irreversible injury (i.e., NOX) while others being involved only in the triggering phase of protection (i.e., connexin 43) or the mediator phase of irreversible injury (i.e., p66shc). The cardioprotection afforded by small increases in ROS is believed to involve in the activation of protein kinase pathways (such as the RISK or SAFE pathway) (for review, see Ferdinandy et al. 2014).

Similar to ROS, NO contributes to the initiation of endogenous or pharmacological cardioprotection. Here, NO might be derived from NOS or generated from nitrite or nitrate (see Fig. 10.1). Protection of NO is thought to be mediated through transient nitros(yl)ation of proteins, most importantly within mitochondria, thereby directly altering their function but also attenuating irreversible posttranslational modification, for example, through ROS-induced protein oxidation (Murphy et al. 2014; Soetkamp et al. 2014).

10.8 Aging

More than five decades ago, Harman et al. proposed the free radical theory of aging, postulating that the production of ROS is a major determinant of life span (Harman 1956). Many studies subsequently showed that aging results in an impairment in the ETC during mitochondrial respiration, primarily because of reduced electron transport by complexes I and IV, and could consequently result in leakage of electrons and increased ROS production. Certain key components of complex I and IV are encoded by mitochondrial DNA and may be altered quantitatively, because of inhibition of synthesis by ROS signaling or enzyme modification and, thereby, inhibition of enzyme activity by excess ROS. These findings had led to the mitochondrial variant of the free radical theory of aging (reviewed by Balaban et al. 2005), which proposes that mitochondrial ROS attack mitochondrial constituents, causing mitochondrial DNA damage and mitochondrial dysfunction, which further aggravates ROS production finally leading to loss of function and cell death.

Two distinct populations of mitochondria exist within cardiomyocytes: one right under the plasma membrane, called the subsarcolemmal mitochondria (SSM), and the other are the interfibrillar mitochondria (IFM) located between the myofibrils. The SSM from aged hearts remain functionally similar to the SSM from the hearts of young animals; however, the IFM are altered in that the activity of the electron transport chain is reduced.

Not only the respiratory chain is affected with aging, also mitochondrial expression of Nox4 is upregulated in the senescent heart, contributing to ROS production and development of cardiac abnormalities. P66Shc also accumulates progressively in aged mitochondria, and p66Shc deletion significantly extends life span in mammals under resting conditions, although this finding has recently been challenged.

Furthermore, reduced mitochondrial biogenesis seen in aged hearts has been linked with the development of cardiomyopathy.

On closer examination, aged cardiomyocytes demonstrate an increased GSH/GSSG ratio, indicating oxidative stress, Nox activation, particularly the p47phox subunit, and an increase in advanced glycation end products, which are generated upon exposure to increased ROS, are correlated with cardiovascular disease and heart failure.

Not only ROS production is increased in aged hearts, but also antioxidant gene expression and activity are affected during aging. The capacity for activating manganese superoxide dismutase and catalase expression during oxidative stress and exercise is reduced over time. Indeed, systemic mitochondrial overexpression of catalase extends life span and delays the aging process in mice.

10.9 Conclusive Remarks

Formation of ROS and RNS contributes significantly to all aspects of cardiac signaling and cardiac dysfunction. As an impairment between ROS formation and ROS detoxification, known as oxidative stress, is causally involved in many aspects of cardiac dysfunction, ROS detoxification has got an attractive target for future drug design.

References

Ago T, Kuroda J, Pain J, Fu C, Li H, Sadoshima J (2010) Upregulation of Nox4 by hypertrophic stimuli promotes apoptosis and mitochondrial dysfunction in cardiac myocytes. Circ Res 106:1253–1264

Afanas'ev I (2011) ROS and RNS signaling in heart disorders: could antioxidant treatment be successful? Oxid Med Cell Longev 2011:293769

Anderson EJ, Lustig ME, Boyle KE, Woodlief TL, Kane DA, Lin CT, Price JW 3rd, Kang L, Rabinovitch PS, Szeto HH, Houmard JA, Cortright RN, Wasserman DH, Neufer PD (2009) Mitochondrial H_2O_2 emission and cellular redox state link excess fat intake to insulin resistance in both rodents and humans. J Clin Invest 119:573–581

Andreadou I, Iliodromitis EK, Rassaf T, Schulz R, Papapetropoulos A, Ferdinandy P (2015) The role of gasotransmitters NO, H_2S and CO in myocardial ischaemia/reperfusion injury and cardioprotection by preconditioning, postconditioning and remote conditioning. Br J Pharmacol 172:1587–1606

Balaban RS, Nemoto S, Finkel T (2005) Mitochondria, oxidants, and aging. Cell 120(4):483–495

Berndt C, Lillig CH, Holmgren A (2007) Thiol-based mechanisms of the thioredoxin and glutaredoxin systems: Implications for diseases in the cardiovascular system. Am J Physiol Heart Circ Physiol 292:H1227–H1236

Bolli R, Marbán E (1999) Molecular and cellular mechanisms of myocardial stunning. Physiol Rev 79:609–634

Bresciani G, da Cruz IB, González-Gallego J (2015) Manganese superoxide dismutase and oxidative stress modulation. Adv Clin Chem 68:87–130

Brown DI, Griendling KK (2015) Regulation of signal transduction by reactive oxygen species in the cardiovascular system. Circ Res 116:531–549

Canton M, Menazza S, Sheeran FL, Polverino de Laureto P, Di Lisa F, Pepe S (2011) Oxidation of myofibrillar proteins in human heart failure. J Am Coll Cardiol 57:300–309

Cantu-Medellin N, Kelley EE (2013) Xanthine oxidoreductase-catalyzed reactive species generation: a process in critical need of reevaluation. Redox Biol 1:353–358

Chen YR, Zweier JL (2014) Cardiac mitochondria and reactive oxygen species generation. Circ Res 114:524–537

Cosentino F, Francia P, Camici GG, Pelicci PG, Lüscher TF, Volpe M (2008) Final common molecular pathways of aging and cardiovascular disease: role of the p66Shc protein. Arterioscler Thromb Vasc Biol 28:622–628

Cox AG, Winterbourn CC, Hampton MB (2010) Mitochondrial peroxiredoxin involvement in antioxidant defence and redox signalling. Biochem J 425:313–325

Dai DF, Rabinovitch PS (2009) Cardiac aging in mice and humans: the role of mitochondrial oxidative stress. Trends Cardiovasc Med 19:213–220

Dai DF, Johnson SC, Villarin JJ, Chin MT, Nieves-Cintron M, Chen T, Marcinek DJ, Dorn GW 2nd, Kang YJ, Prolla TA, Santana LF, Rabinovitch PS (2011) Mitochondrial oxidative stress mediates angiotensin ii-induced cardiac hypertrophy and galphaq overexpression-induced heart failure. Circ Res 108:837–846

Di Lisa F, Kaludercic N, Carpi A, Menabò R, Giorgio M (2009) Mitochondria and vascular pathology. Pharmacol Rep 61:123–130

Di Lisa F, Canton M, Carpi A, Kaludercic N, Menabò R, Menazza S, Semenzato M (2011) Mitochondrial injury and protection in ischemic pre- and postconditioning. Antioxid Redox Signal 14:881–891

Ferdinandy P, Hausenloy DJ, Heusch G, Baxter GF, Schulz R (2014) Interaction of risk factors, comorbidities, and comedications with ischemia/reperfusion injury and cardioprotection by preconditioning, postconditioning, and remote conditioning. Pharmacol Rev 66:1142–1174

Forman HJ, Ursini F, Maiorino M (2014) An overview of mechanisms of redox signaling. J Mol Cell Cardiol 73:2–9

Giorgio M, Migliaccio E, Orsini F, Paolucci D, Moroni M, Contursi C, Pelliccia G, Luzi L, Minucci S, Marcaccio M, Pinton P, Rizzuto R, Bernardi P, Paolucci F, Pelicci PG (2005)

Electron transfer between cytochrome c and p66Shc generates reactive oxygen species that trigger mitochondrial apoptosis. Cell 122:221–233

Granger DN, Kvietys PR (2015) Reperfusion injury and reactive oxygen species: the evolution of a concept. Redox Biol 6:524–551

Harman D (1956) Aging: a theory based on free radical and radiation chemistry. J Gerontol 11:298–300

Hayes JD, Dinkova-Kostova AT (2014) The nrf2 regulatory network provides an interface between redox and intermediary metabolism. Trends Biochem Sci 39:199–218

Hirschhäuser C, Bornbaum J, Reis A, Böhme S, Kaludercic N, Menabò R, Di Lisa F, Boengler K, Shah AM, Schulz R, Schmidt HH (2015) NOX4 in mitochondria: yeast two-hybrid-based interaction with complex I without relevance for basal reactive oxygen species? Antioxid Redox Signal 23(14):1106–1112

Kaludercic N, Mialet-Perez J, Paolocci N, Parini A, Di Lisa F (2014a) Monoamine oxidases as sources of oxidants in the heart. J Mol Cell Cardiol 73:34–42

Kaludercic N, Deshwal S, Di Lisa F (2014b) Reactive oxygen species and redox compartmentalization. Front Physiol 5:285

Kirca M, Kleinbongard P, Soetkamp D, Heger J, Csonka C, Ferdinandy P, Schulz R (2015) Interaction between connexin 43 and nitric oxide synthase in mice heart mitochondria. J Cell Mol Med 19:815–825

Korshunov SS, Skulachev VP, Starkov AA (1997) High protonic potential actuates a mechanism of production of reactive oxygen species in mitochondria. FEBS Lett 416:15–18

Kumar V, Kitaeff N, Hampton MB, Cannell MB, Winterbourn CC (2009) Reversible oxidation of mitochondrial peroxiredoxin 3 in mouse heart subjected to ischemia and reperfusion. FEBS Lett 583:997–1000

Luo S, Lei H, Qin H, Xia Y (2014) Molecular mechanisms of endothelial NO synthase uncoupling. Curr Pharm Des 20:3548–3553

Matsushima S, Ide T, Yamato M, Matsusaka H, Hattori F, Ikeuchi M, Kubota T, Sunagawa K, Hasegawa Y, Kurihara T, Oikawa S, Kinugawa S, Tsutsui H (2006) Overexpression of mitochondrial peroxiredoxin-3 prevents left ventricular remodeling and failure after myocardial infarction in mice. Circulation 113:1779–1786

Matsushima S, Tsutsui H, Sadoshima J (2014) Physiological and pathological functions of NADPH oxidases during myocardial ischemia-reperfusion. Trends Cardiovasc Med 24:202–205

Milsom AB, Fernandez BO, Garcia-Saura MF, Rodriguez J, Feelisch M (2012) Contributions of nitric oxide synthases, dietary nitrite/nitrate, and other sources to the formation of NO signaling products. Antioxid Redox Signal 17:422–432

Murphy E, Kohr M, Menazza S, Nguyen T, Evangelista A, Sun J, Steenbergen C (2014) Signaling by S-nitrosylation in the heart. J Mol Cell Cardiol 73:18–25

Murphy MP (2009) How mitochondria produce reactive oxygen species. Biochem J 417:1–13

Murphy MP (2012) Mitochondrial thiols in antioxidant protection and redox signaling: distinct roles for glutathionylation and other thiol modifications. Antioxid Redox Signal 16:476–495

Nickel A, Kohlhaas M, Maack C (2014) Mitochondrial reactive oxygen species production and elimination. J Mol Cell Cardiol 73:26–33

Orsini F, Moroni M, Contursi C, Pelicci PG, Giorgio M, Migliaccio E (2006) Mitochondrial regulation of p66Shc mitochondrial function. Biol Chem 387:1405–1410

Rassaf T, Ferdinandy P, Schulz R (2014) Nitrite in organ protection. Br J Pharmacol 171:1–11

Schriner SE, Linford NJ, Martin GM, Treuting P, Ogburn CE, Emond M, Coskun PE, Ladiges W, Wolf N, Van Remmen H, Wallace DC, Rabinovitch PS (2005) Extension of murine life span by overexpression of catalase targeted to mitochondria. Science 308:1909–1911

Schulz R, Ferdinandy P (2013) Does nitric oxide signaling differ in pre- and post-conditioning? Importance of S-nitrosylation vs. protein kinase G activation. Free Radic Biol Med 54:113–115

Schulz R, Görge PM, Görbe A, Ferdinandy P, Lampe PD, Leybaert L (2015) Connexin 43 is an emerging therapeutic target in ischemia/reperfusion injury, cardioprotection and neuroprotection. Pharmacol Ther 153:90–106

Semenza GL (2014) Hypoxia-inducible factor 1 and cardiovascular disease. Annual Rev Physiol 76:39–56

Sies H (2014) Role of metabolic H_2O_2 generation: redox signaling and oxidative stress. J Biol Chem 289:8735–8741

Soetkamp D, Nguyen TT, Menazza S, Hirschhäuser C, Hendgen-Cotta UB, Rassaf T, Schlüter KD, Boengler K, Murphy E, Schulz R (2014) S-nitrosation of mitochondrial connexin 43 regulates mitochondrial function. Basic Res Cardiol 109:433

Trinei M, Migliaccio E, Bernardi P, Paolucci F, Pelicci P, Giorgio M (2013) Mitochondria, and the generation of reactive oxygen species. p66Shc. Methods Enzymol 528:99–110

Vanden Hoek TL, Li C, Shao Z, Schumacker PT, Becker LB (1997) Significant levels of oxidants are generated by isolated cardiomyocytes during ischemia prior to reperfusion. J Mol Cell Cardiol 29:2571–2583

Victorino VJ, Mencalha AL, Panis C (2015) Post-translational modifications disclose a dual role for redox stress in cardiovascular pathophysiology. Life Sci 129:42–47

Yun J, Finkel T (2014) Mitohormesis. Cell Metab 19:757–766

Zhang YS, Liu B, Luo XJ, Zhang JJ, Li NS, Ma QL, Jiang JL, Li YJ, Li Q, Peng J (2015) A novel function of nuclear nonmuscle myosin regulatory light chain in promotion of xanthine oxidase transcription after myocardial ischemia/reperfusion. Free Radic Biol Med 83:115–128

Zorov DB, Juhaszova M, Sollott SJ (2014) Mitochondrial reactive oxygen species (ros) and ros-induced ros release. Physiol Rev 94:909–950

Zweier JL (1988) Measurement of superoxide-derived free radicals in the reperfused heart. Evidence for a free radical mechanism of reperfusion injury. J Biol Chem 263:1353–1357

The manufacturer's authorised representative in the EU is Springer Nature Customer Service Centre GmbH, Europaplatz 3, 69115 Heidelberg, Germany. If you have any concerns regarding our products, please contact ProductSafety@springernature.com

Printed and bound by CPI Group (UK) Ltd, Croydon, CR0 4YY
30/09/2026
02225454-0006